Arab Settlements

About Access Archaeology

Access Archaeology offers a different publishing model for specialist academic material that might traditionally prove commercially unviable, perhaps due to its sheer extent or volume of colour content, or simply due to its relatively niche field of interest. This could apply, for example, to a PhD dissertation or a catalogue of archaeological data.

All *Access Archaeology* publications are available as a free-to-download pdf eBook and in print format. The free pdf download model supports dissemination in areas of the world where budgets are more severely limited, and also allows individual academics from all over the world the opportunity to access the material privately, rather than relying solely on their university or public library. Print copies, nevertheless, remain available to individuals and institutions who need or prefer them.

The material is refereed and/or peer reviewed. Copy-editing takes place prior to submission of the work for publication and is the responsibility of the author. Academics who are able to supply print-ready material are not charged any fee to publish (including making the material available as a free-to-download pdf). In some instances the material is type-set in-house and in these cases a small charge is passed on for layout work.

Our principal effort goes into promoting the material, both the free-to-download pdf and print edition, where *Access Archaeology* books get the same level of attention as all of our publications which are marketed through e-alerts, print catalogues, displays at academic conferences, and are supported by professional distribution worldwide.

The free pdf download allows for greater dissemination of academic work than traditional print models could ever hope to support. It is common for a free-to-download pdf to be downloaded hundreds or sometimes thousands of times when it first appears on our website. Print sales of such specialist material would take years to match this figure, if indeed they ever would.

This model may well evolve over time, but its ambition will always remain to publish archaeological material that would prove commercially unviable in traditional publishing models, without passing the expense on to the academic (author or reader).

Arab Settlements

Tribal structures and spatial organizations in the Middle East between Hellenistic and Early Islamic periods

Nicolò Pini

Access Archaeology

ARCHAEOPRESS PUBLISHING LTD
Summertown Pavilion
18-24 Middle Way
Summertown
Oxford OX2 7LG

www.archaeopress.com

ISBN 978-1-78969-361-4
ISBN 978-1-78969-362-1 (e-Pdf)

Cover image: APAAME_20020929_RHB-0097, Photographer: Robert Bewley, courtesy of APAAME.

This book is available direct from Archaeopress or from our website www.archaeopress.com

To my parents – Paola and Daniele – and Silvia

Table of contents

List of figures

Pictures by APAAME (Aerial Photographic Archive for Archaeology in the Middle East) available on http://www.apaame.org and http://www.flickr.com/apaame/collections.

List of plates

Preface

What can the built environment tell about the society living in the settlement? This is the main question of the present work. In the final years of my studies at the University of Siena, I became intrigued by the relationship between frequent patterns of organisation in the urban space and specific social structures that are normally described as "tribal". Furthermore, the second aspect of interest was the perdurance of such patterns from the Byzantine into Early Islamic times, and possibly even later into the Middle Islamic period. Following these two broad topics and revisiting some of the several questions left unanswered by my Master's dissertation, my doctoral research investigated the possible social triggers for specific forms of the built environment and its development from Hellenism to the Early medieval (1st – 8th centuries AD).

Archaeological disciplines have demonstrated an increasing interest in "non-material" aspects of ancient societies, especially following the well-known debate between processualists and post-processualists. Inevitably a discipline such as archaeology, which heavily relies on tangible materials to formulate their interpretations, poses several difficulties once the topic of interests falls also into the immaterial side of material culture. Being that the topic of the present work incorporates such sort of inquiries, one pre-requirement urgently needed to be fulfilled: that is taking some important anthropological notions into consideration, such as "tribalism" (or more precisely segmented societies) and "nomadism" (i.e. pastoralism). They are indeed central in the interpretation of the archaeological data and architectural remains for the study region and time range. Nonetheless, their use in the current historical and archaeological scholarly discourse is often misleading, if not (potentially) misused (see Chapter 2). In fact, despite an increased attention to these theoretical concepts during the last years, in most cases a fully coherent and consistent use of such terms is still lacking, and heavily affects the reliability of some reconstructions.

Therefore, the present work will provide a theoretical overview on the anthropological debate covering the two issues of segmented societies and pastoralism. It will also briefly outline the debate on the term "Arab", where its very definition is closely interconnected historically and socially with these two central concepts. Finally in this discussion, I justify my preference for the term the "Arab settlement", rather than other more commonly used terms, such as the "Islamic settlement" and the "Oriental settlement".

Two subsequent parts of the research are built upon this theoretical and anthropological basis. In the first section (Chapters 4-6) the data are yielded and gathered from different disciplines and then are progressively put together into a general analysis of settlements in the Near East. This is accomplished through thoroughly comparing eight case studies coming from different regions in the Levant (the Negev, the central region of today Jordan, and the central and southern part of the Hauran). Despite the strong regionalism and diversity in some aspects of their historical development, the feature that ultimately allows for comparison between them is that they are situated on the border between the desert and the "settled" areas. The comparative analysis of these eight sites ultimately aims to identify a set of common elements in the built environments and at the same time highlight and explain the regional peculiarities, wherever possible.

The last section (Chapters 7-8) brings together the theoretical framework of the study and the results of the comparative analysis, while also taking the general historical context into consideration. By showing some patterns and behaviours related to the built environment in the longue durée, it will be argued that society, especially family groups and the (potential) presence of nomadic pastoralists, played a central role in the organisation and continuity of settlements, especially between the 5th and 8th centuries. Investigating the relationship between social dynamics and phenomena and architectural forms is indeed crucial in order to understand the creation, maintenance or change, and finally the abandonment of determined spatial behaviour. Furthermore, it allows us to consider the following questions: What was the social structure determining them? To what extent were the identity patterns reflected by the built environment? Which role did the sedentarisation of nomads play? This role might ultimately partially explain the recurrence of some ways of organising the built environment in such a broad geographical area.

Another aim of this work is addressing the long-debated topic of the "Oriental" or "Islamic city" from the archaeological perspective. Preferring to use the term "Arab", while also carefully considering its potential constraints, is an attempt to bypass some limitations of the previous two terms. In my opinion the concept of the "Oriental city" as described by Eugen Wirth, among others, is preferable to the "Islamic city", which evokes a discontinuity in urban forms that cannot be supported by evidence from the archaeological record. At the same time, the "Arab city" is preferable to the "Oriental city" because it frames the analysis into a more specific chronological, geographical, and possibly social context. These too terms, however, are not to be seen as antithetical, the first being simply a specific case in the wide spectrum of possible Oriental settlements.

It should be noted that although the present work is not focused on properly "urban" case studies; namely, none of the settlements considered in this study was a proper Hellenistic polis. In fact, the so-called "rural settlements" can explain some of the phenomena that take place in the larger cities of the former Roman Near East. Furthermore, creating a typology of the settlements by using the available administrative terminology presents its own challenges, especially for the Late Antique period.

While it is beyond the scope of the present study to offer a definitive or universal model of the spatial organisation of settlements, applying this new methodology to the range of case studies and material evidence revealed some significant trends for the study region, including the role of family groups in shaping and maintaining the built environment. It is the hope of this study to offer this new methodology for future archaeological studies, encouraging others to approach architecture and urban planning in ancient settlements in the Near East alongside the many possible social factors that contributed to turning them into a built reality. In this case the contribution of multiple disciplines can indeed result in a more complex and realistic explanation of such a fascinating phenomenon that is the built environment; it might also possibly allow us to glean some of the non-material features determining it, like identity.

I would like to extend my gratitude and sincere thanks to many people, professors, colleagues, friends and most importantly my family, for their help in the realisation of my doctoral thesis, which has culminated in the publication of this book. Even though it is not possible to list them all here, I would like to mention those who most directly helped me through the various stages of my work and by different means: my supervisors Prof. Dr Michael Heinzelmann (University of Cologne, Germany), Prof.

Dr Bethany J. Walker (University of Bonn, Germany) and Prof. Dr Hugh N. Kennedy (SOAS, London, UK); Prof. Roberto Parenti (University of Siena, Italy) and Prof. Emanuele Papi (University of Siena, Italy and SAIA, Scuola Archeologica Italiana di Atene); Prof. Dr Bert de Vries, Prof. Dr Øystein LaBianca, Dr Elisabeth Osinga, and Dr Piero Gilento; Dr Barbara A. Porter and the team of the American Center of Oriental Studies in Amman. I am also particularly grateful to Annette Hansen for her suggestions and for reviewing and editing the text.

I would also like to thank the Deutscher Akademische Austausch Dienst (DAAD) and the Annemarie Schimmel Kolleg at the University of Bonn for their financial support through the three-year "Research Grants for Doctoral Candidates and Young Academics and Scientists" and the one-year "Post-doctoral Junior Fellowship", which helped make my doctoral and post-doctoral research possible, as well as the a.r.t.e.s Doctoral School at the University of Cologne for the administrative help.

Chronology for the region (based on David Kennedy, The Roman Army in Jordan, Council for British Research in the Levant, London 2004)

Hellenistic	332 – 64 BC
Nabatean	300 BC – AD 106
Roman	64 BC – AD 324
Early Roman	64 BC – AD 135
Late Roman	135 – 324
Byzantine	324 – 640
Early Byzantine	324 – 491
Late Byzantine	491 – 640
Early Islamic	640 – 1174
Umayyad	640 – 750
Abbasid	750 – 969
Fatimid	969 – 1071
Seljuk-Zenjid	1071 – 1174
Middle Islamic (see "Late" in Kennedy 2004)	1174 – 1516
Ayyubid	1174 – 1263
Mamluk	1263 – 1516

Part 1: Methodology and Theory

Chapter 1: Introduction

The transition between the Byzantine and the Early Islamic periods is an intriguing phase in the history of the Middle East. Recent studies (for instance Magness 2003 and Avni 2014) have finally challenged, and arguably surpassed, the old tradition associating the arrival of Arab tribes from the Arabian Peninsula with a complete destruction of the classical Greco-Roman heritage (the idea of the "Arab conquest", Donner 1981). In fact, more features of continuity between the two periods have been progressively recognised in many aspects of political, social and economic life.[1] This new attitude and awareness has been applied to studies on the Late Antique period across the Mediterranean in general and is to a large extent due to both the broadening of the spectrum of sources used by scholars for their interpretations, which are no longer limited to historical sources, and to a considerable refinement of archaeological methodology and dating tools. The evolving settlement pattern is one of the first topics that underwent a critical analysis: in contrast with the previous idea of a complete abandonment of several villages, towns and even cities across the Middle East following the Arab invasion, archaeological surveys clearly demonstrated that life continued, often with no macroscopic changes. The very fate of the urban phenomenon: cities did not come to an end with the Arab conquest. On the contrary, several Hellenistic urban features survived and were reused despite the evident changes in the morphology of the cities from the Classical through the Byzantine into the Early Islamic periods (Sauvaget 1934; Sauvaget 1941; Kennedy 1985; Di Segni 1995; Di Segni 1999; Wirth 2000; Walmsley 2011; and Avni 2014 among others).

Here lies one major problem: the detailed analysis of the built environment, in its diachronic dimension, concentrated almost exclusively on the urban context, namely the former classical *poleis*. However, at the same time, the settlements dotting the countryside did not attract the same attention, and they were mainly documented in order to determine whether they survived the Byzantine period. Though as a result of intensive surveys, a considerable number were dated to the 5th and 6th centuries (Avni 2014), which necessitates not only the reconsideration of the chronology of the settlement patterns in this region but also a deeper understanding of this unpreceded development.

The present work aims to shed some light on the countryside and its respective chronology of settlement patterns and the built environment, and more specifically attempts to trace the development of eight case studies spread throughout the Roman and Byzantine Near East, dating from the 1st to the 8th/9th AD. Although, I will particularly focus on the latter phases of this period, which constitute the so-called Late Byzantine-Early Islamic transition. There are several reasons for zooming in on this transitional phase. First of all, the overwhelming predominance of material remains date to these later centuries and have heavily affected our knowledge of the initial developments of most sites. Secondly, the apex in the development of these rural settlements takes place during this phase, starting from the 4th century and increasing until at least the 7th century, not only in terms of the number of sites and their increased dimensions, but also, and more interestingly, the expanded diversity of functions that can be associated with such rural settlements. This is a phase in which the dichotomy between city and countryside becomes blurrier, with the development of an intermediate level of settlement that if maintaining a rural connotation, seem to function as an "urban hub". This can clearly

[1] For instance, the unearthing of a series of papyri from Nessana, in the Negev, and Petra, in modern day Jordan, opened an important window into the societies and the administration during this transitional period, showing a strong continuity from the Byzantine well into the Early Islamic period in social and administrative aspects, including fiscal practices, family structure, and religion (see Chapter 7.4).

connect with general political and economic changes taking place in this phase and more even more closely with the evolution in the morphology and role of the cities in the region. It is therefore difficult to consider these two phenomena as independent from one another.

The former Roman provinces of *Siria*, *Arabia* and *Palaestina* have always been peculiar in terms of settlement dynamics: the presence of diverse environments encouraged humans to develop various adaptation strategies already from the earliest phases when they inhabited the region. In particular, living on a "boundary" line between the desert extending to the East, and also to the South in the case of *Palaestina*, and lands always cultivable more or less year-round, led to the co-existence of a plurality of groups with different socio-economic orientations and mobility or settlement patterns. Not all regions were alike, both environmentally and culturally speaking, and historical evidence attests that each region had its own peculiarities. Despite this, when considering several sites across the Near East, a recurring pattern in the general spatial organisation of hamlets, villages and towns is recognisable: blocks and quarters are often occurring as clusters of farmhouses that are entirely closed to the outside, extremely compact in their inner structure, and almost completely self-sufficient as far as water management is concerned. Furthermore, domestic architecture within settlements shows an even more marked homogeneity. Why do these features occur as such? Which factors determine the development and organisation of the built environment of these rural and "semi-urban" settlements, especially during their Byzantine expansion? In my opinion, one of the key influential factors may be the social structures, and in particular different settings of family-groups.[2] Their perdurance through different historical phases, most notably after the arrival of the new Islamic administration, might also offer an explanation for the continuity in the processes behind the built environment.

A settlement can be approached from different scales and in different levels of detail. While a settlement could be approached through its regional context, scholars up until now have mainly studied settlements from two opposite extremes in terms of scale: the "macro" scale, namely the spatial organisation of the entire settlement and, though less frequently used, the "micro" scale, meaning the spatial organisation and evolution within the domestic space. This is particularly true for the study of cities or larger settlements. Likewise, the application of the "macro" and the "micro" scale analyses normally does not take the parallel settlement development dynamics occurring in the countryside into account. With that in mind, little attention had been paid to the structures of quarters in both cities and in settlements in the countryside, focusing mainly on monumental complexes or the more general layout of the site, with only a few exceptions.[3]

The present work attempts to fill these two methodological gaps in the analysis of rural settlements. On the one hand, it aims to provide a descriptive and diachronic investigation on three different levels, from the "macro" to the "micro": the settlement in its entirety, the intermediate level of the quarter/block, and the house. On the other hand, it aims to elucidate some of the possible factors determining the spatial organisation of these different levels of analysis, while paying particular attention to their (potential) social significance. This analytical approach allows the researcher

[2] A study on the so-called Iron Age "Four-room house" (Faust and Bunimovitz 2003) represents a good reference for its social interpretation of the built environment, relating specifically different typologies of houses with different types of family structures and identity patterns.

[3] Particularly noticeable are the studies on the Southwest and Northwest quarters in Jerash (Lichtenberger and Raja 2015; Blanke 2018; Lichtenberger and Raja 2018; Kalaitzoglou 2018).

simultaneously to disentangle and clarify the multiple connections between built environment and social structures, possibly at different scales.

In order to test this approach, this study considers four different regions on the margins of the desert across the Near East. A set of eight case studies was chosen, because of their level of preservation and the availability of archaeological data from excavations or architectural surveys (Figure 1):

- Mseikeh and Sharah in the central Hauran, in modern Syria;
- Umm es-Surab and Umm el-Jimal in the southern Hauran, on the border between modern Jordan and Syria;
- Umm er-Rasas and Tell Hisban in the central Jordanian Plateau;[4]
- Mampsis and Shivta in the Negev desert in modern Israel.

Alongside the multi-scalar approach to the settlement, I consider social implications of material evidence, which by itself presents its own set of methodological challenges. Investigations on the social stratification and composition of a site are quite common in several archaeological projects (among others: Faust and Bunimovitz 2003; Desreumaux et al. 2011; Walker 2013; Walker et al. 2014) and their reliability mainly depends on the accuracy of the excavation data and materials. However, connecting material evidence to "immaterial" elements, like identity and ethnicity, is extremely complicated, especially since the nature and criteria to define such terms is a hotly debated topic unto itself in archaeology and in other fields, most noteworthy between the contraposing Primoridalists and Instrumentalists.[5] Generally speaking, dealing with issues that go beyond the materiality of the archaeological documentation has long been a challenge for archaeologists, with noteworthy examples being the emergence of New Archaeology in the 1970s and the following movements of the post-processualism (Renfrew and Bahn 2004: 23–27), and the extent which immaterial aspects can be described and explained is a recurring topic in debates.

While archaeological and architectural data represent the core dataset of the study, the contribution of data and methods from other disciplines is central for the interpretation of the former dataset to help answer the questions set out in this study. For the purposes of this study, the dataset was gathered from the available publications on the sites, and to a certain extent also from autoptic observations made in the field by the author, with the only two exceptions being the Syrian sites that could not be visited due to the current political climate. The epigraphical record represents particularly important source material, especially for the social organisations of segmented societies explored in this study. Mentions of family ties and "tribal" affiliations are widespread across the regions examined and cover a wide chronological spectrum. If not coming directly from a specific site like Umm el-Jimal, the evidence comes from immediate surroundings of these sites or nearby settlements. It is clear that the nature and destination of the inscriptions is also extremely diversified, from funerary to dedicatory, from carved blocks to stone-graffiti. It consists of a variety of evidence that offers multiple perspectives on social aspects of the local communities, which would have been otherwise missed by other types of evidence. This is especially true for the rural context, where other historical sources might be not as helpful, since contemporary sources tend to consider the urban perspective more frequently, also when writing on

[4] These two case studies will be considered separately at the end of the chapter because of the nature of their poor conservation. Not all three levels of analysis could be easily applied to these sites; nonetheless they will provide good case studies to test the method I propose here for non-optimal material contexts.

[5] For an overview, see: Hall 1997; Jones 1997. A sort of compromise between these two positions is Bourdieu's "theory of Practice" (Bourdieu 1977).

Figure 1 - Location of the case studies (Satellite image: Google Earth)

the "countryside". This is not to deny any possible use of historical sources within the research; though, this is simply to suggest that extra care in how this information is applied to the rural context. Lastly, difficulties are also often reflected in problematic terminology, whether they are used by ancient sources or applied by modern scholars. Here, cultural anthropology and ethnoanthropological comparisons play a fundamental role, despite the challenges involved in creating parallels between ancient societies and contemporary (or at least more recent) ones. Therefore, the anthropological debate greatly contributes in helping to establish the theoretical background on which the interpretations of the archaeological data are built within this study.

Chapter 2: Anthropological background and terminology

Investigating the social dimension of archaeological data is not a new endeavour in the field. Recently, an increasing interest had been developed concerning social identities, and more precisely on the material markers that allow researchers to reconstruct past identity patterns (Dever 1995; Jones 1997; Faust and Bunimovitz 2003; Desreumaux *et al.* 2011). These inquiries also often involve the idea of ethnicity, which nevertheless is only one of the potential identities that one individual can assume and express. Drawing from the wider debate on this subject involving several disciplines including but not limited to anthropology, archaeology and social history, archaeologists have for the most part also managed to reach some degree of agreement, at least on a theoretical level. The most important result was an increased awareness of the complexity behind these phenomena. In particular, there was a general consensus in recognising the danger of denoting a single feature (such as language or religion) to define an identity-marker, representing alone an entire identity-paradigm or an ethnic group. The great disparity between the way an individual or the group define themselves and the way they are defined and perceived by an outside perspective is possibly the main challenges to confront. Furthermore, there are different contexts (either social, geographical and chronological) that may have a direct impact on how and when multiple or different "identities" paradigms could be activated (Hall 1997: 30).

In a seminal paper by Macdonald (1993), he argues that the lack of direct mentions of nomadic and settled segments of the same tribe in some Greek and Safaitic epigraphical sources from the late Hellenistic and Roman periods in the Hauran does not imply the absence of such social groups. In fact, he explains that such a void in documentation is possible through the "general principle in onomastics that the closer you are to home the smaller group by which you need to identify yourself" (Macdonald 1993: 367). Since most of the inscriptions were found in a context where it was most likely the case that all of the individuals knew each other personally (if they did not *also* belong to the same tribe) and there was no need to refer to a "higher identity level", i.e. the "tribe", individuals (may have) preferred personal names or at least the single family-group name. This choice results in the inability to establish the connection between different groups, nomadic and settled, which in fact could belong to the same larger entity adhering to the same identity.

This is just an example of how different contexts might determine different identity patterns: generally speaking, different conditions (i.e. geographical, political, and economic) inevitably can have consequences for a given identity on both the individual and collective levels, determining changes in the identity paradigm too, especially in the features that determine belonging to the group and/or in the ways in which that group expresses their identity (Fabietti 2011: 129-130). Therefore, an "identity-making" process can be triggered by the political and economic stress that a group is exposed to. The challenge for the archaeologist lies in recognising and interpreting the marks that these "identity-making" processes might have left behind in the material culture.

For the purposes of the present work, these marks will be sought out in the built environment, where creation, modification, and maintenance of structures, and even settlements, might reflect also similar processes in the social identities of the local communities. This study will aim to identify three phenomena that appears to be related to the development of social identities patterns, having direct consequences on the built environment: tribalism (i.e. segmented societies), nomadism (pastoralism), and the emergence of an Arab identity. Though first it is necessary to unpack the complex meaning behind these terms and the history of their use in scholarly and popular literature. These terms are all quite well-known and are hotly debated in academia, though they are also widely discussed in non-academic contexts. Nevertheless, these terms are often oversimplified and turned into *clichés* that in themselves are deeply rooted in stereotypes and can be extremely misleading. This is especially the case in scholarly debates.

Moreover, "Western" imagery had often wrongly overlapped the meaning of all these three concepts. Tribalism and nomadism, in particular, often appear in contexts where they are used as the primary descriptors for the Arabs in general both before and after the arrival of Islam. Furthermore, especially in the recent times, the Arab culture is simultaneously subject to one further dangerous simplification: it is equated to Islamic culture, with the terms 'Arab' and 'Islamic' often being used interchangeably. These "essentialisations"[6] could be easily read as the consequences of the romantic idea of the "wonderful Orient". According to such an idea the Arabs "had to be" nomads, merchants, or brigands, where they were sometimes models of freedom and in upholding authenticity of customs, and at other times they were primitives who were constantly living outside the law and devoid of the basic rules of a 'civilized' lifestyle. In describing their way of life, the term "nomadic" is strongly linked to the term "tribal", which likewise expresses a wavering feeling of admiration for the respect of familiar bonds on the one hand, and on the other hand a sense of superiority in relation to a culture unable to create complex political and social structures like the ones in the modern Europe (Fabietti 2011: 24-47; Said 2003: 7). This situation is recurring in the case of "external" commentators that imposed labels on groups that they perceived as "strange" or distant. In the following sections, I aim to summarize and highlight the most relevant aspects of this debate while also pointing out and rejecting some of these stereotypes, especially in their use in the discourse of the built environment. Furthermore, I will offer some solutions where terminology may be adapted to more precisely discuss these topics in future studies.

2.1: Nomadism and pastoral societies

The concept of a total separation between settled and nomadic peoples is dominant in many historical reconstructions: the perennial contrast between these two societies has long been affirmed by ancient historical sources, and thereafter was often uncritically taken up by modern scholars. There are several examples (Pouillon 1996; Weiß 2007), although the renown 14th century historian Ibn Khaldun is particularly relevant here, not only because he separates these two "entities", but also suggests a sort of "evolutionary" pattern of civilisation[7]: the "nomadic" condition is a sort of natural and original state from which human civilisations starts, satisfying progressively more and more elaborate needs, acquiring richness, and adopting more stable ways of life at the same time (Ibn Khaldūn: 2); the very end of the process, that is described as a never-ending circle, is the development of a state or an empire (Ibn Khaldūn: 3).

[6] "To the 'essentialisation' of a complex reality, result from the application of these gatekeeping concepts (or 'theoretical metonymies', because of their power to qualify on their own a whole), follows indeed the concealment of these elements which are as much significant for the comprehension of the context". Speaking of "gatekeeping concepts" (Fabietti 2011: 43; translation by the author). Fabietti directly refers to Arjun Appadurai, who defines them as "concepts, that is, that seem to limit anthropological theorizing about the place in question, and that define the quintessential and dominant questions of interest in the region" (Appadurai 1986: 357).

[7] As well-explained by Rapoport (2004: 3-4) and Irwin (2018: 51-52), one should be careful in reading Ibn Khaldun for his usage of the terms normally translated as "Bedouin" (*badawi*) and "Arab" (*'arab* and *'urbān*, and *a'rāb*). *Badawi* in particular does not necessarily describe nomads, but also sedentary people living in the countryside. Similarly, the term "Arab" might assume different connotations according to the context (see Chapter 2.3).

Only from the second half of the 19th century, anthropologists started questioning the validity of this dichotomy. The debate was built upon two main approaches: one looked mainly at economic aspects, while the other focused more on socio-cultural characteristics (Salzman 1996a; Schlee 2005: 26-28; Szuchman 2009: 2). One of the first results that came from this debate was the recognition of the variety of groups that fell under the generic term of "nomadism", which eventually resulted in the abandonment of this term. The term "pastoral societies" started to replace it to describe all societies where pastoralism was predominant in their economic specialisation, without any reference to a specific social organisation of the community.

The economic approach was particularly successful in identifying various different societies, which had come into being due to diversified "economic adaptations" (Salzman 1996b; Schlee 2005; Szuchman 2009: 2). Clearly, despite the possible differences, all of these societies do share some features that are mainly connected to their most valuable resource: their livestock. The type of livestock a group has and the size of the population it can support generally depends on five main parameters: self-reproduction (meaning the possibility to increase the size of the herds), flexibility, mobility, vulnerability, and the availability and accessibility of pasture and water. These elements, and in particular the last two, determine a "fixed set of imperatives" the pastoral societies needs to meet to survive: "security" in order to protect the value of the herds, "nurture" in order to sustain the livestock, "mobility" in order to assure security and nurture, "distribution and exchange" both for maintaining security and nurture issues and to reach "social goals" (for instance, welfare, alliances, status, income), and lastly "production" of products for the purpose of consumption or trade. Ultimately, the different balance between these requirements result in different types of pastoralism (Salzman 1996b: 150-153; Schlee 2005: 17-30).

More specifically, Salzman (1996b: 151-153) identifies three aspects to classify pastoral societies. The first one is "the extent to which a population is specialized in pastoral production or diversified in a variety of economic activities" (Salzman 1996b: 151). Specialisation is usually connected to the kind – or kinds – of livestock that is reared: goats, bovines and camels are the most frequent examples, but are not the only options (i.e. reindeers or horses). Still, animals are not the only resource at hand and economic diversification is indeed an important factor to consider. Marx (2005), for instance, underlines the relevance of seasonal wage-work in urban centres as observed in some nomadic pastoral communities.

The second aspect is the destination of the product, meaning if the production is oriented either towards meeting the level of subsistence or for market exchange. In particular, Salzman (1996b: 152) sees a correspondence between a higher degree of specialisation and heavier involvement in trade with a more market-orientated economy. If a pastoral group develops a higher degree of specialisation, it will need to exchange its production in order to access a larger variety of items needed for its subsistence. On the contrary, if a group produces a wider variety of goods, it would be consequently less depended on trading (Marx 1996: 104). Still, and it is particularly true for the more mobile groups, it is hard to imagine a complete independence from the market, in light of their higher economic vulnerability (Schlee 2005: 28-30). Generally speaking, "the variation is not all or nothing, but is often a matter of degree, the extend of direct consumption versus the extend of market exchange" (Salzman 1996b: 152).

The third and last aspect is the degree of mobility, normally determined by the accessibility to water for the herds and by the need to protect it and guarantee its security. Salzman (1996b: 152-153) identifies two possible strategies: one requires that only part of a group follows the livestock, while the remaining and major part of the population continues to live in a fixed location; inversely, the "nomadic strategy" implies a contemporary movement of both the animals and the entire community. Intermediate strategies might also be possible as well as movement specifically for political reasons. In addition to this general schematisation, one addenda is important which was initially suggested by Schlee (2005: 22-23): instead of using recurrent labels like nomadism, semi-nomadism and sedentarism, it would be more representative to consider "respective modes of production, as well as the types of mobility (large-scale/small-scale, fixed seasonal routes/free, opportunistic choices of sites, frequent/rare, coordinated/on the basis of small groups, etc.) for each case". In summary the different pastoral societies and their adaptation strategies are only understandable in light of their context (in its most general meaning including environmental, socio-economic and political) and are in no way to be understood as static and ever-lasting entities. On the contrary, the tight connection between general context and adaptation strategies implies that if the first changes, the seconds might follow as well.

Economically-driven dynamics are not the only focus of the anthropological investigation. Other anthropologists had preferred a more cultural and social perspective, that in turn allowed to point out in particular two opposing social dynamics, one the counterpart of the other: sedentarisation and nomadisation. There can be several factors influencing each process, occurring simultaneously and fluctuating multiple times within a given period, and clearly their modalities highly depend on mobility-related features of the pastoral groups involved. Nonetheless, one major distinction should be made in accordance to their degree of "spontaneity": whether or not these processes are community-driven or imposed by other forces (environmental or anthropogenic). Once the main influence(s) upon which the group's mobility is determined, one can further look into these causes, which may be closely related to the historical or political context, to a change in the socioeconomic pattern, changing environmental conditions, or a combination of these factors (Finkelstein and Perevolotsky 1990: 70). Moreover, it is important to consider the wider context in order to truly understand social dynamics and phenomena and choices groups have done in relation to their habitation. For instance, one major factor that likely leads to sedentarisation processes is a desire for improved security: securing a region might bring together a series of positive consequences, such as the presence of new economic "agents" (new settlements or garrisons), the development of new trade routes, or major exploitation of resources. Such a positive political and economic situation might represent a good opportunity for some nomadic groups to choose less mobile living conditions. Similarly, the presence of religious centres or prosperous "neighbouring sedentary population[s]", and in general some positive "demographic processes" may also have stimulate such processes (Finkelstein and Perevolotsky 1990: 70-71). All these factors are more likely to trigger spontaneous sedentarisation. However, impositions by an external force, for instance a state, may also lead to sedentarisation. For instance, a larger entity like a state may be concerned with the difficulties in controlling the movement of more "nomadic" pastoral groups, and as a result these administrations might colonise determined marginal areas for "defensive" purposes. Changes in degree of nomadisation or sedentarisation may also be determined by environmental or climate changes: the availability of pasture and water is highly dependent on climate, especially in arid or semi-arid regions. For instance, dryer periods might force some groups to seek for new pastures or water in other regions, expanding or completely changing their normal mobility patterns. On the other hand, wetter periods could lead to more confined mobility strategies or limited only to a section of a community, if not to a complete sedentarisation, having more resources at hand in a restricted geographical area.

The integration of more nomadic groups within an already settled community, especially as a consequence of sedentarisation processes, is a further consideration in this discussion, which can also be seen from both economic and cultural perspectives (Szuchman 2009: 3). The economic connection between permanently settled groups and pastoral nomads, which is evident through shared commercial and productive activities, helps in creating social ties and reinforcing respective identities of both settled and nomadic groups, that begin to develop well before a complete sedentarisation of the latter (Marx 2005: 4-5). The most problematic issue to sort out is what precise changes have occurred in respect to these groups identities once this has process takes place. Economic relationships might help in establishing social and cultural bonds, yet at the same time can give way to "ethnogenesis" processes, establishing or reasserting criteria to better distinguish a "We" and "The Others". In this circumstance, the so-called "tribal" identity might play a central role in defining personal or group identities (Schlee 2005: 34-49; Szuchman 2009: 3).

This distinction between economic and social bonds, however, tends to remain more at a more theoretical level, since they are deeply intertwined and difficult to clearly distinguish on the practical level (Szuchman 2009: 3). This is due to the extremely variable nature and frequency of ties between nomadic pastoralist groups and settled communities, as a whole or within their respective sub-groups (Marx 1996: 111; Kressel 1996: 137). To help model this complex social arrangement, Kressel suggests the idea of a "triadic relationship" between nomads, rural communities, and urban communities, all connected within a single socio-economic system, where bonds can change over time but are rarely broken. The existing interconnection between the components of 'triad' is ultimately the most important features the triadic system, making almost impossible to isolate nomads, rural communities, and urban communities from one another and consider each of them on their own (Kressel 1996: 137; Fabietti 2011: 38).

From a more archaeological perspective, these anthropological theories about nomadic pastoralism have some direct consequences for the interpretation of the built environment. The first directly involves the organisation of settlements: processes of sedentarisation could have determined specific forms of spatial patterns, likely occurring within already existing towns or villages (Avni 2014: 281-283). A possible way to determine the impact of sedentarisation on settlements might be to look at the organisation of seasonal settlements or tent camps, as attested by ethnographic studies (Cribb 1991; Saidel 2009). Consequently, this leads to the issue of determining if a settlement can be identified as a development of a seasonal village, and if this process can be connected to the process of gradual sedentarisation. In this case it is necessary to be aware of the variety of pastoral groups (with different degrees of mobility) and consider how each group potentially impacted the sedentarisation process. Combining archaeological and historical sources together with ethnographic examples offers important clues and helps us recognise and reconstruct these diverse social dynamics, though this should be closely considered alongside their respective regional political and economic backgrounds (see Part 3).

2.2: Tribalism and segmented societies

The diversity of pastoral societies lies also in diverse social structures, as mentioned briefly above. For the Near East in general, and the Arab world more specifically, the idea of "nomadism" was tightly connected with "tribalism", which is highly problematic. Using this label, anthropologists normally refer to societies "formed by several lineages which consider themselves as descending from a common ancestor" (Fabietti 2010: 349). The debate around tribalism underwent a similar development as with nomadism: starting with Morgan's study on the Iroquois in 1851, the "primitivist nature" of the so-called tribal society was radically considered the opposite to the "advanced" European society, a dichotomy that characterised nearly all anthropological studies from the 19th and early 20th centuries. As a consequence, the word "tribal" started being used as synonym for "primitive" not only in common vernacular, but also in scientific discourse, starting to work as a "conceptual instrument, according to which the ethnographical data could be ordered for comparisons and classifications of the 'primitive' social organisation" (Fabietti 2011: 86). All societies perceived to be different from European societies were collectively classified as more primitive "others", without considering the existing differences between them. The most common aspect that attracted the attention of early scholars is the idea of a total absence of centralised political institutions, normally seen as a mark for advanced social formations. On the contrary, "tribal societies" were seen as victim to the "continuous processes of fusion and fission of the nuclear components" (Fabietti 2011: 87), making it impossible for them to develop a proper modern state. Contextually, the recognition of these "nuclear components" or "segments" led to the formulation of the so-called "segmented society" models in the late 19th century (Meeker 2005: 81-82; Bonte 2006; Fabietti 2011: 87). In particular, Durkheim's idea of the "links of a chain" (Durkheim 1893) was connected with Morgan's theory of the "descent groups" (Morgan 1871). This gave way to a long tradition in British anthropology in the early 20th century, which saw the affiliation of an individual to a group determined by the linage.

Only during second half of the 20th century, scholars became increasingly aware of the limits of these interpretations and of the plurality of social structures formerly described as "tribal": strong regional and local differences came to light with new data and demanded a shift in methodology when studying these groups (Fabietti 2011: 85-86). Along these lines, Evans-Pritchard's work "The Nuers" (Evans-Pritchard 1940) was a trail-blazing study, showing how the groups forming the segmented systems were not the simple result of akin relations, but also from concrete political and territorial segments. Two elements are particularly relevant from his theory for this study: the "structural relativity" and the dynamicity of the segmented societies. These "fusion and fission" processes are ultimately seen as the basis of the social cohesion, since during these oppositions is "affirmed the relative identity of the groups and of the entire Nuer society" (Bonte 2006: 731). Despite being widely accepted amongst anthropologists, Evans-Pritchard's theory underwent some important critiques (Meeker 2005: 85-92), in particular regarding the significance of the "fusion/fission" of lineages as expressed by genealogies, which could be seen not as a trustworthy reflection of social ties rather as purely ideological constructions (Peters 1967). Furthermore the exclusive role of the lineage, through patterns of solidarity/opposition, is essential in determining the dynamics in the system ("Allegiance theory", Dumont 1971).

To this effort, the discourse on the genealogies of these societies became an important field of investigation. Starting with Wolf's observation that lineage in Mecca preceding Islam could have been fictitious, expressing ideological patterns (Wolf 1951: 335), scholars began to appreciate the social and political implications of genealogies, for both the pre-Islamic and Islamic periods. From being considered as true lists and descriptions of the different linages of tribes populating the Arabian Peninsula, they were finally understood as tool to trace and consolidate the political equilibria between the different "segments" of a group, not only in its nomadic component, but also in respect to settled communities. Ultimately, a genealogy is the expression of the inner fluidity and dynamic of segmented societies. Their constant manipulation on various levels reflects the changes in the relationships between them; the fact that the knowledge of genealogy is prevalently transmitted orally can be seen as a consequence of its dynamicity and the need for flexibility, allowing simple modification in the face of new political circumstances (Meeker 2005: 88; Szuchman 2009: 4; Fabietti 2011: 93). As a result of the political nature of this knowledge it is clear that the genealogical expertise is not widely dispersed among all individuals within a group, but normally in trust of the *shuiukh* (the "tribal" chiefs). Manipulating a genealogy may also imply the adding or removing groups and requires a deep awareness of all the political connections and of the possible consequences (Fabietti 2011: 98).

This dynamic dimension can be also understood in relation to territoriality, defined as the "product of particular social relations and also a determinant of social relations" (Meir 1996: 187-188). Salzman (1978a: 68) defines segmented systems as an ideological framework that could preserve the social link between the different components of a society even when faced with radical changes of different natures, either internally or externally. This applies both to 'more settled' and nomadic sections of a group, since "tribal groups have both nomadic and settled sections which maintain ties with the tribal framework and between which there is a two-way flow of population" (Salzman 1978b: 548). The context is again the most important factor that simultaneously determines and requires the dynamicity of a group, and this may be reflected genealogically as well as territoriality[8]: the appearance, mixing or disappearing of different segments of a group both reflects a change in the social ties (either real or fictitious) *and* in their control and in the modalities in managing specific areas. The "tribal" framework then offers the ideological tool to keep the group united and furthermore support the social structure during these territorial "fluctuations".

This dynamicity in managing territoriality is also one of the reasons that ultimately leads to the juxtaposition between segmented societies and state entities in scholarly discourse, even though it is mainly reduced in terms of a "nomad-state" dualism (Szuchman 2009: 4). This however has further consequences for the wider debate surrounding tribes and state formation in the Middle East. Under a more ideological point of view, the "tribal" social and political organisation is assumed to be "primitive" due to its almost egalitarian structure, with the inner dynamics that prevent an unbalanced centralisation of power. This type of organisation had been long contrasted with the "more civilised" stratified urban society. Two main theories have developed on how these supposedly contrasting social entities came into being: the first seeks to understand how and why tribal formations could develop into states through an "evolutionary" perspective, and features Ibn Khaldun's work as a key reference (Lapidus 1990), while the second underlines the "coexistence" of the two entities and analyses the paradigms ruling their relationship (Tapper 1990). Franz (2005: 61-62) offers an interesting alternative point of view to the previous two theories and suggests that segmented societies could develop or "activate" statehood structures under particular circumstances, such as defending against external aggressors or in accessing resources. The accessibility to resources, often translated into a group's participation in the market, is an element where differences between pastoral segmented groups and states are most apparent (Franz 2005: 56-61).

With that in mind, the mutual relationship between "the State" and "tribes", in particular if composed by nomadic pastoralists, is much more complex than a simple, continuous opposition between these two entities. Several examples of empires or kingdoms relied on segmented societies (either in their mobile or sedentarised sections) to control rural areas in their dominion and also identify specific tribal

[8] "Territoriality assumes several functions. Briefly, it provides domination over an area so that the individual has some freedom of behavioural choice, and control over access to it [...] as mechanism to protect privacy, or to determine the distinction between the person who controls it and the others [...] may also reflect one's social status and social relations [...]. Finally, it provides spatial permanence with psychological benefits" (Meir 1996: 188).

chiefs that would help them achieve such a goal in new territories. This is particularly relevant for the region and the period considered in the present work, in which the progressively increased importance of the Jafnids within the Byzantine imperial system might have played a crucial role in the development of the rural built environment and social dynamics (see Part 3). Their role may have been crucial to both the settlement patterns in this region, meaning the establishment or re-occupation of towns and villages in the countryside, and to the organisation of these sites themselves.

This brings us to another element of primary importance while studying settlements: the sharing of the same social "tribal" structures by several settled and pastoral communities. Considering that it is most likely for similarly structured societies to develop similar ways of organising the built environment, there are cases in which it can be extremely problematic to establish whether the spatial organisation is determined by a permanently settled community or by semi-nomadic pastoral groups (Whitcomb 2009: 243-244). Archaeologically, the potential of recognising sedentarisation processes is a major challenge, at least when relying exclusively on the urbanistic and architectural features of a settlement. This is not only due to the complex nature of the aforementioned sedentarisation processes, but also in light of the frequent social and economic ties linking settled and nomadic pastoralist groups (assuming that they are indeed distinguishable, and not different sections of the same society). Here, the concept of "spatial interpenetration" offered by Salzman is helpful (Salzman 1978a: 67, note 191): cities and permanently settled villages often functioned as magnets for pastoral populations, that – whether forced to do so or spontaneously – tended to settle near or within already existing centres. This tendency may have accelerated the rate of sedentarisation itself and triggered a fusion between the two groups (settled and more mobile pastoral) that shared the same social structure. Given this context, distinguishing the so-called "extended families" of the originally settled population from those of the newly sedentarised pastoral community in the built environment is indeed an extremely difficult task.

An aspect surely connected to sedentarisation processes, segmented societies, and the built environment is the correspondence between the different social units and organisational patterns, which should be carefully considered on all the possible different levels of analysis of a settlement (i.e. the whole settlement, the block/quarter, and the house). The absence of a precise terminology, or ambiguous application of certain terms, describing the different social segments from written sources makes this aspect more difficult to explore. However, to ameliorate this gap in information, I propose the adoption of a flexible terminology to help describe these social segments as well as using a multi-scale perspective in the description of the sites to glean insights on how they potentially impacted the built environment. Moreover, different groups and social formations should be considered as a continuum, rather than clearly distinguishable social units. As argued by Antoun in one study on modern Arab villages in Transjordan (specifically Kufr al-Ma, near Ajlun, in the Late Ottoman-Early modern period), "flexibility and diversity [are] due to the fact that patrilineal descent groups reflect a continuum, structurally, functionally, and ethically rather than a set of discrete and easily identifiable social units" (Antoun 1972: 144). Therefore, since no clear borders are set between the units, the use of vague terminology is inevitable. Nonetheless, Antoun differentiates two levels, the ideological and the practical, on which the subdivision of the social units may be better understood.

The ideological level offers the frame the entire social structure is based on and is ultimately expressed by the patrilineal kinship – real or presumed. This sphere is formed by the *sib* – the largest unit that was generally defined as a "dispersed group of patrilineal kinsmen" and included 9 to 11 generations – and its subunits, the *subsib* (Antoun 1972: 44ff.). Belonging to the same *sib* offered the basis for a communal identity among scattered groups, shaping extra-settlement political ties. As mentioned, the genealogies are rarely an "accurate statement of the consanguine relationships" and therefore the *sib* was the level on which manipulations could take place in order to accommodate changes in the political and social context. The *subsib* are not clearly described but could be seen as the purely ideological equivalent of the clan and the lineage.

The "clan" is probably the interface between the ideological level and the practical level (Antoun 1972: 81-91). In a practical sense, a clan is constituted by a series of lineages that can be ultimately described as the daily application of the ideological and theoretical paradigms expressed through patrilineal ties. If the *sib* and the *subsibs* are ultimately only the theoretical framework on which the social relationships among different groups are understood and established, the clan can be considered the largest proper "unit of political administration", operating on a settlement level. The patrilineal genealogy, and its social implications, again acts as a sort of general "law" for the internal social dynamics and daily life, both economically and politically. It also establishes a degree of hierarchy among the different lines of descent, differentiating which are more "notable". The political power and the size of the group are ultimately the two most important distinguishing factors for the clans, deeply influencing the different patterns of social relationships between the groups, expressed in particular through visiting patterns. Generally speaking, "while patrilineality is the dominant ideology of intergroup and intragroup relations as well as the foundation for critical functions (e.g. authority in the household, inheritance, blood-money compensation), this ideology is adjusted to accommodate groups or individuals who would not qualify for membership or the exertion of authority if that principle were applied strictly" (Antoun 1972: 142). Here, a major role is played by the matrilateral relationships, acting within the patrilineal ties but across the several units. It is ultimately the aspect ruling the praxis in the social and political life of a community,[9] for instance establishing and keeping ties across different lineages and determining alliances to help when social crises arise (Antoun 1972: 77).

Within this practical level, terms as "lineage", *luzum*, and "household" actually correspond to specific socio-economic and even political units. The lineage includes all of the descendants of a single ancestor and becomes identified by that ancestor's nickname (Antoun 1972: 69-77). This lineage includes brothers, paternal uncles and paternal cousins, plus their wives and children, up to a maximum of five generations. Even if this unit is involved in additional aspects of social relations, such as marriage and visiting, their role in the political sphere is the predominant characteristic that distinguishes them in respect to others in the village. In fact, a lineage represents "the minimal social unit seeking political representation within the village" and it is exactly its "political function rather than its structural attributes that is critical for its identity" (Antoun 1972: 77).

The lineage is the highest level where the importance of the matrilineal section of the social ties is clearly manifested. Marriage and social control are in fact the most important tool to prevent the community from desegregating and to create a communal sense of belonging to a settlement. This brings together one further aspect that in the present study has a direct consequence for the built environment on a settlement. According to Antoun, belonging to a lineage or a clan plays a crucial role in determining the settlements' layout, especially to accommodate the social control and marriage patterns established in the village. Members of the same lineage are normally neighbours as well, and lineages of a single clan tend to stay close one to the other. Likewise, the houses belonging to the lineages of the same clan are normally clustered, while independent lineages, on the other hand, appear

[9] "Villagers formally recognized one general principle, the patrilineal descent, and explained their actions on its basis, while operating on the basis of another, matrilaterality" (Antoun 1972: 87).

to be located on the outskirts of the settled area. These same patterns are also present in the land possession (Antoun 1972: 91).

The largest lineages are divided into *luzum*, that can be defined as a "close consultation group [either economic, political or social] [...] smaller than a lineage but larger than a household" (Antoun 1972: 58). The definition of a *luzum* is quite problematic and also the size of this unit can vary considerably (Antoun 1972: 58-69): its members live in the same village, though do not share the same house, and differs from the lineage in that it may also include maternal relatives and affines. Functionally, it appears to overlap to a certain extent with the lineage, though operating on a reduced scale, namely within the single settlement where the *luzum*'s members are living.

Finally, the "household" is defined by Antoun as the "smallest permanent economic unit in the village", symbolised by the single purse of the eldest male member (Antoun 1972: 49-69). It is more a consumption than a production unit, since an occupational specialisation is often encountered among its members. The guiding rule for the household is the so-called "organic solidarity", according to which each member of the group contributes to maintaining the household. These social units normally consist of one to twenty people and are mostly described as "nuclear" rather than "extended" families: they include a man with his wife or wives, their children and son's wives and grandchildren. Brothers or other patrilineal relatives normally form their own households. Furthermore, Antoun noticed that (especially for the Late Antique/Early Islamic, though not as a definitive rule) there was a correlation between the household and dwelling (Antoun 1972: 58). The size and form of the house, for instance, could be particularly informative to this extent. A one-room house can be "almost inevitably occupied by a nuclear family"; a three-room house may reflect an extended family; and finally, a house with five or six rooms and an inner courtyard can probably connected to a polygynous family (Antoun 1972: 58). While keeping this in mind, a direct equation between the house and household should be avoided as it can lead to erroneous interpretations; for example the "coresidence of fathers and sons, like the coresidence of brothers and sisters is not indicative of the presence of a single household" (Antoun 1972: 55), and similarly one can find cases of renting living spaces to people that do not share any social ties with the original owner of the dwelling. While if this could be a praxis more common in urban contexts, as shown by some documents from the Haram as-Sharif (Little 1984), we should also consider these possible organisations in households in rural villages as well.

In summary, "segmented societies", or "tribalism" as often found in the scholarly historical and archaeological productions, is extremely challenging to conceptualize and translate into the built environment. Nevertheless, carefully considering the multiple dimensions of "segmented societies" makes up an important component of the theoretical framework of the present study. The possible correspondence between built environment and social structures is one of the most crucial aspects of this framework (Antoun 1972: 91; Whitcomb 2009: 243). Another key theme, proximity, will be explored to help understand the organisational pattern in settlements through a social perspective and will play a central role in the description and interpretation of the case studies.[10]

[10] "The critical importance of propinquity for village life [...] is obscured by the fact that the tie of the particular individual to his neighbourhood, to his village, or to his district is very often expressed in an idiom of kinship or descent" (Antoun 1972: 143).

2.3: "Arabs": the challenge of distinguishing between external labels and self-definitions

One final word is necessary to clarify the choice for the term "Arab settlements" within the title of this work. While the final explanation of this choice will be reserved for the last part of the volume (see Part 3), some key terms should be introduced in order to avoid misunderstandings further on in the argumentation.

At the outset it is difficult to find one single definition of the term "Arab", since it is highly dependent on the context in which the term appears. There are several reasons why this is the case. The most important relates to the evolution of the meaning this term has undergone throughout its history: different temporal, political or geographical contexts created a constantly changing and increasing "stratification" of possible meanings of the term (Lewis 1993: 2). For a long time, a strict linguistic definition prevailed, identifying the "Arabs" as those who were "Arabic speaking". Nonetheless, the term often appears to describe a more social aspect, leading to its association with "nomads". This application probably had the most consequences for historical sources, where the peoples defined as Arabs are most of the times identified with "Bedouins". With the spread of Islam, the religious aspect progressively assumed a central role in the identity value of the term "Arab". This process to a certain extent appears to be connected to a parallel definition of a proper *ethnicon*, at least during the early stages of Islamic history. This is indeed the first, and maybe unique, moment during which some "criteria" of self-determination are traceable, since there are no other (confidently) identifiable cases earlier than the 6th century AD (Macdonald 2009b: 319; Webb 2016). Only when the Caliphate embraced a wider number of cultures and ethnic groups, the need for a more precise self-determination emerged as did the definition and respective application of "Arabs" to refer to the ruling "Arab" élite. Consequently, a precise set of criteria developed in order to distinguish between "we" and "the others", justifying a particular social and political status. The tool that most helped reinforce this status was modifying the genealogies according to the political needs (see Chapter 2.2). Interestingly, the religious aspect did not directly trigger the differentiation between "Arabs" and other groups, but on the contrary the criteria were chosen within the tradition of the *Jāhiliyya*, the time before the Islamic revelation. Given this brief overview on some problematic issues related to the definition and use of the term "Arab", it is important to focus on some of the key stages of its evolution, stressing some chronological, social and political aspects.[11]

The lack of precise and reliable historical sources or epigraphical finds makes it difficult to unpack the exact meaning and application of the term "Arab" prior to the 6th century AD.[12] After its first proper appearance of the term is in an Assyrian inscription dated to 853 BC[13], ancient Greek and Latin sources

[11] Without going into too much detail into the origin of the word, its etymology is debated as well. Lewis (1993: 2-3), among others (Retsö 2003: 107-108; Webb 2016: 24-26), mentions three plausible theories, all connected with the concept of nomadism: the Hebrew *'Arabhā*, "dark land or steppe land"; the Hebrew *'Erebh*, "mixed and hence unorganized", in that opposing the "organised" way of life of the settled communities to the not organised one of the nomads; the root *'Ābhar*, "to move or pass".

[12] For a complete overview on the mentions of "Arabs" before the Greek and Roman periods, several works are available, among which include: Lapidus 1988, Lewis 1993, Hoyland 2001, Retsö 2003, Fisher 2015, Webb 2016.

[13] The inscription recounts the defeat of the Assyrian Army lead by Shalmaneser III of Kurkh at the Qarqar battle, against a confederation of rebel princes (Lewis 1993: 3; Macdonald 2009b: 285, note 49). Other mentions of "*aribi*", "*arabu*" and "*urbi*" appear in other Assyrian and Babylonian documents until the 6th century BC, normally recording payment of tributes or punitive expeditions against them at least. Similar references can be found in the Old Testament (II Chronicles 9:13-14; Isaiah 13:19-20; Jeremiah 3:2, 25:24; Ezekiel 27:21). Around 530 BC, the term "*Arabaya*" first appears in the Babylonian sources (Lewis 1993: 4).

- including the ethnographic ones - do not offer much useful information, and use the term ambiguously. Herodotus was the first renown author to spread the use of the term "*Arabia*" - already attested in some previous sources[14] - and "Arab", where he referred to the entire peninsula and its respective inhabitants.

Nonetheless, the use of the toponym is actually more problematic than it may seems at first, also in later Latin sources, since it is difficult even to understand which geographical area is really meant under such label. As Macdonald pointed out (2009a), the term *'arab*, already in its first appearances, relates more to a group of people than associated with a specific geographic area itself, initiating a process that defines *Arabia* as every region where "Arabians" were found, not taking into account how individual populations actually defined themselves. Unsurprisingly, this identification is common mainly in the written sources, most importantly - after Herodotus - in Strabo and Diodorus Siculus.[15] These authors, like their predecessor, had surely borrowed the term from sources they referred to, without verifying their accuracy. Therefore, at the beginning of the Hellenistic period the term "Arabias" may actually describe six different regions, and then a total of eight in the following century (2nd century BC): eastern Egypt, Sinai and southern Palestine; Anti-Lebanon; northern Syria; central Mesopotamia; southern Mesopotamia and the "head of the Persian Gulf"; the Arabic peninsula. Afterwards Ammon and part of the Gilead, in the northern Transjordan, and the southern Transjordan would also be considered as a part of "Arabia" (Macdonald 2009a: 15-17).

To add to this geographical uncertainty surrounding this term, frequent fantastic and exotic character descriptions of "Arabs" coupled together with a recurring taste for "strange" behaviours, costumes and animals invalidates most of the information offered by such accounts. These descriptions were likely intended more as a way to strike the reader's imagination than to offer an accurate depiction, let alone be considered "scientific" (Macdonald 2009a: 21-22). The use of the term "Arab" to describe the nomads is one of the most recurring *topoi* in the sources, always offering similar characterisations, in which the "Arabs nomads" are depicted as tent dwellers and shepherds. These commonplaces are often associated with the description of the Arabs as nomad marauders - or outlaws in general - and camel-breeders, with an almost proverbial love for freedom and pride. As Macdonald underlines (2009b: 279), these "characterisations" were limited to the perception the authors and their societies had of the peoples that they were describing, rather than being a realistic picture of the context in question. This is the consequence of the phenomenon of "essentialisation", which reduces a more complex reality into dualistic confrontations between "sedentary vs nomad, civilization vs barbarism, the rule of law vs lawlessness, etc., in positive as well as negative terms" (Macdonald 2009b: 279; see also Honigman 2002: 44-45).

Moreover, the sources are layered with another a certain terminological ambiguity, especially from the 3rd century AD onwards, when also the term *Sarakenoi* ("Saracens") came into use and was spread. It first appeared in 2nd century AD Roman administrative documents (Webb 2016: 45) and partially overlaps the meaning of Arabs, in particular in its earlier use, and most interestingly seems to substitute

[14] The most ancient use of the adjective "Arab" in the Greek literature is in Aeschylus - end of the 6th / beginning of the 5th century BC - in his work *Prometheus*.

[15] Diodorus Siculus - writing in the 1st century BC - is one of the most interesting sources, since he dedicated a whole section of the second book of his *Bibl. Hist.* to *Arabia* and *Arabia Felix* (Chapters 48 - 54), though he borrowed materials written by previous authors, for instance Posidonius of Apamea of Syria and Strabo. He is also the first to unequivocally mention the Nabateans (perhaps mentioned already in *Genesis* 25:13) as "Arabs" (Diodorus Siculus, *Bibl. Hist.*, II 48, 1-2).

the term "Arab" in the historical source.[16] In fact, there is no mention of "Arabs" more or less until the 5th – 6th centuries AD, not even in inscriptions. Nonetheless, the appearance and later predominant usage of the label *Sarakenoi* may not necessarily mean the disappearance of "Arabs", but could testify a change in the external perception of the Arabs. According to Millar (1998), if "Saracen" appears to clearly identify a nomad, the term *Arabes* or *Arabioi* in Byzantine sources may refer more frequently to the settled population of the province of Arabia. A possible explanation for this shift was an increased integration of the Arab communities into the Empire. The Byzantine period likely oversaw a considerable increase in contacts with these communities, especially through a more consistent employment of members of pastoral groups or "tribes" as *foederati*, a strategy already started by the Romans. The fortunes and political role played by the Jafnids from the 5th century onwards are a good example of this process (see Chapter 7). At the same time, the integration was most likely also encouraged by the adhesion to the Christian religion and the spread of the monastic movements further into the more peripherical regions of the Eastern provinces. The additional contact with the monks or hermits could have offered these groups a possible "device" to facilitate their integration in the Empire, also on a social level (Millar 1998: 313). Interestingly though, 5th century Christian sources like Origen and Eusebius never mention Arabs, while they refer to the Saracens as descendants of Abraham (Millar 1998: 301-311).

Another term that has consequences for the term "Arabs" is "Ishmaelites". This term initially was used to describe nomadic groups, and later incorporated all other stereotypes listed above for the Arabs.[17] Far from being a simple philological cavil, the appearance of the term "Ishmaelites", also in some way correlated to the Arabs, could reflect an important phenomenon: the development of a first original myths initiated by these groups themselves. Specifically, it could be seen as a form of manipulation of the genealogies and a creation of descent stories as frequently described by anthropologists also for contemporary "tribal" societies, that resort to such "legerdemain" in order to define, consolidate and reinforce social and political ties (see Chapter 2.2).

All the aforementioned examples are definitions imposed by external observers and not entirely respectful to the true identity of the groups described. As mentioned, only in the 7th century AD with the establishment of the Umayyad and Abbasid Caliphates, it is possible to detect clear indications on how the groups later described as "Arabs" defined themselves, leaving only a few remarks possible for the previous periods. Some studies have tried to analyse epigraphical finds, written in different languages and alphabets and coming from several parts of the Mediterranean, where the word "Arab" appears.[18] This kind of source is generally considered a direct expression of its client, and is therefore less filtered by stereotypes as in the case of historical texts. Nonetheless, it is still extremely difficult to identify definitive examples of proper ethnic self-designation.

---

[16] One possible etymology is connected with the Arab *šarqiye*, "Those who migrate each year to the inner desert" (Macdonald 2009a: 20). Macdonald also underlines that this word may be found in the Ancient North Arabian, where *'s2rq* means "to migrate to the inner desert [regardless of the direction]" (Macdonald 2009a: note 120). According to another hypothesis (Lewis 1993: 4), the term indicated originally a specific nomadic tribe living in the Sinai. Afterwards, "Saracen" would have been extended to describe the nomads in general in Greek, Latin, as well as in Talmudic sources, without any particular reference to a single tribe. With the spread of Islam, the word underwent one further evolution, starting to identify all the Muslims.

[17] In his rewriting of the Old Testament, Josephus mentioned too some Arabs groups as descendant of Abraham. In this circumstance, he referred probably to the Nabatean, seen as the "principal 'Arab' people" (Millar 1998: 301).

[18] Some notable examples include: Hoyland 2007, Macdonald 2009b, Fisher 2011, Fisher 2015, and Webb 2016.

Given the actually available evidence, it is particularly interesting to underscore that on sixteen well identified examples, only three come from outside Egypt as noted by Macdonald (2009b: 285-290). In addition to these cases, thirteen more are still unsure. The Egyptian samples seem to confirm the general tendency of an ambiguous and diversified significance for the term "Arab", at least in Ptolemaic and Roman Egypt. Surprisingly though, clear references to nomadic style of life are quite uncommon.[19] Only some cases document geographical references, and not even a clear professional connotation seems distinguishable, as suggested by other sources of the period as well.[20]

Among the other non-Egyptian evidence[21], an epitaph found in Namara in southern Syria is particularly relevant (Figure 2).[22] This inscription is in honour of Imru' al-Qays, apparently chief of the Lakhmid

©bpk / RMN - Grand Palais / Les frères Chuzeville)

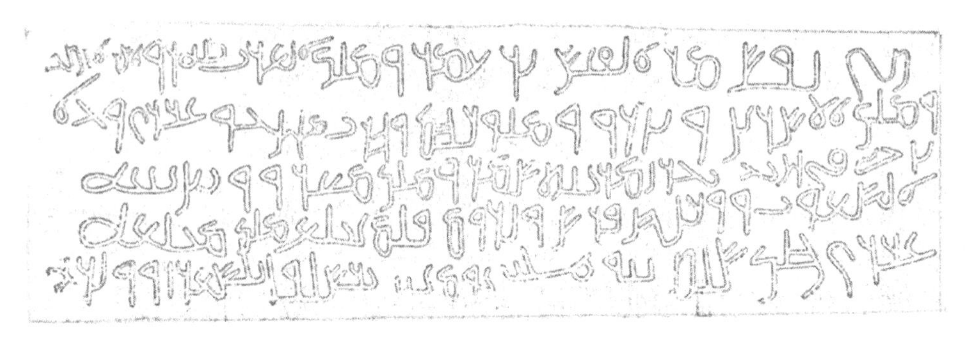

Figure 2 - Imru' al-Qays Inscription from Namara, Syria (©bpk / RMN - Grand Palais / Les frères Chuzeville) with transcription (by Dussaud and Macler 1903: 314)

---

[19] A direct equalisation between "nomads" and "Arabs" is suggested by Gawlikowski (1995), which at the same time denies any possible ethnic characterisation. Macdonald strongly opposes this theory, since it does not take in consideration the changes that the concept underwent though time and also takes the step from the wrong assumption that the nomadic social structure is necessarily different from the one of the settled communities.

[20] Nonetheless, Honigman (2002) singles out a recurring link between "Arabs" and watch, patrolling or military professions in general, mainly in connection with the desert. According to Macdonald though, the professional spectrum described by the mentions of "Arabs" is not so clearly characterised but shows a much broader variability (2009b: 283-285). He suggests though a possible fiscal significance, connected with a particular tax-status. Nonetheless, clear references are still missing (Macdonald 2009b: 292).

[21] The other two examples come from the Greek island of Thasos, dated to the third century AD, and from al-'Uqlah, in Yemen, probably contemporary to the Greek one but dated still uncertainly.

[22] The inscription had been the focus of a huge number of articles and mentioned in several volumes, among which: Lewis 1993: 4-5; Bowersock 1994: 138-147; Ball 2000: 96-98; Retsö 2003: 467-480; Sartre 2005: 361-363; Shahîd 2006: 31-53; Macdonald 2009b: 290; Fisher 2011: 143-144.

tribe, dated to AD 328 and written in *Old Arabic* using the Nabatean alphabet.[23] He is mentioned as "the King of all the Arabs" (*mlk ' l- 'rb kl-h*), even if it is not clear whether he defined himself or was defined as such by his respective social group. Known also through other sources,[24] Imru' al-Qays is indeed a fascinating character. Born in al-Ḥīra and initially reported to be under Persian influence, he became a confederate of the Byzantine Empire, which is the reason why he was buried in Namara, in the province of *Arabia*. Although the epitaph most likely exaggerates several aspects of his status and his achievements, it is of primary importance the political recognition he appears to have gained officially by the Byzantine imperial power. Nonetheless, there are several problematic issues in the inscription's interpretation, especially in terms of its political implications. In fact, this epitaph could represent the first evidence for the phenomenon of integrating a local chief into the Byzantine administrative framework that would further develop in the following centuries, particularly with the Jafnids (see Chapter 7).

The significance of the term "Arab" is also quite unclear, since the available information do not allow to distinguish between an ethnic, geographic or social meaning. At the same time, however, it does not seem to refer univocally to "nomads" and diverges from the "classical" social significance of the term. Ultimately, it is impossible to state either if the "Arabs" that Imru' al-Qays ruled over were a distinguished ethnic group, either internally or externally defined, or if he himself saw himself as belonging to it.

The discussion around the inscription and its significance in the emergence of the "Arab" identity continues to be important in the wider debate on the "self-determination" of subsequent "Arab" groups. Macdonald does not list it among his examples of "self-determination" (Macdonald 2009b: 290), while other scholars like Hoyland and Fisher consider it an important sign of a growing Arab self-consciousness. To them, whether Imru' al-Qays considered himself an Arab or not, does not change the fact that a group of people decided to erect a monument written in the Arabic language (Fisher 2011: 144).

Some other applications could offer a slightly more precise valence of the term "Arab", and thence of "Arabia", namely the term's use in bureaucratic/administrative contexts. A good example is one of the three "non-Egyptian" inscriptions mentioned above, found in the Greek island of Thasos[25], despite some debate surrounding its dating and the inclusion of *Septimia Canotha* (the town of the inscription's client) into *Arabia* from *Syria* (Bowersock 1994: 115-116; Macdonald 2009b: 302-303). Clearly, this administrative application of the term follows the creation of the Roman province of Arabia in AD 106, and therefore

[23] The lintel is conserved at the Louvre Museum in Paris (Inv. AO 4083). One of the possible transcriptions of the inscription reads: "This is the monument of Imru' l-Qays b. ' Amr, king [*mlk*] of all the Arabs; who sent his troops to Thaj and ruled both sections of al-Azd, and Nizar, and their kings; and chastised Madhhig, so that he successfully smote, in the irrigated land of Najran, the realm of Shammar; and ruled Ma'add; and handed over to his sons the settled communities, when he had been given authority over the latter on behalf of Persia and of Rome. And no king had matched his achievement up to the time when he died, in prosperity, in the year 223, the 7th day of Kislul" (Fisher 2011: 77)

[24] Among others, Hishām al-Kalbī. For a complete examination of the sources, see: Retsö 2003: 473-480 and Shahîd 2006: 31-53.

[25] *IG* XII 8, 528: Ῥουφε[ῖ]νος Γερμανοῦ οἰωνοσκόπος Ἄραψ πόλεως Σεπτιμίας Κάνωθα Γερμνῷ τῷ ὑῷ ζήσαντι ἔτη κβ′ μνήμης χάριν. According to Macdonald, Rufinus's statement of being "Arab" could depend on his extraneousness from the context the inscription was found in, where the precise location of the city of *Septimia Canotha* might be not clear, needing an "administrative" indication. However, a more identity related interpretation is not entirely excluded (Macdonald 2009b: 302-303).

it falls again under the type of labels imposed upon this area by "outsiders". It is possible that the appellative of "Arab" for this new province in some way mirrored the perception that the Romans had about the ethnic belonging of the inhabitants, or at least their majority. However, this application of the term likely did not take the way the groups actually identify themselves into too much account.

An inscription from Palmyra, dated to AD 132, shows how this administrative label could have possibly been felt as imposed by the local population.[26] The client describes himself through three different perspectives: the first is political, as "Nabatean"; the second is social, as "the Rwḥite", referring to his tribal belonging; and the last one is military, "mounted soldier at Ḥirtā in the 'Ānā camp". For the present argumentation, the most important aspect of this statement is that the individual described himself as being "Nabatean", since the province of Arabia was just created and the former Nabatean state was completely integrated into the new Roman Empire. This detail is also interesting if compared with the Namara inscription, which is dated approximately two centuries later. Macdonald suggests that through the inscription the client "was making a political statement about his independence from Roman rule, in a city which was itself independent of Rome. He acknowledges his Palmyrene patron (*gyr*) but states his own and his family's origins" (Macdonald 2009b: 302; see also Al-Otaibi 2015: 297ff).

The blurriness of the term "Arab" can be explained by the fact that until the Islamic period the individual identity was expressed more frequently in terms of tribal membership, but scholars are divided in how each of them interprets the occurrences in which it appears. According to some (Hoyland 2007; Macdonald 2009b; Fisher 2011), its use betrays the recognition of cultural and linguistic features that several peoples (may) have in common, while not belonging to the same tribe. Of course, it is still unclear if this recognition follows a developing conscious of the "Arab" identity already in the Late Antique. For instance, Hoyland (2007: 227-228) suggests that the increasing intensity of the struggle between Rome and Persia, and later between Byzance and Sasanians, worked as a perfect trigger for a self-designation process. In particular, the more frequent interaction with steppe tribes and the influence of Christianism could represent good examples of "stress situations" giving way to ethnogenesis phenomena. The linguistic aspect and the birth of Arabic poetry are seen as clear indicators of this growing identity consciousness. Webb, however, expresses a completely different opinion to those previously mentioned (2016: 60-109) and considers all the recurrences of the term "Arab" as over-imposed labels, which do not reflect the identity of the individuals described as such. By suggesting the term "Arab-less", he argues that there is no ethnic or social structure before the Islamic period that can be described as Arab.[27]

What seems to be clear is that the rise of Islam and the conquest undertaken by the first Caliphs during the 7th century AD do represent a crucial moment for the development of an Arab identity, and specifically for its standardisation. With the Umayyads, the term already presents a quite clear ethnic connotation, marking a consistent change from its meaning in the Quran itself, where it seems to describe mostly Bedouins (Lewis 1993: 5) or be an indirect semantic derivative of the term *a'rāb* (Webb 2016: 115-126).[28] This ethnogenesis can be explained in light of the necessity to distinguish the "Arab"

[26] *CIS* II. 3973: *'bydw br 'nmw [br] š' dlt nbṭy[']rwḥy' dy hw' prš [b]ḥyrt' wbmšryt' dy 'n'...*
[27] He considered the Ma'add as the only possible self-conscious ethnic group (later defined as Arab) to be documented in the pre-Islamic period (Webb 2016: 88-89).
[28] According to Webb, the adjective *'arabī* connects the "Arab-ness" to the linguistic sphere, namely the pure language of the revelation, not specifically defining nomads. In this sense, the term does not have to be considered as a plural of the term *a'rāb*, but an indirect derivative: "Hence the Qur'an's sense of *'arabī* may have indirectly

ruling elite clearly from the other ethnic and cultural groups who became their subjects. Consequently, tribal individualities might have lost importance in defining the ruling group in favour of a more common and wider self-definition (Lewis 1993: 5-6; Macdonald 2009b: 306).[29] In this context, the two words *'Arab* e *A'rāb* were distinguished for the first time, describing the new political and potentially ethnical entity, while also keeping the original meaning "Bedouin" (Lewis 1993: 5-6; Webb 2016: 115-126). At the same time, the tribal genealogy underwent a new and more precise codification, in particular because they started to appear also in written form, no longer being exclusively orally transmitted. Clearly, the new fruition and distribution of this knowledge changed radically, but still they continued to mirror social relations and existing political assets like in earlier periods. Beginning with the first Caliphs, the genealogies justified the belonging of singles families, clans or even entire tribes to the ruling Arab elite, in light of their Arab origin and/or genealogical proximity to Muhammed and his closest companions (Webb 2016: 197-208).

If the later Umayyad and Early Abbasid rules realised the highest point in the Arab ethnic consciousness (Webb 2016: 240ff), the mid-9th century marked an important change, with the loss of power by the Arab elites. Therefore, the original meaning of the term lost the importance it had previously in association with the ruling group (Lewis 1993: 86-89). Especially by the end of the Abbasid period onwards, with the arrival of Turks in the 10th century, the term "Arab" once again lost its association with a specific ethnic group, and instead was used to broadly describe social characteristics of a group. Those designated as "Arabs" once again became "nomads" ("Bedouinisation of a Memory", as defined by Webb 2016: 337ff)[30], and this application of the term became widespread in western chronicles as well, although in which case the term "Saracens" was generally preferred to refer to Muslims. Lewis noted that from this period one can also find the expression "Abnā' al-ʿArab" or "Awlād al-ʿArab" ("sons or Children of the Arabs"), "applied to the Arabic-speaking townspeople and peasantry to distinguish them from the Turkish ruling class on the one hand and the nomads or Arabs proper on the other" (Lewis 1993: 8).

Along these lines it is worth considering Rapoport's study where he analysed the terminology employed by the historical sources (especially *badw*, *ḥaḍar*, *'arab* and *'urbān*, and *a'rāb*) to describe the actors of some rebellions in 13th and 14th century Egypt, who opposed the Mamluk administration and were described as "Arab tribes" (Rapoport 2004). He argues that the key to properly understand the broader political and social contexts within these rebellions would be to establish "a clear distinction between pastoral nomadism as an economic option, tribalism as a form of social organization, and bedouin-ness as a cultural identity" (Rapoport 2004: 5). Accordingly, he established a differentiation based on the kind of crops cultivated and irrigation methods: *badw*, either nomads or settled peoples, specialized in grain cultivation and dependent on the seasonal inundation; *ḥaḍar* specialized in cash-crops (e.g. vegetables, fruits, and sugar cane) and heavily investing in the local irrigation system. One further difference between the *badw* and the *ḥaḍar* is that the formers are mainly Muslim and can bear arms, while the latter are predominantly Christian. Most importantly, Rapoport underlines that "al-Nābulusī 's dichotomy of Christian ḥaḍar and Muslim badw meant that not only were all the badw Muslims, but, more surprisingly, all the Muslims were badw" (Rapoport 2004: 12). This is explained in the light of the

descended from the Bedouin/*a'rāb* semantic universe, developed from centuries of pre-Islamic desert ritual practice which the Qur'an then Islamised and exalted for itself alone." (Webb 2016: 125).

[29] "During this first period in Islamic history, when Islam was an Arab religion and the Caliphate an Arab kingdom, the term Arab came to be applied to those who spoke Arabic, were full members by descent of an Arab tribe, and who, either in person or through their ancestors, had originated in Arabia" (Lewis 1993: 5-6).

[30] The term "Arab" starts to describe even Kurdish or Turkoman nomads, which are no Arab from an ethnic perspective (Lewis 1993: 8)

Islamization and Bedouinisation processes in the Fayyum, where "the conversion to Islam was accompanied by the assumption of a Bedouin tribal identity" (Rapoport 2004: 13).[31] Here, the term *'urbān* is particularly relevant, since it distinguishes the Arab tribes in Mamluk Egypt and Syria from the "pure Arabs" (*'arab*) form the Arabian Peninsula.

In summary, there is no simple definition for "Arab" in the past, as its application and how it was understood by various groups entirely depended on the social structures and political climate of different regions at different times. Still today, it continues to be a label that is difficult to define, since it neither has a juridical significance, despite its presence in the official name of some countries, nor characterises ethnically the Arab speaking peoples.[32] Similarly, the term's association with "nomads" is still present in the colloquial use of Arabic, though it lacks any clear connotation of a specific ethnicity or identity. Within such a blurred framework, religion might play an important role, since "many Arabs still exclude those who, though they speak Arabic, reject the Arabian faith and therefore much of the civilization that it fostered" (Lewis 1993: 9). As far as the present thesis is concerned, the term "Arab" will be used according to the context described as much as possible. Following Macdonald's suggestions (Macdonald 2009b: 1) and avoiding an ethnic or entirely religious connotation of the term, the cultural and linguistic dimensions of the term are probably the most representative, though one should be aware of the dynamic and (to a certain degree) controversial nature of such features. At the same time, a certain level of breadth and vagueness in the definition needs to be accepted, even if it is likely that the population was not entirely ethnically "Arab", at least until the Byzantine period, and the defining features of "Arabs" became apparent in different phases.[33] In addition, it would be limiting to consider an identity significance of the "Arabs" exclusively in terms of ethnicity: admitting that such an ethnic identity was only acquired in the Islamic period does not exclude that it might have had another usage, perhaps that was not recorded by historical sources. Some groups could have perhaps recognised themselves as something socially different from the rest of the population, either being mobile pastoral or entirely settled communities. Most likely, the integration in the Roman and Byzantine army and administration at first and later the Christianisation of the Near East played a crucial role in forming such a social identity, which led to the development of a distinguishable ethnic group in a later moment and under specific political and social conditions.

It is under such a set of circumstances that a group (or set of groups) identifying as "Arab" peoples come into being, and as a consequence so did the "Arab settlement". The intention of the present work is to offer one additional material element for reconstructing identity formations through the analysis of the built environment. The deliberate choice for the label "Arab" is also informed by the general debate around settlements (and mainly urban centres) in the Middle East, polarising into two opposed

[31] "By adopting an ideology of lineage or of bedouin-ness, the rural communities found a sense of superiority in Islam and nostalgia for the autonomy of a nomadic past" (Rapoport 2004: 20).

[32] "Is the Arabic-speaking Jew from Iraq or the Yemen or the Arabic-speaking Christian of Egypt or Lebanon an Arab?" (Lewis 1993: 1). The problem behind this question, according to Lewis, is that the answer may be different from both the concerned people and by Muslims in general.

[33] "Inevitably, categorization simplifies, selecting features [...] as defining and diacritical, and leaving others aside. This simplification may be fruitful, if the feature selected is determining or diagnostic. But the dangers are clear: an oversimplifying reduction of a multi-dimensional reality to a single selected dimension, or a stereotypic ideal type in which many features are baselessly assumed to follow once the defining one is identified" (Salzman 1996a: 21, note 1).

positions: "Islamic" vs "Oriental" cities. In my opinion, "Islamic" might confer an excessive emphasis of the religious component to the argumentation (even if on a lower degree than "Muslim", which is a label sometimes applied to settlements). This is also due to its improper use in common vernacular and by scholars of various disciplines. Furthermore, as far as archaeological, urban studies and historical disciplines are concerned, the label can be also misleading either in terms of chronology and culture, considering Islam's millenarian history, geographical extension, and cultural variety. The risk in its use is that it implicitly underestimates strong traditions throughout the region that continued to play an important role in the local communities and with which the "Islamic" tradition continued to interact. The term "Arab", which should be understood here as a geographic (Middle Eastern) specification within the more general concept of "Oriental" settlements, is more holistic in that it respects the continuity between the different phases examined in the present research and is more neutral and hopefully generates fewer misunderstandings than the term "Islamic".

Chapter 3: Typology

In the previous chapters, the interdisciplinary nature of the research was underlined, offering the conceptual framework within which the archaeological data will be considered and filtered. It was also shown how several aspects of an anthropological approach can be highly problematic; these difficulties are moreover shared by other issues specifically connected with archaeological approaches and evidence. In both cases, in fact, a researcher is often confronted with improper or imprecise terminology, which leaves their interpretation open to ambiguities and misunderstandings and potentially leads to their misapplication in a discussion. Similarly, when approaching studies concerning archaeological or historical settlements, one will immediately encounter a plethora of terms used to describe sites. City, town, large village, and village are the most common, but also the various "local" or "ancient" equivalents, such as *polis*, *kome*, *medina*, and *qasr*, must be taken into consideration (among others: Safrai 1994: 17-103; Hirschfeld 1997; Seligman 2011: 422-429; Pini 2019). Nonetheless, the problem is not the wealth of terms in itself, but what they precisely describe or described. There are two sides of the problem: on the one hand, each society – chronologically and geographically intended – has a different concept of what each type of settlement should be, especially in the "middle field" of the semi-urban or semi-rural towns or large settlements; on the other hand, there's a clear linguistic problem in accurately translating some terms that do not have an exact parallel in other languages, often leading to misunderstandings and potentially misapplication of those terms. As for the anthropological concepts mentioned earlier, a final and definitive word is not presented here, as it is beyond the scope of the present research. However further explanation (and a certain degree of standardisation) is required in order to offer a common basis for discussion for the following analysis, seeing the great differences identifiable among historical sources and modern scholars.

Despite the terminological challenges, the chapter aims to offer a broad overview of the different spatial solutions identified on the various analysis levels – settlement in general, quarter, block, house – in particular for the Late Byzantine and Early Islamic periods, from the 5th to the 8th centuries AD. This part is not to be considered as a typology, since it does not systematically analyse all the possible variants of the different types, nor their specific diachronic development. However, it provides a necessary categorisation and graphic representation of the criteria on which the single sites will be investigated in the following section (Chapters 4-6); in so doing these criteria make up a shared set of features on which the final comparison will be made. In fact, despite the different environmental and socio-economic contexts and possibly several individual aspects of historical development, the eight case studies share several features not only from a chronological point of view – with an extremely noticeable expansion during the Byzantine period, continuing into the early Islamic – but also culturally. Most notably is the close connection between nomads and settled communities and the presence of "Arab" (or those described as such) groups. As mentioned above, the specificities of each site or context will be considered within a common framework offered by the first "typological" section. In fact, with only a basis of a shared number of criteria, the eventual different variants, e.g. in shape and size, could be clearly identified and possibly explained.

Furthermore, most sites also share an exceptional state of conservation, allowing the detailed study of their spatial organisation without extensive excavations. In addition, with the exception of the two Syrian settlements, I was able to personally visit the sites and directly verify some features and aspects for their comparative analysis. It is not infrequent to have structures still almost entirely standing, in most cases with only few marginal modifications taking place in later periods than the one considered here. The case studies cover a wide spectrum of settlements' types, having at least one confirmable rural and one semi-urban example for all different regions, and will help address two major questions: what do the labels usually applied to these types exactly describe, and on which basis is a site defined a large village or a town? By considering the answers to these questions for each site, this can perhaps offer more concrete associations with specific terms and their related settlement type as well as encompass the 'spectrum' of what each settlement type can contain.

3.1: The "Settlement" level

The most general level of analysis is the settlement in its entirety. It is difficult to graphically identify and show a possible "typological" categorization, at least for two reasons: the first is the aforementioned difficulty with highly abstract and imprecise terminology. Closely connected to this is the large number of variables influencing the urban organisation in its entirety, determining a multitude of potential subtypes. One such variable influencing the physical structure of the settlements is the topography of a regions, which is likely to have the most direct and evident influence on the built environment. Furthermore, this situation can be convoluted by overlapping series of settlement types. The size of the settlement has often been used as decisive criteria to solve the impasse and to try offer a typology.[34] In my opinion, this is an extremely misleading criterion, neglecting several features that are ultimately more decisive. To demonstrate this point, I will consider the case studies in this study (Plate 1). Based exclusively on their size, these sites would potentially consist of at least one *polis* - Umm el-Jimal, with an extension of *c.* 33 ha - some towns - Sharah (20 ha) and possibly Umm es-Surab (*c.* 12 ha) - and the rest villages - ranging between the 4.5 ha of Mseikeh to the 8 ha of Shivta. This designation is both too simplistic and could not be further from the truth. While leaving the detailed description of the individual case studies for discussion later in the text, suffice it to say that Umm el-Jimal can be hardly defined as a proper *polis* - not even a Byzantine one. Furthermore, the label "village" does not apply to Mampsis, which on the contrary shows marked urban features. Although the size of a settlement can offer some indications on its role and perhaps its status, it is necessarily to take it into consideration with other organisational and functional features. These features ultimately allow us to distinguish between the different types of settlement.

The first element that will be considered is how the settlement is physically delimited from the outside. Even if the ways used to mark this delimitation may vary considerably - a proper wall or continuous and blind face formed by the most external dwellings, topographic factors like wadis, or an orographic drop - the most general distinction is between "open" and "closed" settlements.[35] The former normally describes a less urban site, in which one may identify a major continuity and spatial fluidity between the inner areas and the proper countryside. On the contrary, closed settlements usually present a much clearer separation of the two areas. In this regard, it is important to stress the fact that this separation between the inner urban space and the countryside may be operated in a variety of ways.

[34] For instance, Hirschfeld (1997: 61) affirms that the extension of 8 - 10 ha is distinguishing the villages in south Palestine in confront with the rest of the province, where rural settlements rarely reached such size. This allowed me to imagine, that if a site is larger than 10 ha, has to be considered necessarily something different from a village.

[35] Hirschfeld define them "introverted villages" or "dispersed villages" (Hirschfeld 1997: 60). Seligman adds one third type of "orthogonal villages", where the houses are "arranged in parallel and perpendicular lines" (Seligman 2011: 426): none of the case studies considered for the present study belongs to this last type.

Apart from the *poleis*,[36] which clearly show a demarcation thanks to the presence of defensive structures, namely the wall, with towers and monumental gates providing access to the internal infrastructure network, in general closed settlements appear to be less structured in the other contexts. Clear defensive structures enclosing settlements are rarely identified and are often simply the outer faces of the most external blocks that function as an established border. Generally speaking – and of course some exceptions can be found, like in Mampsis and Umm el-Jimal – the access to the settlements is more fluid and less "regimented". The narrow ways or paths running between the different houses function as throughway to the inner part of the settlement, sometimes marking the entrance with simple doorways (Figure 3). Needless to say, there are also some "mixed" situations, presenting both

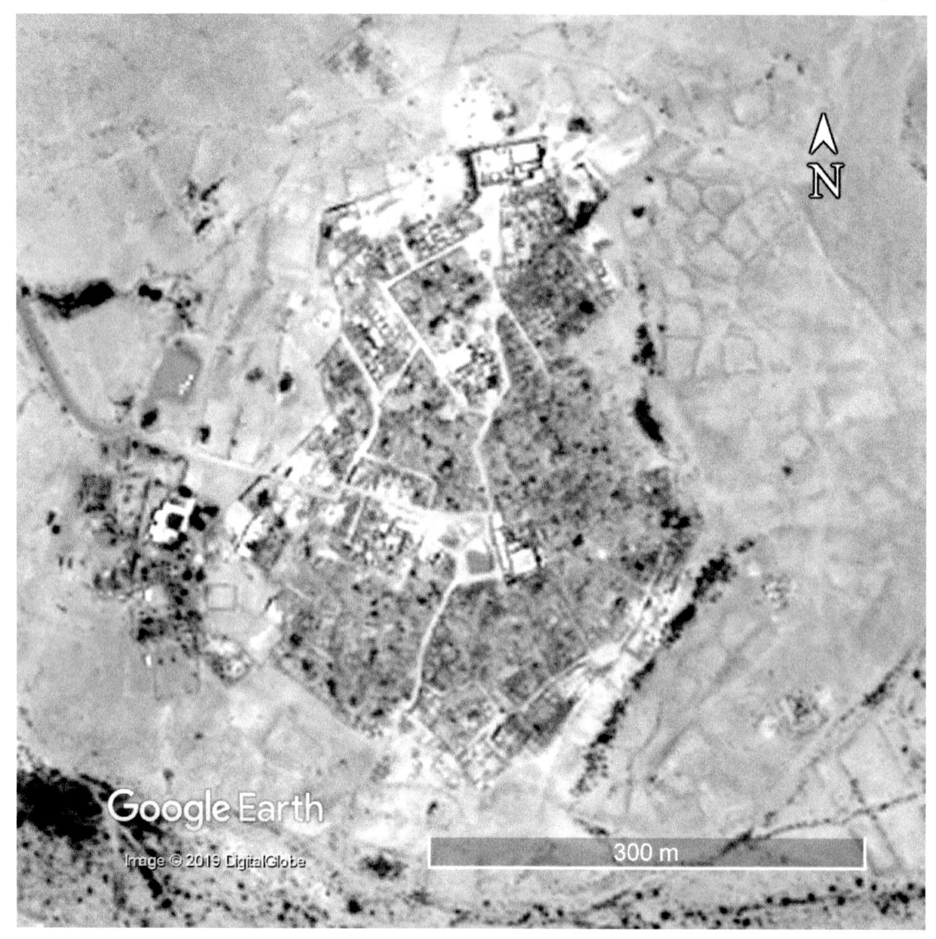

Figure 3 - Shivta, an example of "close" settlement (Satellite image: Google Earth)

[36] The Hellenised *poleis,* being likely the only proper type of settlement that can confidently identified, escape this extremely general categorisation, despite some quite evident differences among the different sites. Without going into too much detail and considering the same aforementioned criteria, it is obvious that the *poleis* are the only settlement showing most of the time a clearly planned layout. Their design consists of a clear organisation in quarters, delimited by the main thoroughfares, namely the *cardo maximus* and the *decumanus maximus* (or *decumani maximi* if there is more than one), which ultimately define the spatial pattern as well as the monumental frame of the city. Often, the crossroads are monumentally marked, for instance by *tetrapyla* (i.e. the four-way arches). Additionally, a more developed hydraulic infrastructure follows the street frame, with aqueducts supplying major reservoirs or cisterns, which then redistribute water to the public buildings and decorations (like *nympheia* and fountains) and also to some private domestic cisterns. Moreover, these are typically urban features, which entail the continuity of the spatial pattern, the compactness and the density of the urban fabric. This overview of poleis seems to be the case at least until the beginning of the Byzantine era, if not the last decades of the Late Roman period. From that point, some important changes underwent in the conception itself of city (see Chapter 7.5).

"structured" single gates and narrow paths between buildings or even more opened and wider tracks – this latter feature being normally more typical for an "open" settlement (Figure 4). Finally, common to both "open" and "closed" settlements is the presence of an intermediate zone before the proper countryside: a sort of buffer area constituted by enclosures, orchards or fields surrounding the entire built area.

The distribution of the quarters/blocks is the second major feature that will be considered. In this regard, the street pattern is one of the most important aspects of this feature. Generally speaking, one can find a more or less geometric and regular system, or a more winding and irregular one (e.g. "le chemine de l'ane", the "track of the mule" as defined by Le Corbusier). Theoretically, both of them can be either hierarchical – with a distinction in major throughways and secondary alleys – or undifferentiated. What complicates matters is that it cannot be always determined if the quarter distribution determines the street network or the way around: this is probably one of the major differences between proper urban centre and the other (not entirely or entirely) rural settlements,

Figure 4 - Umm es-Surab, an example of "open" settlement (detail of APAAME_19531031_HAS-58-032, Hunting Air Survey 1953, courtesy of APAAME)

since the planning of a street pattern is the formally most important feature in determining the distribution of the urban fabric.

However, there are clear differences between "compact" and "scattered" settlements. In "compact" settlements, it is ultimately impossible to distinguish between quarters, and the different paths in the settlement simply define a series of blocks, often characterised by an extremely complex inner structure. The more "scattered" settlements, on the contrary, offer a clearer demarcation of the different quarter, normally being separated from one another by "blank spaces". These areas, however, are far from empty. Often lines of enclosures or also communal facilities like cisterns or reservoirs can be identified, showing possibly some kind of "public" use of these spaces: places for public events and gathering of people, like a marketplace, or simply areas with facilities for keeping animals belonging to sellers, caravans, or to the local inhabitants themselves. In both "compact" and "scattered" settlements, it appears that no clear planning effort was made (if not limited to extremely small portions of the urban area), and the paths or the rarely paved streets simply followed spontaneous routes within the settlements.

As far as the hydraulic infrastructures are concerned, few differences can be detected between the different kinds of settlements and only appear to be centralised and systematically planned in few circumstances. Only the larger sites are supplied by aqueducts, bringing water to the main public cisterns and reservoirs present in the open spaces inside the settlements or immediately outside them. These structures rarely appear to be connected to domestic private cisterns in the quarters, which were the major source of water for the population. If they were not connected to a central water-infrastructure, this leaves few remaining possibilities for their supply: while the collection of water from the winter rainfalls was the most vital resource, the manual pouring of water carried to the public reservoir is not to be underestimated, as suggested for Mampsis (Chapter 4.3.2).

One of the most problematic aspects of analysing whole settlements as a whole is approaching them through a functional perspective, since the data required to do so are extremely rare: this makes the spatial distribution of activities – or functional zoning – often difficult to outline. Moreover, the most common feature is a functional mix, where separating productive, residential, religious and administrative areas is almost impossible. Sometimes only the main religious or military complexes appear to be isolated or at least marked in the settlement frame, creating a self-standing functional zone (Figure 5). Major productive areas, identified by the presence of massive installations, are rare in settlements and are often connected also to residential structures. In general, it is extremely difficult to find a clear distinction between productive or public areas and residential zones, documenting instead a high degree of functional mixing, as far as the productive and residential functions are concerned. The religious and administrative building may even appear completely integrated and connected to both productive or commercial and residential areas.

The aforementioned "blank spaces" of the scattered settlement could offer an interesting perspective on the function of a site, since they possibly could have been a distinguished zone that was dedicated to commercial or social goals (Figure 6). They could ultimately represent a functional zoning, as the *agora* and the colonnaded streets with their *tabernae* in the poleis.

Figure 5 - The large reservoir in Umm el-Jimal, fed by an underground aqueduct, close to the "Diocletian Fort" (APAAME_20091019_DLK-0248. Photographer: David Kennedy, courtesy of APAAME)

a) From the ground looking southwards with the Commodus Gate on the right (by the author); b): Satellite picture showing the same area (Satellite image: Google Earth)

Figure 6 - The "Blank space" in front of the so-called "Commodus Gate" in Umm el-Jimal.

A further possible functional indicator for the site is the presence of an acropolis, a topographically distinguished quarter and zone in the settlement (Figure 7). The three more-frequent designations of such areas are military (thanks to the highly defensible position they normally offer), religious (with a more ideological valence, underlining their predominance through the dominant position, visible from every point in the settlement), or administrative (for various practical and ideological reasons). Mixed or combined functions of the acropolis, however, are not infrequent and the function might also change through time. The presence of a military camp or a monastery, not necessarily founded on a topographically prominent position, but still appearing as a distinct area within the settlement, can also have an interesting and similar spatial relationship as an acropolis to the rest of the settlement (Figure 8). Sometimes these two parts, though clearly marked and distinguished, appear to be tightly connected

Figure 7 - The Acropolis of Avdad, in the Negev (photo by the author)

Figure 8 - The Site of Umm er-Rasas, where the former Roman fort is clearly distinguishable on the right (Jordan) (APAAME_19980520_DLK-0018. Photographer: David Kennedy, courtesy of APAAME)

to one another: open spaces or monumental links (such as monumental stairs) may indicate this connection. In other cases, the acropolis/military camp occupies a more marginal or isolated position, sharing only a few spatial connections with the rest of the settlement. The typical spatial organisation of the Roman forts, with well-defined blocks and the regular frame of streets, seems to be only rarely preserved, showing in many examples radical changes in the disposition of the buildings. Nonetheless, the outer outline of the walls of the earlier camp – as the entrance gate – are often kept, with buildings abutting to the inner face. A further distinction has to be made for the old *castra* that sometimes, during or at the end of the Byzantine period, appear to cease their defensive functions and be converted to residential quarters, with productive and religious structures also present (see for instance Umm el-Jimal and Umm er-Rasas, Chapters 4.1.1 and 6.2).

3.2: The "Quarter" level

Some considerations made for the analysis of an entire settlement can also be extended to the residential quarters. Morphologically speaking, it is difficult to find a recurring type of quarter, since many variables could modify the form. However, one can generally distinguish two different kinds of quarters: the compact quarter and the more open quarter. The more compact one, where the buildings and the structures are extremely close to the other, leaves almost no empty spaces and its several blocks are defined by narrow pathways (Plate 2a and 2b). The more open quarter type, where the structures are more discontinuous, replicates the same spatial pattern seen for the "settlement" though on a smaller scale. In most cases, both these types of quarters may appear like clusters of blocks that if occupying a more peripherical area can also function as "boundary" for the settlement (Plate 3). Despite these two different types of organisation, the access to and the inner circulation of the quarters is similarly structured, with pathways guaranteeing the going in and out, without monumental entries or clear linking point(s) connecting the quarter with the rest of the settlement. The same paths spontaneously develop inside, defining the different blocks without any sort of initial planning, in contrast with what normally seen in the Hellenistic organisation of the *poleis*, where a frame of secondary streets, starting and leading to the main thoroughfare, outline the different *insulae*.

A final type of quarter, though extremely difficult to find, is an enclosed quarter, which appears to be clearly separated from the rest of the settlement without being topographically separated like in the case of an acropolis (Plate 4). It is also sometimes difficult to determine if this separation can be somehow explained by the development of the site; namely, if like in the case of earlier forts, these quarters are indeed the original nucleus of the settlement.

On the quarter level of analysis too, the functional zoning is the most problematic aspect, showing the same sort of difficulties as investigating the settlement as a whole. Furthermore, it is also difficult to determine the economic stratification among the quarters, i.e. a socio-economic zoning in most settlement types, even in larger quarters. By contrast, the situation in the major *poleis* is different, since both administrative compounds and the major trends in socio-economic stratification are more clearly differentiated among the quarters. Quarters generally tend to exhibit a mixed signal for this aspect. The features normally linked with more aristocratic quarters – like a concentration of larger and richer houses – are often difficult or nearly impossible to detect. Similarly, one cannot locate particular inner patterns, like the presence of larger and wealthier complexes at the centre of the quarter or block surrounded by smaller and poorer buildings.

3.3: The "Block" level

Although few generalisations could be made for the general morphology of the settlement and for the physical features of the quarters considering the wide span of local variables, the block, on an even more detailed level, appears to be more standardised and recurring with few variations in the several case studies. It is then possible to identify the different spatial solutions that might have been adopted. One can distinguish three major types of blocks, with subtypes resulting from particular conditions, mainly linked to their location in the settlement.

The first type of block is actually an isolated building, like a courtyard house or a public building. Needless to say, if it is a residential complex, the spatial features vary according to the typology of the house, which will be dealt with in more detail at a later point. The second type of block is the "cluster": the outline is normally irregular and not planned, and the dimension can vary according to the number of buildings attached one to the other. What differentiate the various subtypes is actually the inner circulation and the spatial solution adopted. The simplest solution is a series of edifices sharing some walls, each having an opening to the public pathway and rarely allowing a direct inner communication through all the different complexes (Plate 5). In the case of residential blocks, the separation of the different houses is reflected by the distribution of the courtyards and of the water facilities – namely the cisterns, at least one for each unit. The second and more complex alternative is a group of buildings that not only share walls but also one or more private *cul-de-sacs* – or dead-end alleys – sometimes closed off by a gate from the public pathway. The access to the single units is guaranteed by the same alley, even if the inner circulation between the structures depends on their typology. Increasing the degree of complexity, the third subtype of the "cluster" block is a group of structures sharing one large, central courtyard, closed off from the outside of the block, to which each building opens. It is not rare, in for example residential units, to have additional courtyards belonging to the single house separated from the communal one and opening to it. The last subtype is actually a combination of the second and third sub-types: a common private alley leading to a central courtyard (Plate 6). Finally, there is a particular variant that can be found normally in the most external reaches of the settlement. Since most of the sites, despite their dimensions, do not have a wall surrounding and protecting the settlement, the outer blocks to a certain extent take on this function. Normally, the buildings constituting these complexes offer a continuous blind face to the outside of the settlement, with no opening with the exception of some small windows mainly for ventilation purposes. The impression from this standpoint is that an incomer entering the settlement comes face to face with a wall incorporated into the settlement texture.

The third type of block is the *insula*-type block (Plate 7), characterised by a more geometric and regular outline and closely reminiscent of a "Hellenistic" *insula*. Like in the first "Cluster" subtype, the buildings following this spatial organisation share their side walls, often resulting in a row or a double row of structures. As a result, each unit is therefore independent from one other, with private facilities and an independent opening to the public pathway.

Despite these several types and subtypes of blocks, they do nevertheless share some common features: the most important one is the functional polyvalence of the spaces. As seen for the quarters and the entire city, this is one of the most problematic issues of this research, in light of the lack of detailed and broad reports for the sites from archaeological excavations. Despite this, it is quite rare to find particular concentrations of productive installations and purely residential areas within the same block. More frequently, in fact, the two functions often occurred next to the other, likely on a family scale. Under this aspect, along with small wine and olive presses that can be found in many settlements, the courtyards played a particularly important role. They were indeed not only a designated space for the inner circulation through the block and the different units, but also the preferred location for productive and daily-life activities, such as cooking. Subdivisions of the courtyard, perhaps to separate the different kind of animals or to designate areas for small installations like tabun ovens, are often found in the archaeological reports and provide extremely important information on the proximity of the residential and more productive spheres within these settlements. One need to keep in mind that economically speaking most of these sites seemed to be more oriented towards subsistence, or at least to a local market trade.[37] For sure, it is rare to find productive installations on a proper industrial scale not only inside the settlements, but also in their immediate surroundings. Religious and administrative buildings, however, can often be quite integrated in the functionally mixed blocks. Though this is generally not the case of the major monastic or ecclesiastical centres, which are normally isolated within the settlements' frame, small churches or chapels tend to be more frequently tightly incorporated in the block, while also having some preferred inner communication's ways with some of the other buildings (Figure 9). It is also common to have religious complexes accessible only through private courtyards of a cluster or even through a residential unit.

Figure 9 - A block in Umm el-Jimal, where a small church is clearly visible amid the block (APAAME_20091019_DLK-0232. Photographer: David Kennedy, courtesy of APAAME)

[37] And here is probably one of the most important difference between these rural or semi-urban settlement with properly urban centres: the first reflects an economy mainly oriented toward the subsistence, while the latter have structural features connected with the distribution of the surplus.

3.4: The "house" level

As far as single residential units are concerned, i.e. the "house level", categorisations have been proposed by several scholars, though none made a systematic typology. That being said some types and their variants, including the "Single", "Complex" and "Courtyard" houses, had been analysed in detail and diachronically. These types are the most frequent examples that are found in the settlements over time and space. It is important to underline that, as was the case for blocks, the designation of this level of analysis as "houses" or "residential units" can be highly imprecise, meeting more a need of generalisation than an effective scientific fact. The functional classification of these complexes is in fact far from being univocal. Generally speaking, however, a house will contain areas that cater to the following needs: private family life, the reception of guests, cooking, productive activities, and sanitation (Wirth 2000: 370). It is fairly common, for example, to find productive installations (though certainly not for an industrial scale) integrated in the single unit structure, near the properly residential spaces. Sometimes these functional spaces can result in physically distinguishable – or at least clearly separated – quarters or rooms. This integration of different areas or even the existence of polyvalent spaces within the single unit is common in rural, subsistence-based settlements that participate in a local or regional market and can be explained by the fact that most social units within the settlements aim to maximise their autonomy in supporting these everyday needs. Of course, this is a trend and not a general rule, since great variation can occur and reflect the different role of the settlements both socio-economically and administratively. "Houses" or "residential units" in the *poleis,* by contrast, are extremely different from those in the village, in light of the major socio-economic stratification of the population and the diverse economic specialisation. Nonetheless, especially for the Byzantine and Early Islamic periods, some characteristics of rural residential complexes, such as the development of irregular courtyards (spared ground leftover from building activities) and a more accentuated polyvalence of the spaces are shared also with urban context, as suggested by some residences in Jerash (Gawlikowski 1986; Blanke 2018: 44; Lichtenberger and Raja 2018: 158-161; Walmsley 2018: 251-252).

Another feature that may prove difficult to determine through archaeological work, especially if the conservation status of the structures is not extremely good, is the number of stories of the different buildings. The lack of information, both from documentary and ethnographic sources as well as well-preserved archaeological examples, likely affects our interpretation of the distribution and functionality of the spaces within "houses". In general, also according to what is observed in the best-preserved settlements (e.g. Umm el-Jimal), there were usually up to three floors. Only particular dwellings, such as tower-houses, may be higher; nonetheless they are rarely documented in this area, being typical for the Red Sea and Southeaster Arabian regions, especially for the Late Byzantine-Early Islamic periods. Although, if they are present, they are normally integrated into a courtyards' systems. Isolated tower-houses, though not comparable to the standards seen in Yemen, appear to be a little more frequent in earlier phases of the settlements, mainly Hellenistic and Early Roman phases, which were afterwards gradually incorporated in the other kind of residential units. In scientific literature "tower-houses" are often connected either to a defensive or at least with land control, even though there are no clear indications from the architecture itself that can unequivocally support their association with this function. Most scholars, in fact, are more favourable to see these kinds of structures more as a display of social power.[38]

Apart from the aforementioned types of housing, scholars that have dealt with the topic, not necessarily from an archaeological background, have quite different opinions on the number of different types that

[38] For a general description of this typology, see Röhl 2010: 99-100.

it is possible to identify. This is also due to the fact they tend to consider specific case studies that are of a very different nature from one another. In fact, despite some of the common features that were previously mentioned, local traditions and socio-economic factors have often determined different degrees of inner complexity of the houses both in terms of number of rooms and their distribution, ultimately resulting in different types. For the purposes of the present categorisation, I will mainly refer to the following studies, which ultimately offer several overlapping aspects despite their differing foci of research: the residential typology by the geographer Wirth,[39] the study of the Negev settlements by Hirschfeld (1995), and the research on the domestic architecture in the Hauran by Clauss-Balty (2008: 41-103). All of the different types of houses share a common feature: the courtyard (Raymond 1994: 3-18; Wirth 2000: 370). Of course, it is not a Roman, Byzantine or Islamic feature. Moreover, many of the different housing types – if not all – are actually the development of older types, already attested during the Iron Age (Wirth 2000: 359ff). Functionally and from an organisational perspective, the courtyard is the real core of the house, but the way it is formed depends on the typology of the house where it is found. The most evident function the courtyard plays is to regulate the inner circulation and the physical distributions of the rooms of the building. Its relative position in the house and the way the rooms are distributed around it (or them, as in some form of dwellings) is ultimately determined by the differentiation of house types, as will soon appear evident. Nonetheless, the most characteristic feature of this space is its polyvalence.

The courtyard firstly guarantees light and fresh air to all rooms within the unit. The importance of this aspect – and therefore, the fundamental role of the courtyards also in social terms – can be better understood when considering the priority privacy need of the families living in the house. In fact, it is one of the more characterising features in segmented societies – as is the case with the Arab culture formalised in the Islamic period, though maintaining its pre-Islamic cultural preferences – and deeply influences many choices on spatial distribution made in the local architecture. One such choice is the absence of windows facing the outside of the building; while the light and the air circulation is provided by the courtyard, the privacy of the house's residents and their activities is simultaneously maintained. Moreover, the courtyard provides an open space where productive activities and cooking processes can occur. It therefore is common to find installations and/or facilities linked with such spheres, like *tabun* ovens or smaller pens for animals. In some cases, they could even answer needs for the reception of guests.

As mentioned above, considering the constant presence of courtyards, it is ultimately the diverse degree of complexity of the spatial organisation of the other rooms that leads to differentiation between the different types of houses. The first and more general division, according to this principle, that can be recognised is to between "*konjunkiven*" and "*injunktiven*" houses. Wirth actually reuses concepts proposed by Koldewey for describing the typology of *Neoassyrian* and *Neobabylonian* dwellings in Mesopotamia, which are direct models for later typologies (Wirth 2000: 265). The two kinds of structures can easily be distinguished by the outline of the general plan of the residential unit. Conjunctive houses, in the first case, have several different and irregular shapes, since the different rooms or groups of structures are spontaneously added according to the particular needs and conditions at that specific moment in time. The courtyards themselves are formed more from the spare ground leftover from building activities than real house-planning; therefore, their resulting forms are also extremely variable, following the process of growth/evolution of the house, and the space leftover in the parcel from the addition of new constructions to the initial core. On the contrary, injunctive houses show a

[39] Wirth (2000: 359-376) mainly refers to proper urban situations, but many of the considerations could be also extended to other contexts.

more regular and geometric plan, with the inner spaces being built starting from the outer walls inwards and the courtyards located in a more or less central position.[40]

This conceptual aspect of conjunctive vs. injunctive houses has to be combined with the complexity and articulation of the complexes, namely the number of structures composing it. According to Hirschfeld, there is a "tripartite" categorisation: the "Simple" or "Single room" house, the "Complex" house, and finally the "Courtyard" ones. The "Simple" type normally has roofed structures on one or two sides of the courtyard (Plate 8). The "Complex house" and the "Courtyard" house have similar numbers of roofed sides (at least three). What really differentiates them is the spatial concept ruling their organisation: ultimately, they can be considered the "conjunctive" and "injunctive" variants of the same house. "Complex" houses are the result of "adjoining dwelling units around a common courtyard" (Hirschfeld 1995: 46),[41] thus a proper conjunctive character (Plates 9a and 9b). On the other hand, in "Courtyard" houses, the central court is considered a properly planned room,[42] determining also a more regular outer shape of the entire complex (Plates 10a and 10b). The difference between these two last types is therefore theoretically quite clear, even if not always easy to recognise on the field. The indications offered by the shape of the courtyard can be helpful to a certain extent. Complex houses' courtyards are ultimately the result of leftover ground from the roofed areas and are delimited by the wall. They do not occupy any particular position in the house, as one for instance may see for the "patios" of the urban houses or in the so-called "central court" houses. In any case, for both typologies, courtyards continue to be the most important functional space of the house, where the extreme versatility of each house is shown and the major part of the daily activities took place.

The case studies of the southern Hauran and northern Jordan, which share all the features just mentioned, present an extreme homogeneity in the recurrence of a particular subtype of "Complex House", observed and described in several sites in Syria (Clauss-Balty 2008: 57-59). All Hauranian domestic architecture seems to follow an almost standardised model of "Complex House", which development is probably the same as the "normal" aggregation process that characterises the "conjunctive" Houses. The dwellings are modularly structured, as already discussed by Bulter (1930) in his analysis of the domestic architecture for several sites in the Hauran (Figure 10). On the ground floor,

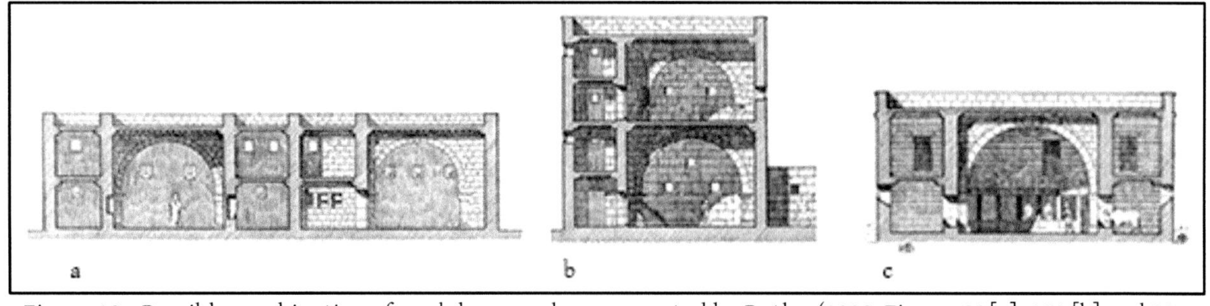

Figure 10 - Possible combination of modular parcels as suggested by Butler (1930, Figure 66 [a], 174 [b] and 88 [c]): a and b have both the larger hall functioning as "reception hall", while c shows the possible destination of the larger room and the accessory spaces of the aisles as stables.

[40] Because of this spatial organisation, these kinds of houses are also referred to as "Central Courtyard houses". In the Islamic period, this courtyard is defined as the "*wast ed-dar*", the centre of the house. As aforementioned, the centrality is not only spatial, but also functional.

[41] Hirschfeld (1995: 46-47) differentiates also two further subgroups: the "urban apartment house" and the "rural farmhouse". They nonetheless share the same spatial organisation.

[42] Also in this case, Hirschfeld distinguishes two subtypes: the one with a columned inner courtyard (also referred to as "atrium" or "peristyle" house), and the simple type without columns (Hirschfeld 1995: 57). According to the scholar, the first type has to be connected more with the Greco-Roman influence, while the second one is the result of a more local tradition.

a central role is played by a large room, whose roof can even be double the height of the room and normally flanks it at least on one side. This large room is generally interpreted as the reception hall, according to a series of recurring elements: the aforementioned dimensions, the presence of decorations, otherwise quite rare in other parts of the house, and finally, a direct and wider access opening to the courtyard that is also decorated.

On the aisle of this "hall", some smaller rooms are also often found and are usually interpreted as service areas or storerooms. Rooms similar to the reception hall can also be found, but are related to other functions. Despite presenting similar dimensions and also a wide opening to the courtyard, they could be used as stables. In that circumstance, instead of finding decorations (even though in some cases decorations may be found on reused materials used for the construction of these rooms), some kinds of installations, especially mangers, can be found. It is important to note that the inner circulation may vary according to the different designate of the "hall": for instance, sometimes the services room open directly to the audience hall, while in other occasions they are accessible only through other side areas. In some cases, the service room on the aisles are still present, but likely are not related to a reception room; they instead open directly on the central space and are separated from the larger room only by the mangers. They were therefore used also to accommodate animals and possibly separate the different types of livestock. For this particular aspect, these rooms share many features with similar stables in the Negev area (see Chapter 4.3). The upper storey of the service rooms also could have been directly accessible from the lower floor and had been often interpreted as storage areas. Otherwise, the second floor was normally reachable only using the stairs build on the outer face of the wall facing the courtyard,[43] and the rooms here are normally smaller and less decorated. For this reason, they are interpreted as the living quarters of the family of the house. In some cases, even a third floor may be found, but the function is normally considered the same as the lower one, answering to a need of more living space for the inhabitants. One further element is that often common for the roofed areas of dwellings assume a "L" or a "U" around a courtyard, that can be interpreted in my opinion as intermediate stages of development between the "Simple house" and the entirely enclosed "Complex house", that is slightly less recurrent than the "L" and "U" subtypes (Clauss-Balty 2008: 61, 63, 66, 69; 2010: 202-204) (Figure 11).

Another observation has to be made in order to complete this categorisation. For all the different kind of dwellings mentioned, there is a further distinction regarding the interface between the inside and outside of the house, namely the way the entrance of the building is articulated. In general, there are two different types encountered: in the first case there is a direct access to the courtyard that opens without obstacles apart the fencing wall on the path or street. In the second case there is a complex variant, where a room, functioning as a sort of *vestibulum* though often appearing like a sort of corridor, mediates the access from the external sphere to the inner courtyard.

One must also be aware of, especially in the case of multi-storied houses, is the high variability of the roofed structures. As mentioned before, normally there were one or two stories in houses, with some rarer case of three storied. The system of accessing these upper rooms is one more way to differentiate between subtypes of houses and can also be indicative of regional differentiation. In the Hauran area, the most frequent subtype is to have stairways leading directly from the courtyard, so external and normally with the steps inserted toward the outer face of the buildings, and thus always facing the inner side of the complex (Clauss-Balty 2008: 57). The other subtype, found only in the case studies in the Negev, is an inner stairway that normally also starts from the courtyard, but develops internally,

[43] This is ultimately one of the major differences with the Negev settlement for instance, where the stairs are normally internal, constituting a sort of tower-room giving access to the upper floors (see Chapter 4.3).

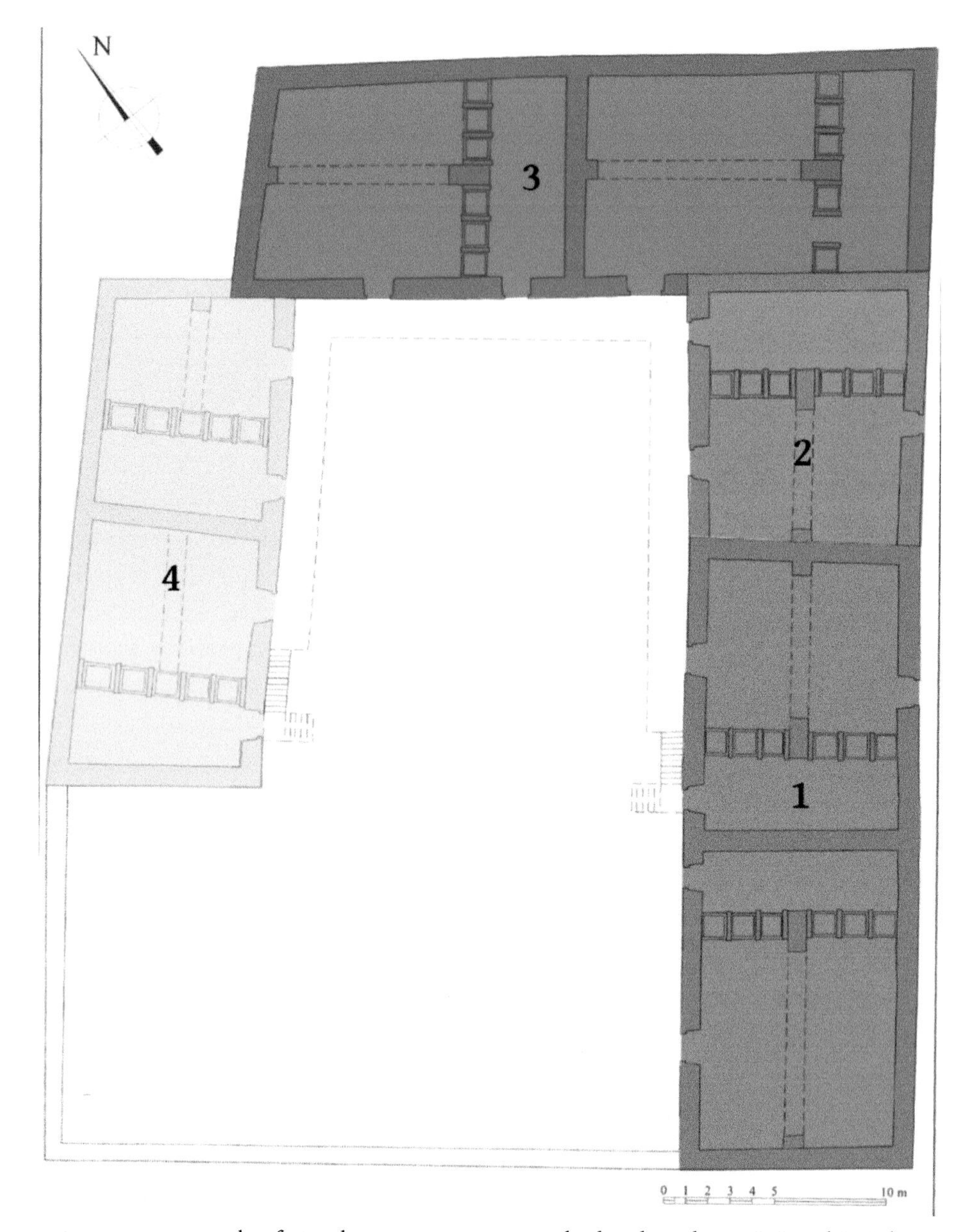

Figure 11 - Example of Simple House, progressively developed into "L" and "U", by adding first the north-eastern aisle and later the north-western (elaboration by the author on Clauss-Balty 2008, 99, Pl. XVIII Figure a)

sometimes in a spiral-form, defined sometimes "staircase-tower", which is extremely well represented in Mampsis. In all these cases, the staircases or the staircase-towers can be also linked with some "*loggia*" structures that cover part of the courtyard and are supported by a colonnade or a partial peristyle on the ground floor.

The functional zoning of the house, as evidenced by the previous discussion, is a very complex matter. It is more complicated for the less articulated dwellings, such as the "Simple houses" with only ground floor structures. If no installations such as wine or olive presses or mangers can be found, then the clear distinction between functional zones in the house is almost impossible. The most encountered case is that they have highly polyvalent spaces, with the courtyard playing an even more crucial role in the house economy. On the other hand, for buildings with at least one upper storey, a more general hypothesis can be made on the specialisation of spaces. With the additional consideration of ethnographic examples, it can be deduced that the ground floor level belongs to a more "public" sphere, while the upper storey consists of the more private quarters of the family. Furthermore, the more representative rooms, including the reception hall, are normally easy to identify on the ground level, since they directly open into the courtyard. In most cases these rooms (the "larger rooms" in the module briefly described above; Figure 11) differ in dimension, with a higher roof and wider surface and also decorations (mainly engraved) are also fairly common, especially if the inhabitants belonged to higher socio-economic levels. In addition, in the most complex structures, facilities such as stalls, may be present close to the reception hall.

Moreover, it is fairly common to find inner and (more or less) provisory subdivisions that complicate the functional interpretation of the spaces. This division normally reflects momentary and specific needs, suited to the family living in the house. Apart from all of these subdivisions, in the region and case studies considered there is a substantial modular conception of the rooms, with a tendency of squared or sub-squared particles, aggregated one next to the other in a more or less standardised pattern. A similar phenomenon can be documented also for houses in their entirety in all the regions considered in the present analysis. Specifically, the fusion of different housing units into a single unit is far from rare and often involves all of the aforementioned types of dwellings. The reasons behind these processes of unification of separated houses are again difficult to determine clearly, since different causes – acting simultaneously or not – could lead to such a result. Among other factors, the extension of the family group and the increased socio-economic power are possibly the two more frequent. In general, family dynamics are likely fundamental triggers for changes in spatial organisation, and this aspect will be further considered later in the research.

Part 2: Comparative Analysis

Chapter 4: The case studies

The case studies in this dissertation belong to four different regions: the Southern and Central Hauran, the Negev, and the Central Plateau in Jordan. These areas, though with their clear regional characteristics, do share several aspects in spatial organisation on the three levels of inquiry: the settlement in general, the quarter/block and the house. Some of these morphological trans-regional features had been described in the earlier section.

In this chapter, I will focus mainly on two aspects: the diachronic evolution of the settlements and their spatial pattern specifically in the Late Byzantine-Early Islamic period. Since the analysis considers three different levels, the case studies need to meet some requirements: first of all, the state of conservation must be sufficiently good to investigate each level of analysis; secondly, the available archaeological or architectural reports should allow the reconstruction (even more generally) of the development of each site; last but not least, a Late Byzantine occupation should be clear and not extremely disturbed by later phases. All but two case studies fulfil such requirements: Tall Hisban and Umm er-Rasas. Tall Hisban lacks the conservation status necessary for all three levels and there is low visibility of the Late Byzantine phases, since the Middle Islamic occupation heavily re-settled the same areas. In the case of Umm er-Rasas, the availability of archaeological data is limited, since the main focus of research for this site has been on the churches and the mosaics, which does not offer much chronological information for the rest of the settlement. Therefore, these two case studies are taken under examination following the comparison of the other sites and are to be seen mainly as a "test" of the method offered here. The usefulness of considering these two case studies is not only to verify how helpful a similar inquiry-approach could be in future studies, but also allow to point out some phenomena that are not so evident in the other settlements, and extremely important in deducing the final conclusions of the study. Some general spatial behaviours, for example, are recurrent in the different regions considered during the same chronological framework, which may relate to the fact that these regions are not so distant one to the other and the (contemporary) social context appears to be similar. The proposed observations and conclusions are not defined as "predictive", as in the processual theoretical tradition, since I believe that such a label could lead to dangerous generalisations. Therefore, the comparison between other sites can help balance the lack of data on certain sites and provide a more holistic view of settlements in the region.

4.1: The southern Hauran (Jordan)

The region spreading across the present border between Jordan and Syria is for a large part the administrative region of the Roman *polis* of Bosra (Figure 12). Among several factors – both local and extra-regional – two seem particularly important to me in determining the economic prosperity of this area, particularly visible in the Byzantine period. First of all, the strategic location, close to important trade networks and liminal to the desert steppe, populated by several entirely and partially nomadic pastoral groups. Secondly, though not of secondary importance, the basaltic plateau, guaranteeing a higher fertility of the soil and also sustained by sufficient rainfall provided ideal conditions for agriculture. Some of the features of the settlements from this region are in my opinion direct consequences of these two elements. One further particularity of this region is the employment of basalt as the primary building material, either in the official/monumental and residential architecture.

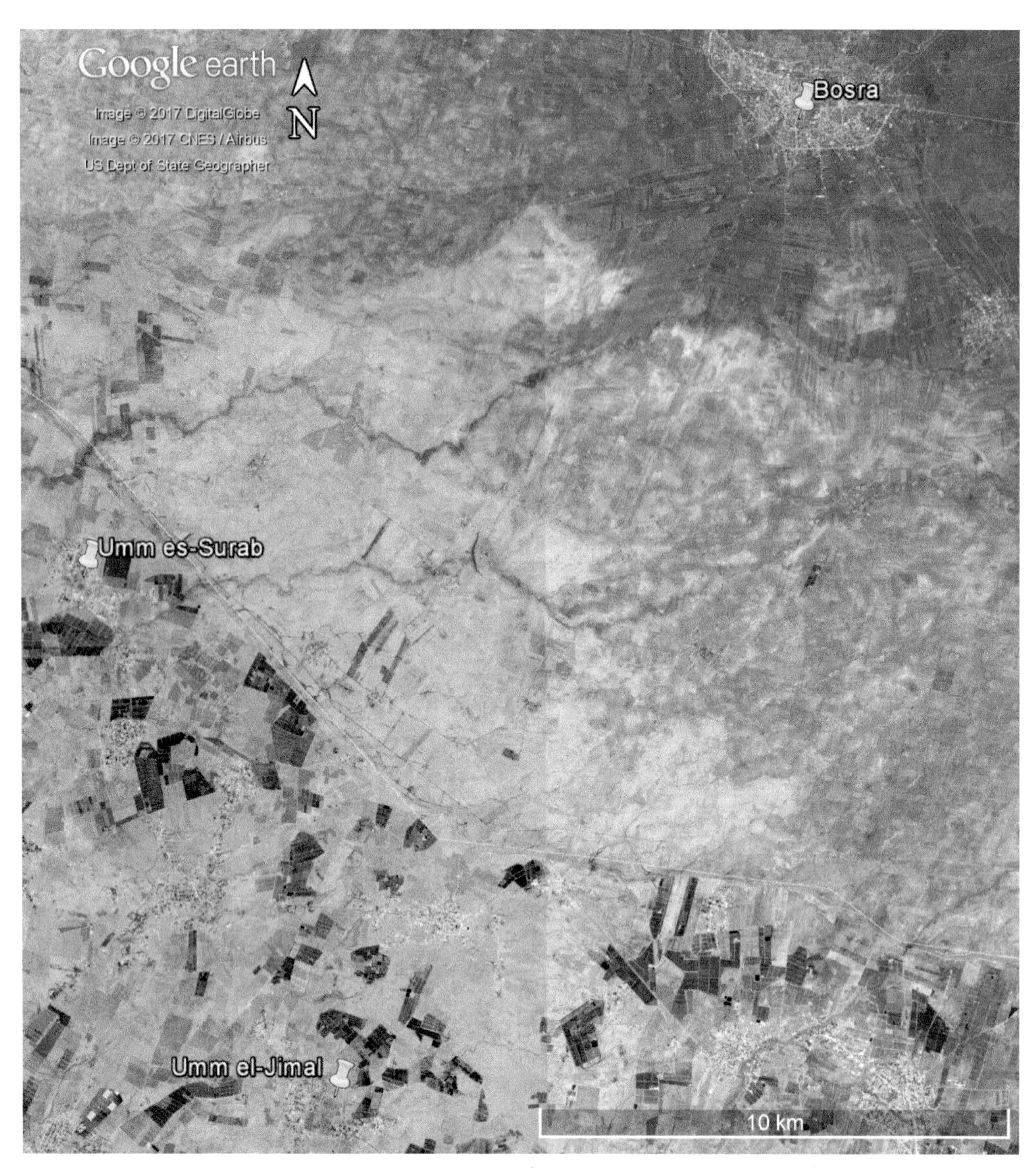

Figure 12 - The Southern Hauran (Satellite image: Google Earth)

4.1.1: Umm el-Jimal

Umm el-Jimal represents, in my opinion, one of the most astonishing case studies available, not only in Jordan but in the entire Near East. Its extremely good preservation is in fact a rarity that can hardly be found elsewhere, especially if considered the large scale of the settlement – *c.* 33 ha. The site did not experience the same level of destruction as other villages and towns in the area. This aspect may be explained in light of two major factors: first of all, the modern village that developed in the second half of the 20th century, a part from the construction of a road damaging the south-eastern corner of the settlement, did not overlay and destroy the ancient site, leaving it in its pristine condition; the second reason is the use of an extremely resistant building material, volcanic basalt, of which the entirety of the structures, including the roofs, in the site are made. Of course, the incredible conservation of the site does not necessarily imply that the interpretation of the remains is free from difficulty. In fact, several questions can be posed, especially in regard to its historical background, where its absence in written sources leave Umm el-Jimal's ancient toponymy[44] as well as the status and administrative functions and the types of institutions present an enigma. Not even the large number of epigraphic finds in the site help counterbalance this gap in knowledge.

Another issue to be dealt with is more directly connected with the dating of the site itself. I would say that the main reason for the lack of an absolute chronology of the single phases of the settlement can be seen as a direct consequence of its incredible material conservation. Because of this, researchers studying the site have mainly focused on the analysis of the standing structures and conservation projects, rather than conducting proper extensive archaeological excavations aside from a few limited probes in some of the major complexes. The only chronological indications are consequently most frequently the relative dating of the development of single houses or complexes, available nonetheless quite homogeneously in all areas across the settlement. But only few absolute dates are available and are normally offered by epigraphic finds. This issue is particularly the case for the pre-Byzantine occupation of the site. In fact, almost all the Roman, and earlier, structures have been completely destroyed or reused by the enormous expansion of the town during the Byzantine period. Therefore, the exact foundation and the earlier stages of Umm el-Jimal's history for the most part are still unknown. In relation to this aspect, the presence of a Hellenistic or Roman settlement on the site of the later Byzantine town has been long debated, especially in light of the presence of the village of el-Herra, a few hundred meters southeast of the site. The structures and materials found here, in fact, are all dated between the Nabatean[45] and Early Roman periods, showing an abrupt abandonment between the Late Roman and Early Byzantine times (3rd-4th century AD)[46], in coincidence with the start of the impetuous expansion of Umm el-Jimal. The Roman fort built during the fourth century, heavily modified in the following centuries but still recognisable from aerial view (Figure 36), had long been considered to coincide with the acceleration of the abandonment of this process, and ultimately with the creation of the original nucleus of the Byzantine town. This theory has since then undergone some radical revisions, mainly because it appears that Umm el-Jimal was settled even before this camp was erected, as some of the excavation probes and epigraphic finds suggest. Moreover, a recent analysis of the ceramic finds, testified that the presence of Nabatean and Roman pottery is widespread throughout the entire surface of the site, even though they are not connected with any precise structure (Osinga 2017: 274). It appears that already in the 2nd century AD a settlement was starting to develop in exactly the same place as the Byzantine settlement.

Furthermore, two Latin inscriptions seem to be particularly important in these two matters, namely the possible foundation of the site and its status, being also some of the few sure chronological indicators for the development of the settlement. The first is the commemorative inscription of the "Commodus Gate" (Inscription nr. 232, Figure 13), found *in situ* and recognised by Butler as an original part of the wall above the arch dominating the town entrance (Butler 1930: 156-157). Dated between AD

[44] Some hypothesis has been proposed, but none of which appeared to be entirely convincing, as reported by de Vries (de Vries 1998c: 36-38). Butler proposed the town of Thantia as identification of Umm el-Jimal, also present on the *Peutinger Table* and the *Notitia Dignitarum* (where we have the name Thainantha). This hypothesis was discarded since in the time this last source was written (AD 408), Umm el-Jimal appears to be without any proper garrison. Another suggestion was made by MacAdam in 1986, seeing as a possible identification the ancient Surattha, mentioned in Ptolemy's map (2nd Century AD). Also, in this circumstance, we cannot be sure without any epigraphical documentation found in the site.

[45] The establishment of such settlement could perfectly correspond to the development policy carried in the region by the Nabataean Kings following the foundation of Bosra during the 1st century AD.

[46] Staring from the 4th century, el-Herra is used as dumping site.

177 and 180, the text mentions the construction of a *vallum* that Littmann, the curator of the publication of the inscriptions found during Butler's expedition, interpreted as a permanent fort.[47] The exact position of such a structure is still unknown, but he suggested the presence of an earlier small *castrum* of which the Commodus Gate would have been the eastern entrance. One further interpretation of the evidence presented by de Vries is particularly interesting: he considers the *vallum* as the city wall surrounding the settlement (De Vries 1998b: 229). The probes made under the defensive system of Umm el-Jimal seem to confirm the presence of a first wall and afterwards a rebuilt wall (Parker 1998b: 143-147): it perhaps belonged to an earlier fort, as suggested by Littmann, or it could have been already an "independent" wall surrounding a smaller portion of the settlement. The second and even more problematic document is the lintel coming from the so-called "Cathedral" (Inscription nr. 233, Figure 14), also discovered by Butler (Littmann, Magie, and Stuart 1930: 132-134). The block was reused for the construction of this church and its original collocation, and the kind of building it belonged to, is still unknown. The inscription is nonetheless confidently dated to AD 371 and the text mentions a *burgus* built by the *equites IX Dalmatae*. The exact meaning of this term is the most problematic issue: after a series of parallels with other provinces, Littman suggested that it should be interpreted as a small

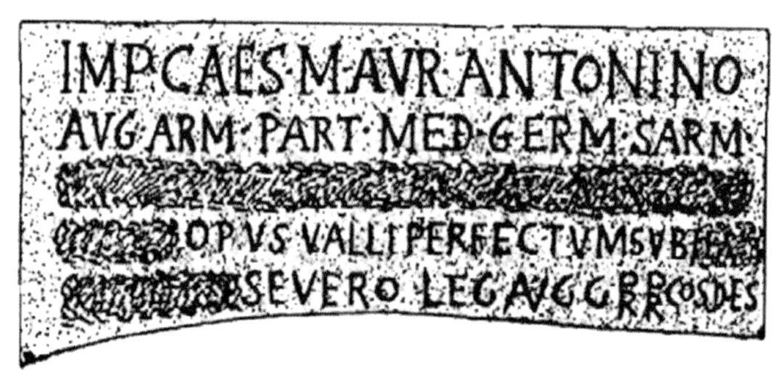

Figure 13 - Inscription nr. 232 from Umm el-Jimal (Littmann, Magie, and Stuart 1930, 131). Transcription: Imp(eratore) Caes(are) M(arco) Aur(elio) Antonino / Aug(usto) Arm(eniaco) Part(hico) Med(ico) Germ(anico) Sarm(atico) / [et Imp(eratore) Caes(are) L(ucio) Aur(elio) Commodo Aug(usto) Germ(anico) / Sarm(atico)] Opus valli perfectum sub ... / ... Severo leg(ato) Aug(ustorum) pr(o) pr(aetore) co(n)s(ule) des(ignato).

Figure 14 - Inscription nr. 233 from Umm el-Jimal (Littmann, Magie, and Stuart 1930, 132). Transcription: Salvis d(ominis) n(ostris) Valentiniano, Valente, et Gratiano, / victoriosissimis, semper Aug(ustis), dispositione Iuli, / v(iri) c(larissimi), com(itis), magistri equitum et peditum, fabri- / c(a)tus est burgu[s] ex fundamento mano devo- / tissi[m]orum equitum IX Dalm(atarum), s(ub) c(ura) Vahali trib(uni), / in consulatum d(omini) n(ostri) Gratiani, perpetui Aug(usti) iterum / et Probi, v(iri) c(larissimi).

[47] Moreover, he interpreted that it was not a properly fortified *limes* as in Britannia, as suggested by other scholars (Littmann, Magie, and Stuart 1930: 132).

watch-tower.[48] On the other hand, Butler offered another hypothesis, on which the curator of the epigraphic survey expressed several doubts, but that remains nonetheless suggestive. He in fact saw the *praetorium* as the original building mentioned in the inscription: if it was the case, Littmann underlines that the term would have been "applied inexactly" in the inscription, finding no direct parallels in other contemporary sources. He was therefore firmly convinced that the most probable hypothesis is that the term refers to an "outpost of the main *castra*" (Littmann, Magie, and Stuart 1930: 134).

The presence of Latin inscriptions seems to demonstrate the importance of the site already in its earlier, less known stages of development. In Jordan and the Near East in general, the presence of inscriptions written in Latin is extremely rare, and the fact this site has at least five of them suggests a certain degree of Umm el-Jimal's importance in the regional context. Moreover, the two inscriptions mentioned above also documented the sure presence of some sort of administrative (the *praetorium*) and defensive structures (the *vallum*, *burgus*, and the different forts) being clues for a certain relevance already in the Roman phases, if not earlier. From these scattered indications, it is clearly difficult to establish the real extent of the settlement during these early phases. One can find clear traces of Roman structures only in three areas, surviving the Byzantine overbuilding of the town: the "Commodus Gate" and the walls directly connecting to it, the large reservoir, and finally, in the central building of the "*praetorium*". As far as the "Commodus Gate" is concerned (Figures 15 and 16), as mentioned before, it is one of the monuments that offer also an almost certain date of construction between AD 177-180. Along with being one of the few chronological indications for the early settlement, this structure also represents the only clearly "defended" entrance to the site, at least among the ones still standing, even if it was better preserved at the time of Princeton's survey. The difference with the other gates is evident, and Butler already did not hesitate in considering it as the main entrance of the town. Flanked by two towers, the arched gate opens to at least two roads connecting the site with the regional street-network: one towards Qasr el-Ba'ij, and intersecting the *Via Nova Traiana* around the XIV mile (*c.* 10 km far from the settlement) and the other in the N-NW direction, towards Khirbet el-Kaum and Umm es-Surab.

Figure 15 - The "Commodus Gate" as preserved today (photo by the author)

[48] Littmann (1930: 133) underlines also that "the *burgi* were utilized in a system of defences extending along the frontier". According to Littmann, the interpretation of the term "burgus" as small watchtower is also supported by other three inscriptions also dated during the reign of Valentian, Valens and Gratian, coming from the Renan-Danubian Limes - namely Noricum, Pannonia and Germania Superior.

Figure 16 - The "Commodus Gate" when the Princeton's expedition visited Umm el-Jimal at the beginning of the 20th century (Butler 1930: Figure 134)

One of the problems of the structure is determining its relationship to the enclosing wall of the town that does not present any proper defensive character, such as the absence of towers and a thickness not sufficient to withstand a proper siege. Some probes effectuated in two sections of the defensive system (one close to the Commodus Gate and the other near to the so-called "Barracks") confirmed that a first wall was constructed during the Late Roman period between the mid-2nd – 3rd century AD with "two faces of coursed masonry and an interior of rubble and soil" (Parker 1998b: 143ff). Apparently, the Commodus Gate was contemporary to this first phase. A second phase, which is clearly distinguished by the earlier one by the use of smaller and irregular basaltic stones and a higher ratio of chick stones, dates to the Late Byzantine period, between the 6th and the 7th century. In both cases, the general appearance of the wall does not change considerably, given the impression of a structure that had always little proper military function. A few other important buildings and installations provide insights on the development of the site between the Roman and the Byzantine periods. Not much can be said about the large reservoir, spread over a surface of *c.* 30x40 m, close to the Diocletian castellum (Figure 17). It represents the largest public water management structure of the town, which is fed by an aqueduct gathering water in some of the wadi north of the site. It likely continues to function also in the Byzantine period, representing the most important urban water source, also connected through channels to some private cisterns of the Byzantine period.

Some more information can be gathered from the so-called "*praetorium*", which is particularly interesting since it could also constitute a model for the "courtyard" organised houses built later on in

Figure 17 - The Large Reservoir of Umm el-Jimal (photo by the author)

the site (Figures 18 and 19).[49] This private residence is clearly different from the other houses in Umm el-Jimal, especially if one looks at the architectural quality of the central building of the complex. The atrium with its *compluvium*, a basilica and the cross-room (interpreted as an audience hall, Figure 20) are indeed the only clear examples of Greco-Roman tradition in the domestic architecture survived in the site. The interpretation of the building as *praetorium* - a term that normally identifies the tent of the general in a military camp or the house of the governor appointed by the central government - is in my opinion nonetheless still uncertain for at least two reasons[50]: the first is that it is unknown if other buildings (later totally destroyed) also had such features; and secondly, even admitting it was an *unicum* in the site, the choice of the choice of building in the Greco-Roman tradition may also had been a simple demonstration of the particular socio-economic status of its owner.

Ultimately, the question is if the presence of a *compluvium* and other architectural peculiarities, together with the high quality of the structures themselves, reflects a more Greco-Roman stylistic fashion or is direct consequence of a well-determined and characterising function of this building. Moreover, it is not certain if the inscription mentioning the *burgus* is really referring to this building; this does not consider that we do not know exactly what this term describes. Brown (1998b: 161) is nonetheless quite sure in affirming that the so-called "*praetorium*" was at least originally connected to some sort of civic. The fact that the first phases of this building are dated to the Roman period is indeed one of the few certainties available. More specifically, the complex composed by the *atrium* with the

[49] De Vries considers the "Barracks" as a better model for the local development of the Byzantine residential architecture, based on the courtyard-system (de Vries 2000: 40). I do not agree with this hypothesis, since the function of the Barracks does not seem to have been in any case entirely residential - eventually only some areas had a domestic use - and the development itself is different from the one we will see in the *praetorium*.

[50] The first to propose such interpretation was Bulter (1930: 160-166): he interpreted the complex as a *praetorium* for two reasons. The first one is the abovementioned inscription nr. 233 found in the so-called "Cathedral" (Figure 15), where a *burgus* - interpreted by Butler as a *praetorium* - is mentioned. The second is the presence of a cross-room, who as a direct parallel in the *praetorium* in Mismiyeh (Phanena), interpreted as an audience room.

Figure 18 - Aerial picture of the main structure of the praetorium of Umm el-Jimal (APAAME_20091019_SES-0145. Photographer: Stafford Smith, courtesy of APAAME)

Figure 19 - The complex of the praetorium (on the left end) (APAAME_20111002_MND-0832. Photographer: Matthew Dalton, courtesy of APAAME)

Figure 20 - Roof decoration in the "Cross-room" of the praetorium (photo by the author)

four ionic columns supporting the *compluvium*, the basilica and the cross-room would have been the original nucleus around which the later Byzantine "Complex House" would have been built.

According to the probes effectuated (Brown 1998b), the earliest features comes from the cruciform room, dating to the Late Roman's later phase, namely the third to the early 4th century AD. This possibly would be the period of construction of this entire nucleus, which was also provided with a channel, coming again to the cross-room, which exact function is still difficult to clarify. With the exception of this nucleus, all the other structures from the Late Roman period had been totally erased by the following building activities, which heavily modified the general organisation of the *praetorium*. If in fact the 4th and 5th centuries are missing in the archaeological record, a widespread activity involved the whole area in the Late Byzantine period, starting with the 5th century. Generally speaking, though, a strong continuity marks the passage from the Byzantine into the Early Islamic period without any evident change in the use of the spaces. The last phase is characterised by clear traces of abandonment, following a change in the use of the rooms, in turn probably consequent of the collapse of some of the structures, perhaps because of the earthquake in AD 749, which heavily hit this region. Like other buildings in the settlement – and plausibly the town in general follows the same chronology – a total abandonment starts in the 9th century and lasts until the Druze occupation in the Late Ottoman – Mandate period.

As perfectly demonstrated by the *praetorium*, the transition to the Byzantine period involved an impressive development and some heavy changes in the settlement, which led to a widespread massive destruction of the earlier structures and left few traces of them in the Byzantine town. The circumstances of this destruction are unclear: some relate them with the Palmyrene rebellion (Osinga 2017: 274), but no elements can be definitively related to the two episodes. What is clear is that already in the early 4th century building materials from earlier buildings are reused, and specifically for the construction of the Tetrarchic *castellum* (de Vries 2000: 40). Of this "legionary camp" inspired construction (Figure 21), few remains are left that can be described: only the external shape is clearly recognisable from the aerial picture, but the original inner organisation was totally disturbed in later periods. It would be logical to think that the abandonment, or more precisely the change in use of this structure, followed the construction of the "later *castellum*" (see infra) as a consequence of a reduction in the size of the garrison presiding the town (Parker 1998c: 131-142). Apparently, its military function ceased at the end of the fifth century, implying the development of some residential areas and at least one new church (de Vries 1998b: 231); however, only limited evidence on the changes taking place is available. Similarly, the so-called "Nabatean temple" (Figure 22) appears to be originally constructed in the Late Roman or Byzantine period, probably being one of the last non-Christian religious buildings built in the region. Being a part of the same quarter as the *praetorium*, this building was interpreted as Nabatean by Butler during its survey at the beginning of the 20th century in light of some decorations found by the participants of the expeditions. The excavations conducted here finally elucidated the chronology of the complex, which sheds light on several questions regarding the dating of the settlement's overall development (Parker 1998a; de Veaux and Parker 1998).

Figure 21 - The Tetrarchic castellum of Umm el-Jimal (APAAME_19980512_DLK-0056. Photographer: David Kennedy, courtesy of APAAME)

Figure 22 - The so-called "Nabatean Temple" of Umm el-Jimal (photo by the author)

According to the records, the area was not occupied before the Late Roman period, between the mid-second and early 4th century AD. The first structure however was only built at the end of this period, in the 4th century, as is quite confidently interpreted by Parker and de Veaux.[51] As seen in other buildings in Umm el-Jimal, the transition to the Late Byzantine period marks a radical change in the organisation and function of the complex. Between the 6th and 7th centuries AD, the reuse of the temple's cell is clearly attested with substantial modifications to the structure, for instance the removal of the pavement that left a simple beaten earth floor. A new domestic use of the spaces is suggested by the presence of at least two hearths. An occupation that - in contrast with what is observed elsewhere in the site - does not seem to continue into the Early Islamic period, for which no activities are documented. Only the Late Ottoman period sees a new phase of activities here, with the conversion of the cell into a stable and a continued domestic use of the other structures.

Another building with a challenging chronology is the "later *castellum*" (Figures 23, 24 and 25), the building that Butler named the "Barracks", for which different hypotheses were proposed. Its military character was immediately recognised by Butler, who also supposed a later conversion of the building into a monastery before its abandonment. According to the evidence revealed from some probes, Parker (1998c) substantially confirms the same periodisation offered by the Princeton team. The area was probably only partially occupied in the Northern sector by the Late Roman period. In the early Byzantine phase, between the 4th and the 5th centuries AD, building activities appear to increase consistently, with the creation at least of an enclosure wall and several installations within it. A *latrina*

[51] He affirms that "the building bears no relation to Nabatean cult buildings, but it does appear to fall within the tradition of Roman temples in the Levant" (de Veaux and Parker 1998: 160). The temple is described as a small complex, "with distyle in *antis* porch and triple doorway", finding direct parallels to the provincial architecture of the region. The temple possibly also had a *temenos* around the main building, later completely erased. The consequences on a purely historical level of such a late construction for a temple are particularly interesting, attesting that temples continued to be built even in a period when Christianism was already spread in the province of Arabia.

was also clearly constructed in this period, which continued to be used through the entire Late Byzantine and Early Islamic periods together with another room of which its function was not able to be determined. Parker ultimately confirms the validity of Butler's hypothesis, not only concerning the date of the construction of the building, but also in its interpretation as the *castellum* built by Flavius Pelagius Antipater mentioned in the inscription (Inscription nr. 237, Figure 26), seeing the impressive architectural similarities with the fort in Qasr el-Ba'ij.[52]

Figure 23 - The "Later Barracks" of Umm el-Jimal, looking south (photo by the author)

Figure 24 - The inner courtyard of the "Later Barracks" of Umm el-Jimal, looking north (photo by the author)

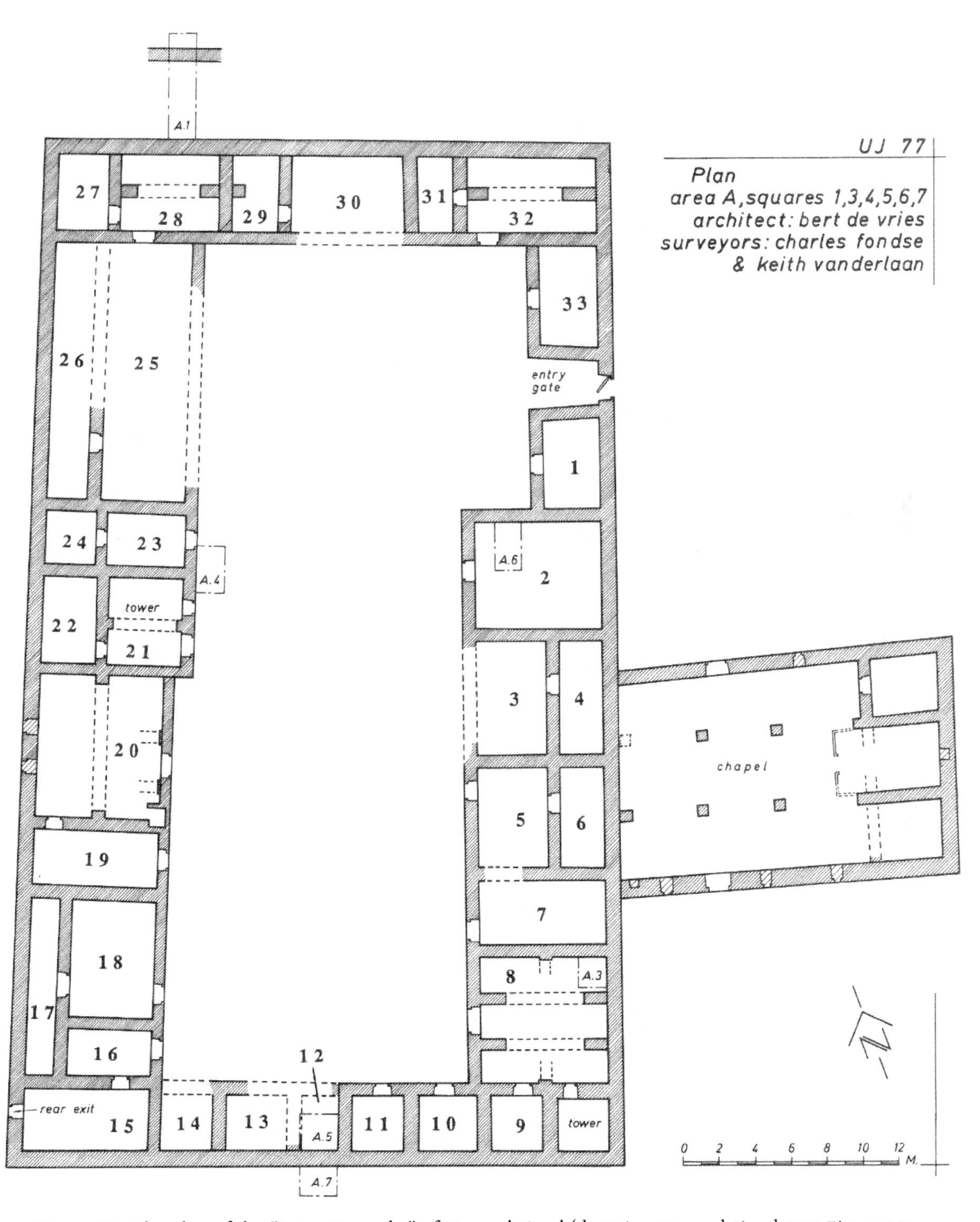

Figure 25 - The plan of the "Later Barracks" of Umm el-Jimal (de Vries, Umm el-Jimal: 133, Figure 83; courtesy of Prof. B. de Vries and the Umm el-Jimal Archaeological Project)

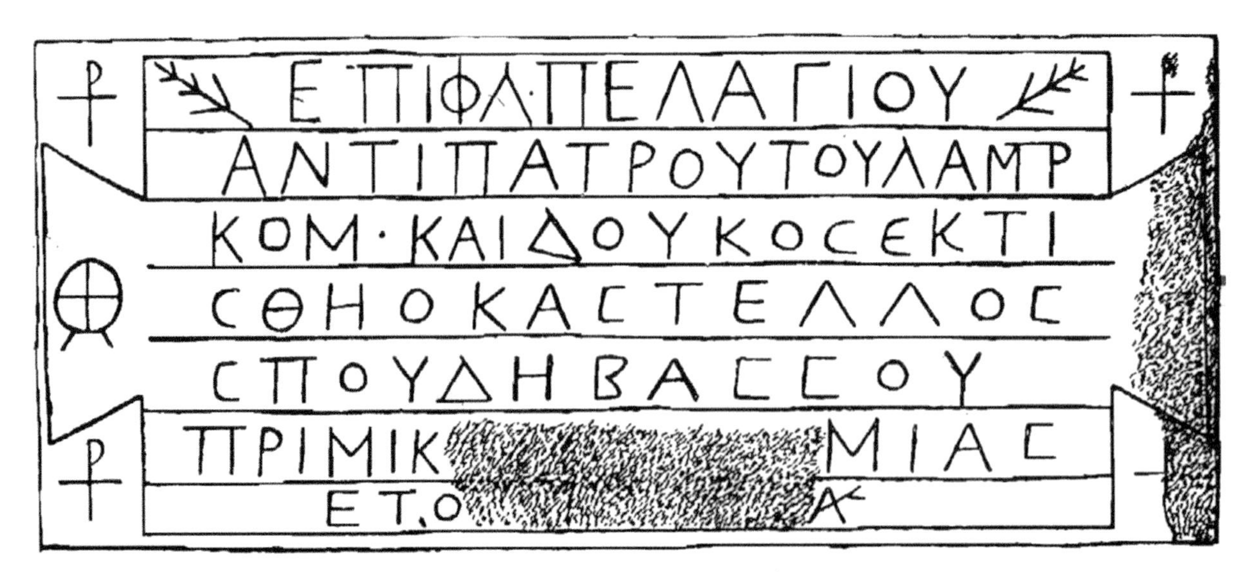

Figure 26 - Inscription nr. 237 from Umm el-Jimal with transcription (Littmann, Magie, and Stuart 1930: 136). Transcription: Ἐπὶ Φλ(αουίου) Πελαγίου / Ἀντιπάτρου τοῦ λαμπρ(οτάτου) / κόμ(ιτος) καὶ δουκὸς ἐκτί- / σθη ὁ κάστελλος / σπουδῇ Βάσσου / πριμικ

All of the structures from the original phase of the *castellum* were heavily modified in the later periods, when the complex underwent a noticeable development (Parker 1998c: 140). As Parker affirms, the Late Byzantine period sees a "substantial remodelling" of the complex with the creation of the tower in the south-eastern corner of the barracks, of the "corral" wall to the north and possibly a paved path between it and the proper barracks. The earlier structures also appear to undergo substantial modifications, such as the court's gate. In a later moment, a chapel was also added on the eastern side of the enclosure, an element that Butler interpreted as a conversion of the barracks into monastery.[53] This hypothesis is not entirely rejected by Parker, who nevertheless counts it simply as a possibility, since other *castella* in the region, most notably the aforementioned Qasr el-Baij, were also provided with such religious structures. Apparently, the structure continued to be used without substantial modifications, at least structurally, in the Early Islamic period when some rooms appear to have a domestic use, and nonetheless continued to be occupied until *c.* the 9th century AD. This moment represents, probably also for the entire settlement, the beginning of a long gap in occupation, which only seems to end with the Druze settlement in the region during the Late Ottoman – Mandate period.[54] These examples make it extremely evident that the crucial moment in the development of the Umm el-Jimal is indeed the Late Byzantine period.

It is especially in the 6th century AD that the development process, which began gradually in the Early Byzantine period, could have reached its apex:[55] an almost systematic dismantling and reoccupation of

[53] The addition of the chapel is summed to the presence of numerous "religious" invocations inscribed on the perimetric wall of the complex and the modern toponym Ed-Deir (meaning Convent): these are ultimately the reasons which Bulter is using in support of its hypothesis. De Vries, on the other hand, proposes a possible alternative use of the Barracks in their later phases as a caravanserai (Vries 1998b: 237; 2000: 3): in fact, he affirms that "if the open spaces [present in the settlement] were used for caravans, this building is the only candidate for the role of hostel or caravanserai".

[54] Nonetheless, some aspects of the reconstruction offered by Parker are particularly convincing and a renewed architectural analysis remains of the Barracks would be necessary, especially as far as the earlier phases of the construction are concerned.

[55] As clearly demonstrated by Osinga, the ceramics finds allow the identification of a general Late Byzantine Transition occupation and activity, and likewise the architecture. The data set, however, does not allow the sub-

the Hellenistic and Roman town of which only few scattered rests are left across the site. Almost all the new courtyard houses appear to be built starting from the 5th century, determining an extraordinary rate of expansion of the entire site, totalling eighteen churches and *c.* 140 "residential" units by the end of the Byzantine period. De Vries deduces a particularly interesting interpretation regarding the change in the settlement organisation in this phase: "the transition is not from planned to unplanned, but from planning from the outside inwards to planning from the inside outwards" (de Vries 1998a: 127). Even if referring to "planning" can be misleading, I agree with this view. The new focus of the building activity is on the private sphere: the new complexes are all organised around courtyards and the fulfilment of the daily life needs of a single family are the primary interest of the inhabitants. This is not to say that no attention is paid to the public sphere, since there is evidence for new public constructions continuing to take place, most notably of some religious complexes. Nonetheless it does not seem to be the first priority of the inhabitants of Umm el-Jimal.

As far as Umm el-Jimal's role in the regional context, the sheer size of the settlement reached by the site, *c.* 33 ha, is not the only factor that shows the settlement's significance during the Byzantine period, at least from an economic point of view. In fact, if one pays attention also to the presence of certain structures, this impression is reinforced. In particular, one of the major churches – the so-called "Cathedral" – played an important role and can be easily identified as the seat of a bishop (Figures 27 and 28). Similarly, Umm el-Jimal's administrative importance seems to be confirmed by the maintenance of military structures, even though on a reduced scale, with the only "later *castellum*" still in use.

A more delicate matter is the settlement's transition into the Early Islamic period. The examples considered until now tend to suggest a continuity in the occupation of many of the structures. In most cases, the function of some spaces appears to change, but normally the structures are kept in place. The

Figure 27 - The apsis of the "Cathedral" of Umm el-Jimal (photo by the author)

phases to be clearly distinguished and the occupations during the 5th, the 6th and 7th centuries to be precisely divided (Osinga 2017: 275-279).

Figure 28 - The entrance and the Nartex of the "Cathedral" in Umm el-Jimal (photo by the author)

only trend that is quite clear throughout the entire settlement is a systematic removal of debris and the laying of new floors.[56] This is not only evident in the Umayyad phases in the *Praetorium*, but also in another particularly interesting complex, namely the largest house in the site (Houses XVII-XVIII, Figures 29 and 30). Recent excavations in the complex have offered new important information, not only on the development of the single complex, but also on the chronology of the entire site (Osinga 2017: 105-141). The original construction of the structure appears to have taken place in the Late Byzantine, though possibly even in the Early Byzantine period (4th century), similarly to the *praetorium* and the other residential units. It is not clear if it reflects the original organisation of the complex, but it is certain that at one point during the Late Byzantine period House XVII and XVIII were unified into a single large complex. It needs to be further stressed that apart from the standing structures, for which their dating is problematic, no clear evidence of the 6th century can be found, since the later refurbishing and reflooring activities heavily disturbed the traces of these earlier phases.

This is a problem shared by the Early Umayyad phase, in which the chronology is further complicated by the problematic distinction between material culture from the transition period and the Late Byzantine period. The major contribution of the new excavation likely concerns the later occupation of the complex. It was believed earlier, as suggested by the excavation of two rooms in 1977, that the Umayyad period marked the end of the occupation of the complex (Brown 1998a: 195-201). More specifically, this period was subdivided into three sub-phases: the first one bearing witness to the removal of all the debris, followed by the creation of the new floor, and finally the abandonment of the room. The phase of abandonment appears to begin with the Abbasid period and last until the Druze occupation in the Late Ottoman – Mandate period. Recent data from excavation, however, offer a radically different chronological framework. It appears that the Late Umayyad-Abbasid periods in fact bore witness to an important activity in the entire complex (Osinga 2017: 137-138), as partially

[56] De Vries (2000: 4) suggests that the "clean-up phenomenon" dictated by hygienic necessity. The debris removed in the Umayyad period were apparently consequence of the accumulation of material "from dense occupation stretching over at least three centuries".

suggested by Brown's excavation. In particular, Osinga sustained the South Gate was entirely an 8th century construction (Osinga 2017: 115-117). I have some doubt regarding this dating for two reasons. The first is that they stopped excavating at a basaltic floor underlying a phase with mixed Roman through 8th century material; in other rooms and complex, it was shown that several refurbishing and

Figure 29 - Plan of House XVII-XVIII (Osinga 2017: 106, Figure 4.2; courtesy of Dr E. Osinga)

Figure 30 - House XVIII of Umm el-Jimal (photo by the author)

reflooring activities tended to erase all traces of the earlier phases. Not excavating until the bedrock ultimately does not offer a dating for the standing structures, and this an important piece of information lacking, especially in light of a purely architectonical analysis of the gate. This is ultimately the second doubt I have in regard to the chronological hypothesis offered. From the interfaces (or line of intersection) indicated in the photographic documentation by Osinga (Figure 31), it is clear that the central building forming the gate underwent several transformations, although it is not in its entirety dated later than all the other structures, as it seems to be suggested. The right wall is abutting against the central building, being clearly a later addition as is also the case for the upper part of the tower. Similarly, the arched central "tower" is later than the wall on the left, which on the other hand, was also partially rebuilt in a later phase, probably at the same time of the upper part of the tower. The situation is more homogeneous on the external face of the building, where the entire eastern part, including the lower part of the central tower, appears to belong to the same phase, as well as the eastern and the lowest section of the western parts of the *vestibulum*. On the contrary, the upper part of the western wall of the *vestibulum* seems to join to the wall of the arched wall. From inside the *vestibulum*, it is also clear that the arched door, also typologically different from the outer entrance to the building, is abutting against the entire eastern wall, being particularly evident also in coincidence with the first corbelling line.

Figure 31 - The "south Gate" of House XVIII in Umm el-Jimal (Osinga 2017: 137, Figure 4.14; courtesy of Dr E. Osinga)

Thus, if the refurbishing of the gate dates to the 8th century, and possibly its restoration after a partial collapse of an older structure, with the construction of the new inner gate, it follows that the eastern part of the gate and more importantly the outer section dates to an earlier phase. Of course, this hypothesis is pending on further excavation data. Anyway, what is certain is that the 8th century sees a continuation of the occupation of the complex, which apparently did not change consistently in its use. Furthermore, new excavations have also attested a Mamluk phase of occupation of the complex, especially visible in the re-plastering of the cistern[57], reflooring of some areas and restoration of some structures, before the modern reuse of some rooms, most notably the stables in House XVII.

Generally speaking, what has emerged from the data confirms that the Early Islamic period did not represent a dramatic change in the settlement in its general appearance, such as the organisation of the spaces in the houses. We simply can observe the maintenance or the continuation of the same processes, or better yet building behaviours, which we have already seen in the Late Byzantine period, if not earlier.[58] It is indeed the Byzantine phase of the site that represents the site's most radical spatial revolution, not only in terms of size, but also in the conceptual spatial organisation ruling all the new building activities. At the same time, one can clearly see a definitive phase of abandonment starting with the Abbasid period: the 9th century represents the beginning of a long gap in Umm el-Jimal's occupation, which was briefly revived during the Mamluk period and then again definitively in the Modern period. The causes of the abandonment are still uncertain: the hypothesis proposed by de Vries is that the devastating earthquake that struck the Levant in AD 749 hit the local communities hard, and they were not able to recover completely (De Vries 1998b: 231-232). Surely, it took place in a span of between AD 749 and 800, since no ninth century material has been found in the site (Osinga 2017: 283-284). As far as the modern occupation is concerned, its building activity normally does not influence our interpretation of the ancient structures very much: the two main trends are the rebuilding of roofing – that nonetheless are clearly distinguishable from the Byzantine ones thanks to a much less carefully applied technique – or the edification of irregular dry-stone enclosures or small rooms.[59]

As mentioned before, the Umm el-Jimal one can visit today is still substantially the site of the 6th century at its maximum extension (Figure 32). The following analysis therefore specifically refers to this phase. Nonetheless, several of the observations are also valid for the following Early Islamic phases, which did not alter the general organisation of the settlement. On the highest level of analysis, Umm el-Jimal shows its most peculiar features, shared on a reduced scale also by other settlements in the area like Umm es-Surab. The settlement area is clearly delimited by a wall, offering only five accesses.[60] The characteristics of these enclosures are themselves interesting and have been mentioned above: a military function of these walls can be quite certainly excluded, considering its extremely thin structure and absence of proper towers along its track. Similarly, the accesses show little in common with proper city gates, with the only true exception being the so-called "Commodus Gate" (see Figures 15 and 16)

---

[57] Dated to AD 1220-1388 by al-Bashaureh (quoted in Osinga 2017: 139).

[58] De Vries mentions the possible conversion of two churches into mosques, namely the Numerianos and the West churches. Even if it represents important cultural information – showing that at least a portion of the population did embrace Islam already in the Umayyad period – it does not represent a significant change from a strictly urban perspective, since the religious function of the building is maintained.

[59] These later phases had been well documented in Umm es-Surab in some of the complexes surveyed by University of Siena (see below). The case of Umm el-Jimal does not seem to differ much from this other site as far as the Druze occupation is concerned.

[60] In the survey made in 1905 by Butler, also a sixth gate is shown, opening to the south (Butler 1930: 158). The gate was already ruined when the Princeton expedition visited the site and the construction of the modern road had destroyed even more heavily this portion of the site.

Figure 32 - General plan of Umm el-Jimal (after de Vries 1998c: 15, Figure 6; courtesy of Prof. B. de Vries and the Umm el-Jimal Archaeological Project)

and partially the East gate (Figure 33).[61] These are in fact the only ones also presenting small towers[62] controlling the entrance. The first one seems to be the main entrance to the settlement, of which its importance is also underlined by the Latin imperial inscriptions. All of the other accesses to the town on the contrary are simple gaps left between the buildings, appearing more like normal house doors than urban entrances (Figure 34); in some cases, one can find arches between two buildings like the Southwestern entrance or small door with inscribed lintels like the western gate.[63] It is therefore clear that whatever the real function of the Byzantine "walls" were, the features hardly fit the description of a proper military structure. Even if it might have provided a certain degree of defence from the raids of the nomadic groups of the area, this was almost surely not its main purpose.

Nonetheless, and most interestingly, the walls clearly divide the inner side of the settlement from the outer countryside. This aspect is even more important in light of the inner spatial organisation of the settlement: in fact, the first and more evident aspect is absence of a compacted settled area, or at least of the built surface. On the contrary, a large part of the site included by the walls appears as a blank area, separating the different quarters. When looking more closely at these "blank spaces" one realises

Figure 33 - The so-called "East Gate" (Butler 1930: Figure 135)

[61] It's simpler and more modest than the "Commodus Gate". It is a single arched gate, without any tower on the sides (Butler 1930: 157).
[62] According to the Butler's survey, also the southern gate should had shown the same structure, with tower flanking the arched passageway. As mentioned, though, the structure was already badly conserved and Butler had not offered any other detail. Today no traces are left.
[63] Both are briefly described by Bulter (1930: 157-158). In general, since the wall is abutted by houses or other buildings in many sections, these "gates" look more like simple openings between the complexes forming the quarters.

Figure 34 - The Western entrance to the settlement (photo by the author)

that, even if not occupied by buildings, they were far from "empty". A series of raw and often single-row enclosures were spread across these areas. Their precise purpose is difficult to establish, and they do not always appear to be directly connected to any particular complex or building. The presence of cisterns among these enclosures may suggest that these areas were to a certain degree designated for public use. Considering the area where Umm el-Jimal is located, it is not difficult to conceive the possibility of a more commercial or trade-oriented facility: the enclosures could therefore answer to the more or less frequent need to accommodate the livestock that would be sold on market days and fairs. They could ultimately represent the only functionally defined area of the town, since the residential and productive functions in general do not appear to be spatially separated in specialised areas. Interestingly, only one gate opens directly to these areas: the "Commodus Gate", coincidentally the main entrance to the town (Figure 6). Moreover, it allows access to the more extended of the blank spaces, where also many of the most rich and noticeable buildings of Umm el-Jimal open to, such as the so-called "Praetorium" (Figure 35b), the Numerianus Church (Figure 35c), the so-called "Cathedral" (Figure 35d) and the West Church (Figure 35e).

These "blank spaces" are ultimately defined by the different quarters; the distribution of such quarters is the other most noticeable spatial feature of the settlement. We can easily identify four different quarters, each one occupying more or less one corner of the settlement. One exception to this rule because of its position within the settlement is the old *castrum*, lying midway between the northern and south-eastern quarters (see Figure 21). The Roman fortress built during the third to fourth centuries follows Diocletian's project of reforming the provinces and, in this particular case, strengthening the so-called *limes Arabicus*. Even if there are not many certainties available for the constructions from the earliest phases of Umm el-Jimal, it is certain that this structure was the main military structure of the site, until the construction of the later *castellum*. As mentioned above, the construction of this building, or better yet the final realisation of it, probably determines the fate of Diocletian's fort: it was converted into a productive and residential area, and in addition one church was erected. This quarter ultimately presents different features than the others in the site (see infra).

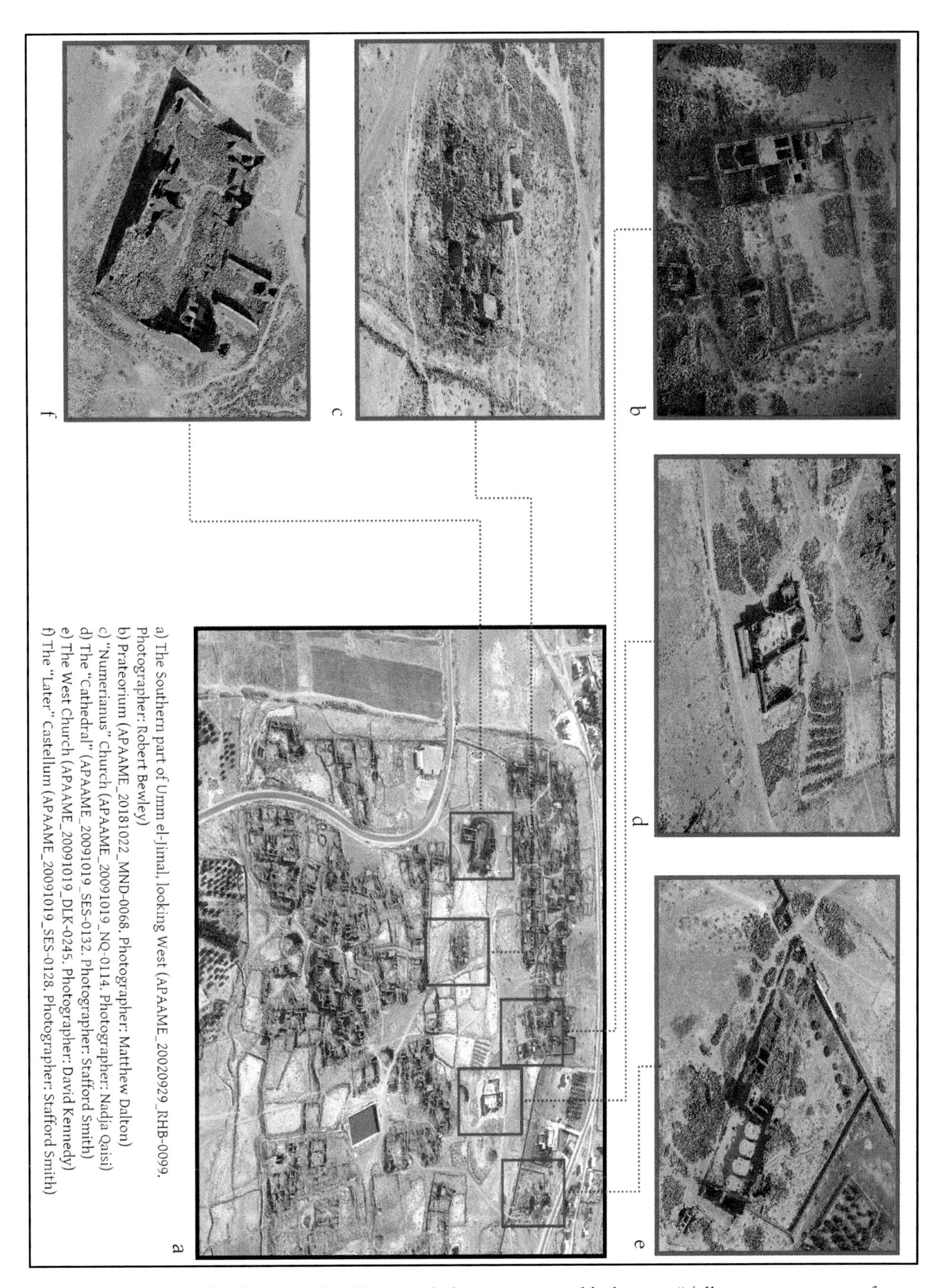

Figure 35 - Isolated buildings and buildings with direct access to "black spaces" (all pictures courtesy of APAAME)

Normally, the disposition of the different areas, including the "blank" ones, does not follow any geometrical plan, potentially being generated by the progressive addition of buildings next to one other. No raster of the streets is detectable; the circulation inside the settlement, also towards the outside, follows more spontaneous pathways. Similarly, no clear central organisation is developed for the hydraulic system. One aqueduct supplies the large reservoir next to the old Roman fort, the only clearly public facility identifiable in the settlement, and apparently four "private" cisterns: one in the Numerianos Church, one in the "Praetorium", and the other two in private dwellings in the southwestern and eastern quarters. One cannot exclude the possibility that other complexes were supplied by this aqueduct or by secondary channels from the reservoir, but we do not have much information on this matter. What is certain is that no systematic hydraulic network is detectable on a complete urban scale and most of the complexes are independent, as far as the provision of water is concerned. In addition to the four quarters, a series of single complexes, completely isolated from the other quarters and surrounded by "blank" areas, can be pointed out. The relative importance indicated by their prominent position is ultimately connected with their function. The first is the only military building of the settlement after the conversion of the old *castrum*. It's uncertain if (and eventually in which terms) this structure changed its function: from barracks into monastery (as suggested by Butler) or caravanserai (as suggested by de Vries); the only certainty is that the later-dated *Castellum* remains in use and stays isolated (Figure 35f). The other three single complexes are churches, the only ones not integrated in a quarter or even directly connected with a single house: the Numerianus Church (Figure 35c), the so-called "Cathedral" (Figure 35d) and the West Church (Figure 35e). They do not present any extraordinary features from an architectural or decorative point of view, nor do they show clear marks for a particular status. For instance, the Julianos Church, embedded in the northern quarter, has similar characteristics and also more or less the same dimensions.

Generally speaking, it is difficult to see a proper recurring typology of churches among the fifteen examples in the site: there is a high variability in forms and size (Butler 1930: 171-194). I would say that there is only one element that makes these three buildings peculiar: their position. Their contrast with the other churches is striking: some clearly appear to be "private" religious buildings, accessible only through private courtyards (Figures 36c, 36d and 37c), while others are likely part of urban monastic complexes (Figure 38d).[64] On the opposite, these three churches are not integrated in any quarter and this isolation seems to suggest that they not only held a major predominance in the urban hierarchies, but also a more accentuated communal dimension. This is particularly true for the "Cathedral" and the West Church.[65] In fact, the first (see Figure 35d) occupies the exact centre of the settlement, just a short distance north from the new barracks. The position appears to be extremely important, not only for its more or less equivalent distance from each quarter but also for its direct visual connection with the "Commodus" gate. The West Church (see Figure 35e), on the other hand, is particular since it is outside the walls, but enclosed by a new wall, similar to the first one, immediately next to the "Commodus" Gate. It appears as an attached area, not completely inside the settlement, since the original wall is maintained and only a smaller gate is opened, but not entirely outside, in light of its enclosure. Generally speaking, on an urban scale, each component, the quarters and the single complexes, appears to be isolated and autonomous from the others, without any direct connection linking them. This separation is clearly neither determined by a functional specialisation of the different parts of the

[64] Not surprisingly the Southwest Church and the Masechos Church and their connected structures are on the fringes of the settlement.

[65] The Numerianos Church is also not directly integrated in the southwestern quarter, although the distance between them is quite short, and a series of enclosures connect the building with the other structures to a certain extent. Nonetheless, it should still be considered a totally autonomous and isolated complex.

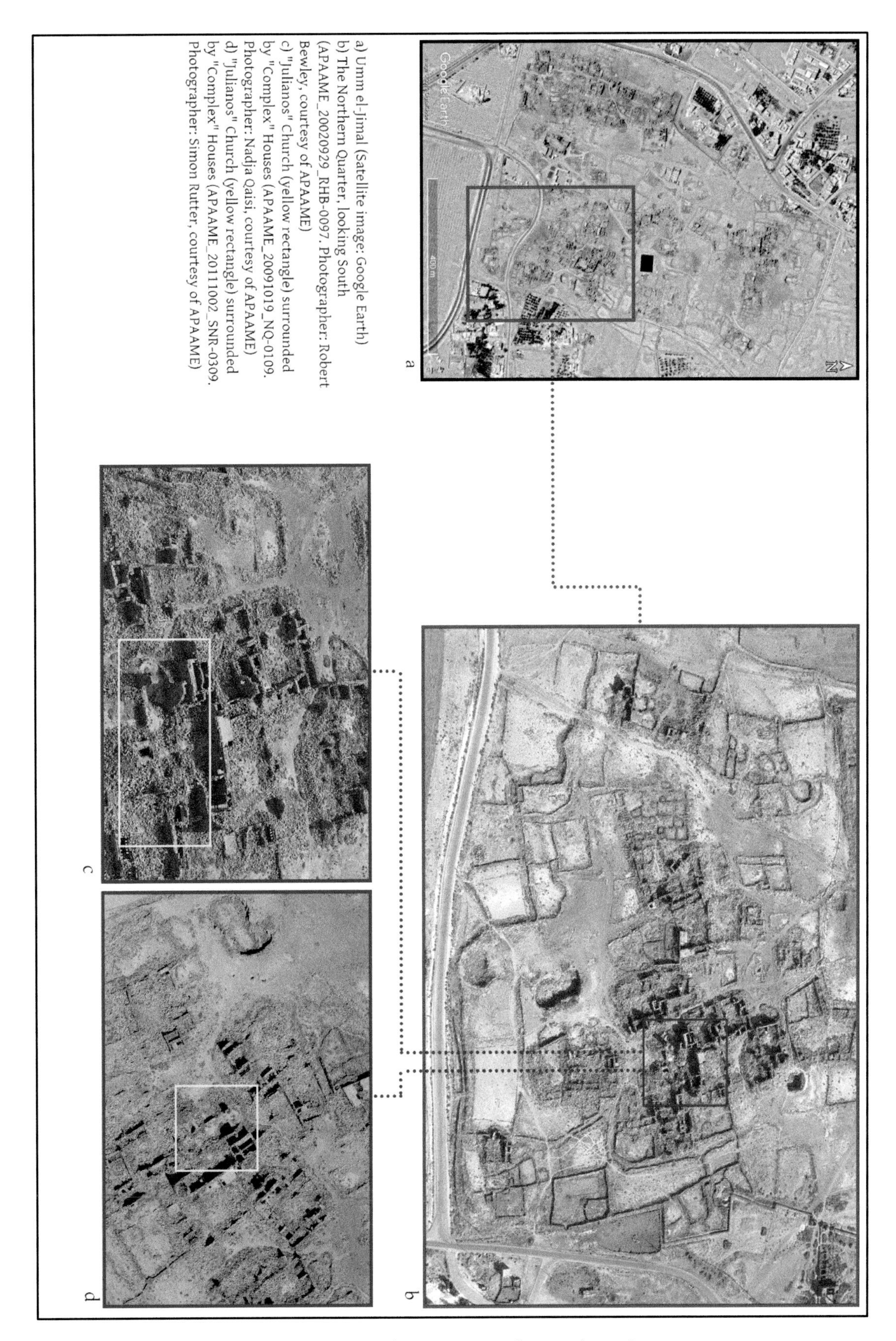

Figure 36 - Northern Quarter of Umm el-Jimal

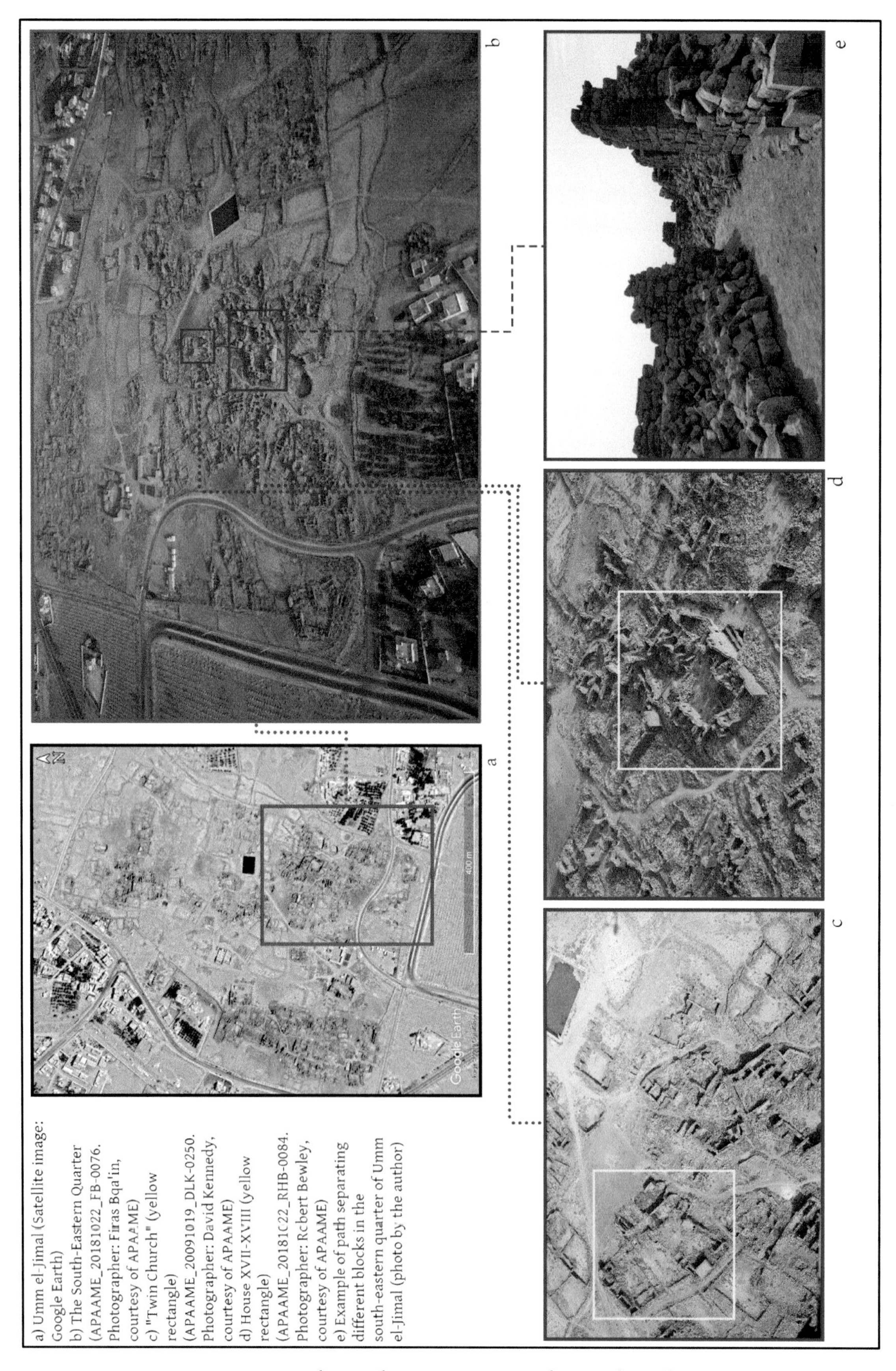

Figure 37 - The South-Eastern Quarter of Umm el-Jimal

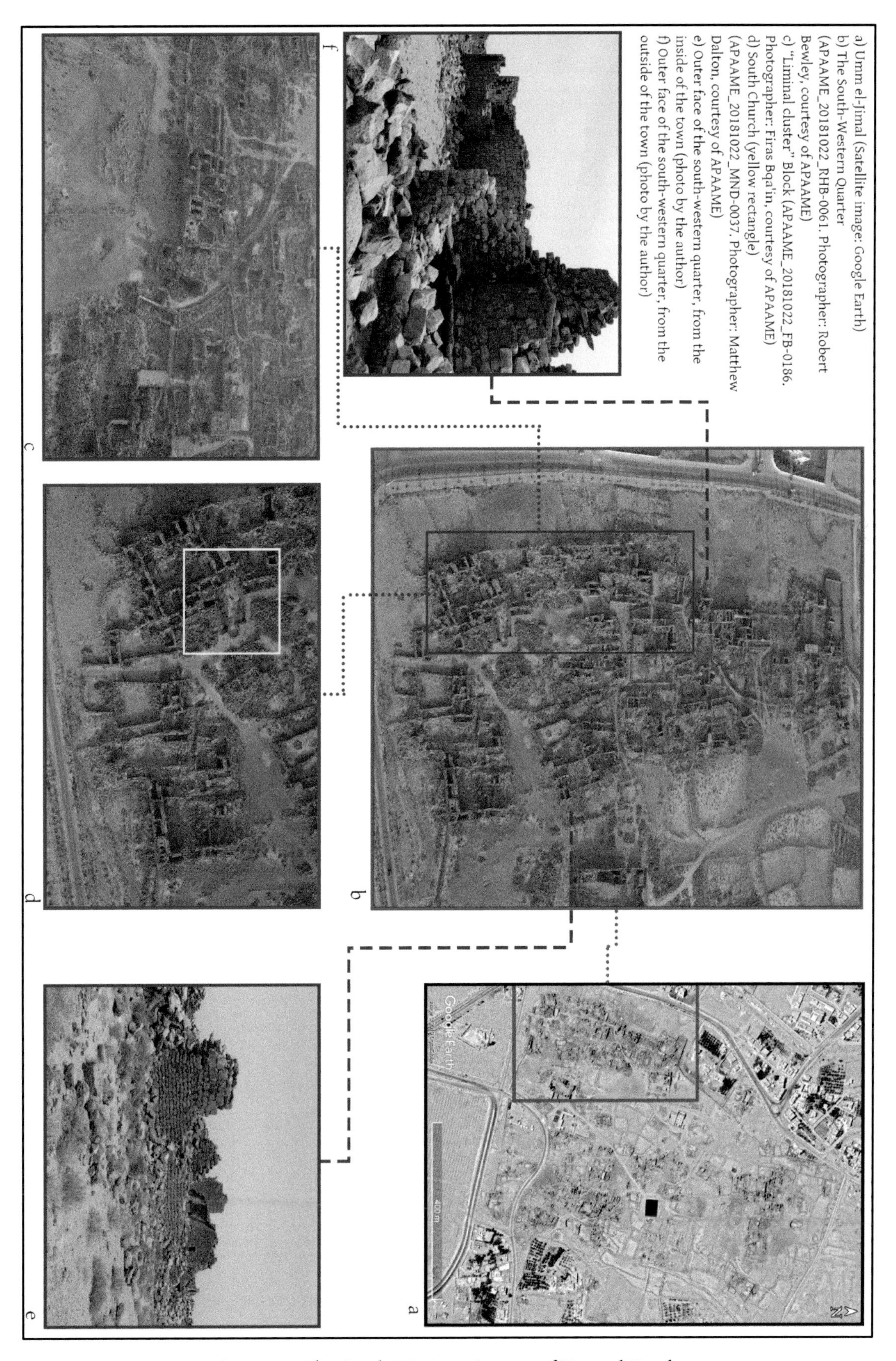

Figure 38 - The South-Western Quarter of Umm el-Jimal

settlement – for instance, there are no traces of an industrial or commercial zone separated from the residential quarters – nor suggests an economic stratification in the community, since the larger houses are present in all the quarters like the smaller ones. The only tie that keeps everything together, offering a sort of unitary appearance to the settlement, is the not military wall. Forcing the interpretation slightly, one can even consider this aspect as the demarcation of the "identity border" within the community living in Umm el-Jimal, isolating and clearly distinguishing it from the outside. The basis of this communal recognition, though at this stage, is far from clearly identifiable.

Therefore, all the quarters are irregular clusters of buildings, although in a slightly more orderly disposition if it was situated in a peripheral position, like the western quarter (Figure 38c). The throughways are actually narrow paths obtained by leaving some space in between the blocks (Figure 37e). The separation from the rest of the settlement is clearly marked through the outer face of the quarter, presenting an almost continuous and blind front to those coming from other parts of the town (Figure 38e).

This same aspect is repeated towards the outside of the town, since the buildings are abutted and, in some sections, are actually forming the aforementioned city-wall (Figures 38c and 38f). No windows or doors – if not some high and small openings for the inner aeration – are present, since they all overlook inwards the quarter.[66] Only few accesses are spared, leading to the inner path-system developed between the blocks. No clear preferential entrance is marked: no proper gate is identifiable and even the ones opening towards the outside of the settlement have a modest appearance. This last aspect – the direct access from the outside – is indeed a common feature to all the quarters. The proximity of these openings of enclosures or even small fields that create a sort of "buffer zone" between the quarters, and ultimately the entire settlement and countryside is indeed interesting. The general impression is that there are smaller settlements within a larger settlement, each quarter having its own development almost independent from the others. As mentioned above, the only element bringing all these scattered quarters into a communal urban dimension is the wall surrounding the town and the few isolated buildings.

As mentioned above, the old *castrum* (see Figure 21) represents the only exception to the rest of the quarters for certain aspects, being the only one that presents a clear geometric shape and a more or less identifiable street system. Nonetheless, the conversion from the military designation determined a deep and radical change in the inner layout. As it may be noticed in other sites even more clearly[67], the typical Roman organisation is not recognisable, with only few of the major streets spared, normally the *via principalis* as in the case of Umm el-Jimal, and most of the blocks, or better yet the former "barracks", unified or completely rebuilt. The result is an inner organisation similar to the one seen in the other quarters, with a series of spontaneous pathways and *cul-de-sac* branching off and determining the different blocks. The tendency of keeping the outer fortification of the fort in place, as margins of the new quarter, also is directly parallel to the need to clearly distinguish the quarter-space from the public sphere.

The self-sufficiency of the several quarters is also marked by one further aspect: the presence of religious facilities. Fifteen churches and chapels have been identified thus far and if one considers that there is a total 140 houses in Umm el-Jimal, this proportion is impressive. Each quarter has at least one church, and in some cases also a quite complex assembly of buildings, with one or more courtyards and

[66] And more specifically, towards the courtyards, that mediate the access to the houses from the street. See below.
[67] For instance, this is also the case in Umm er-Rasas (see Chapter 6.2).

facilities connected to the major building, which possibly identifies urban monasteries.[68] For most examples, there is no separation between these religious centres and the rest of the structures, with whom they often share some walls if not also courtyards. De Vries counts eight churches that "were built into private complexes, some of which, like that of the Double Church [in the eastern quarter], were domestic. Four of these had their entries from the courtyard of those complexes, so that they appear to be entirely private [...] The other four, as well as the remaining seven (that is eleven in total) had entrances accessible from the street)" (de Vries 2000: 3). In any case, the presence of a church does not provide any particular spatial solution, being included in the "urban framework" like the other non-religious buildings. It is a further confirmation that even in the quarters, the functional mix of the spaces is one of the most typical features in these settlements. Thus, the religious function isn't treated differently from other functions, and the buildings conforming to it are normally embedded in the urban texture as the other more domestic structures.

Many of the features described for the quarters are also applicable on a lower level of analysis, namely the block. In Umm el-Jimal there were three substantially different ways of structuring them, to a certain degree connected to their relative position within the quarter and the settlement. The most frequent one, normally occupying the bulk of the quarters, is a cluster of courthouses (Figure 39). The single units are simply juxtaposed, sharing some structures like walls dividing the single courtyards.

Figure 39 - An example of Cluster Block from the South-Eastern quarter in Umm el-Jimal (APAAME_20181022_RHB-0099. Photographer: Robert Bewley, courtesy of APAAME)

[68] It would be interesting to determine the precise chronology of the development of the quarters where urban monasteries can also be found. If the monasteries were in fact the earliest structures to be built, followed by the houses, we could see in a more urban context the same process that we possibly see in some villages like Umm es-Surab: namely, the role of such complexes - exactly as forts - in attracting and aggregating buildings in their surroundings, especially in cases of sedentarisation dynamics.

The dimension may vary considerably depending on the number of units attached one to the other. The other solution is mainly more frequent on the borders of the settlement (see Figure 38c). In these blocks, the single units appear to be organised in a single row, with all the entrances oriented towards the inside of the settlement. This solution can be explained if one thinks that, being the borders of the town, they integrate (if not also to a certain extent constitute) the walls surrounding the entire site; this is most visible in the western and eastern quarters. The last solution is the least recurring and is closer to the "*insula*" of the Hellenised settlements (Figure 40): the geometric shape is normally the most visible feature, but is present in only a very limited number of cases in Umm el-Jimal. In fact, even more geometric single dwellings, like the *Praetorium,* are indeed part of larger irregular clusters, surely not following the Hellenistic model. Apart from the different relative disposition of the units, the various blocks share almost all of the other features. As for the quarter in general, the self-sufficiency of each part seems to be the basis of the organisational solutions that were implemented. The water supply is guaranteed by at least one cistern in each block, normally there is one in each household, often isolated from the central "public" water system. Considering the proximity to the major reservoir and the aqueduct, the eastern quarter seems to have the greatest number of blocks connected to that infrastructure, even if not systematically. One further channel also connects the reservoir with one of the external blocks in the western quarter, but these are the only documented exceptions.

In addition, the "independence" of each unit within the block is emphasized by the organisation of the accesses and the inner circulation of the blocks. In fact, the different complexes normally have their own entrance and furthermore the inner circulation between units is avoided. One can rarely find *cul-de-sacs* or courtyards opening into the entrances, even if there are some important examples like in the northern part of the western quarter (see Plate 6 and Figure 19). The creation of this sort of "buffer

Figure 40 - The Nabatean Complex, in the South-Western quarter, is one of the few examples of "insula" type Block in Umm el-Jimal (APAAME_20181022_MND-0067. Photographer: Matthew Dalton, courtesy of APAAME)

zone" or semi-private areas, mediating the completely private areas to the public paths, appears to be mainly connected with the wider blocks. Moreover, only few major complexes appear to have units connected by a direct passageway, but these are clearly later phases of use of the structures. Originally the complexes were separated houses, which were united after being owned by the same family.

These last remarks on the private and public spheres confront an important issue. As previously mentioned, the need for privacy played an essential role in determining spatial solutions, especially in these regions. The fact that the blocks, like the quarters in general, are blind to the outside can be explained in relationship to this need. The private areas, where the daily life of families took place, had to be separate and removed from the external view as much as possible. Nonetheless, the way the blocks are organised in Umm el-Jimal, in my opinion, betrays a reduced feeling of risk towards an individual family's privacy. It is suggested by the scarcity of devices such as the *cul-de-sac* and implementing solutions that prevented even the slightest intrusion in the private sphere and creating – as mentioned before – a sort of "buffer zone". The courtyards often open directly to the streets and sometimes share dividing walls with opens spaces of the other units of the block. Considering that the buildings were frequently more than one storey high and that roofs were also used, it follows that "vertical" privacy was almost absent. The impression is that the level of intimacy, or of social control, among the people living in both the block and the entire quarter, allowed for less strict privacy devices. One cannot say the same in relation to the people living in the other quarters or outside the settlement, since the way the external face of the quarters was conceived completely precludes outsiders from having direct visual contact with the inner areas.

In regard to the functional specialisation of blocks shows the same trend as the quarters in general, namely the absence of any clear functional zoning of the spaces. The residential and productive functions often share the same spaces and are not easily distinguished in the archaeological reports, since large recognisable installations, such as olive or vine presses or kilns, are absent in the site. Only in few examples, some rooms recalling the form of *tabernae* could be identified: their small dimensions and the direct opening to the major pathways may suggest a sort of commercial use of these spaces. But since no structure or installation has been found, this interpretation is still tentative. It is also important to mention here that several blocks have enclosures attached to the external walls, probably in order to keep livestock. The only clear functional specialisation of complexes, as is the case for the quarter, is in presence of religious buildings, churches or chapels. Nonetheless, they appear to be tightly merged with the other structures of the blocks, normally guaranteeing a simpler inner circulation than in the clusters only constituted by "residential" units. As seen for the major residential units, it could be that some of the largest churches also had some facilities in the immediate proximity, such as residences and stores, but also in this case the material documentation is not sufficient to clearly distinguish an urban monastery from "private" churches build in domestic buildings. It can therefore be stated quite confidently that the functional versatility of the spaces is a general feature of the settlement on all levels of analysis, and it's even more evident on a house level.

Almost all of the houses in Umm el-Jimal share the same spatial concept. The homogeneity in their set of features is incredible and only few exceptions can be pointed out. The "conjunctive" or "complex" house is the standard form: no clear limitations were set per lot, allowing the progressive addition of new structures according to the momentary need or possibilities. The shape of the houses is therefore extremely variable, but regular geometric plans are rare. The courtyard is normally flanked by roofed structures on at least two entire sides, and the single houses are not very frequent. Generally speaking, the domestic architecture in Umm el-Jimal follows the tradition of the Hauran as described by several

French expeditions is Syria. In the southern part of the Hauran, the "L" or "U" shaped roofed areas are also organised on at least two storeys around a courtyard[69], which may ultimately confer a more geometric appearance to the entire complex.

In Umm el-Jimal, there are only a few peculiar houses that present some diversity from the standard Hauranian domestic architecture. The reasons for these differences might be understood in light of particular pre-existing complexes embedded in the new complex, or possibly specific social dynamics. It is therefore not entirely surprising to find the so-called "*praetorium*" among these "exceptions". The aforementioned development of the structure, from its Late-Roman foundation until the Late Byzantine development, and the changes that took place specifically in the Byzantine period have conferred the definitive form to the building. The main unit of the courtyard system is ultimately what is left from the earlier buildings, and this is the aspect that is most surprising, considering the degree of destruction that affected the Roman and Early Byzantine settlement (see Figure 18). The fact that only the assumed reception rooms were maintained can probably be understood in light of their excellent architectural quality and decorations, even if we cannot be sure if the original function of these ambiences was kept. The "*praetorium*" also maintains some of the water infrastructure from its earlier phases: a drain supplying a small basin – just behind the cruciform room – and the cistern. Furthermore, according to the general plan of the settlement made by de Vries, it seems that this cistern was also connected to the public aqueduct, underlining the importance of this complex – one of the few structures surely benefitting from this infrastructure. The addition of the Late Byzantine structures, abutting on the southwestern corner of the Roman structure, confers the more common "L-shape" to the entire complex, again entirely covering two side of the new courtyard. These new structures appear to have only one upper storey, while the original building had two. It is interesting to note – in relation to the privacy-issue – that the new aisle is in a way blocking the view from the neighbouring dwelling to the West. On the other hand, the side of the courtyard facing the inner part of the settlement are simply delimited by an enclosure wall, providing a less efficient visual separation of the inner space. The contrast with the other liminal building in the quarter – and also of the other quarters – is quite striking and it would be interesting to understand with a major degree of certainty if these differences are due to particular architectural development of the complex or to the maintenance of an exceptional function within the settlement.

The other complexes that present some distinctive features actually form a block of their own, and at a certain point are confidently, if not originally, unified into a single unit. House XVII and House XVIII already attracted Butler's attention (Butler 1930: 199-202) in light of the large size of the block – especially of House XVIII – and more specifically their limited correspondence with the "typical" domestic architecture in the Hauran (Figures 37d and 41). Even when considered as two different units - as supposed by Butler – it is indeed without a doubt that both houses hardly recall the typical courtyard farmhouse from this region, but for different reasons. In the case of House XVII, in fact, the plan and the dimension more closely resemble some "Courtyard Houses" in the Negev settlements, while House XVIII recalls the "Complex House" type in a more Hauranian tradition. When looking at the complex in its entirety, its unicity is also striking, both in the site and also on a regional scale. The only direct parallel that can be found thus far is Building XII in Mampsis (Chapter 4.3.2). Therefore, the two parts of the complex in Umm el-Jimal may reflect a different specialisation of the spaces and perhaps an internal functional zoning. The major difference is the presence of stables in House XVII, which on the contrary are absent in House XVIII. This absence would have been particularly striking considering the dimension of the unit, and the fact that the stables are a normally a common feature in

[69] The upper floor is always accessible through stares placed on the outer face of the walls facing the courtyard.

Figure 41 - House XVII-XVIII

the domestic architecture in the region. Being connected with House XVII, House XVIII consequently would have also been equipped with some stables, which were clearly distinguished by an area that we can interpret as the residential one. Though both zones have similar features, the most notable is the presence of former entrances, each given access to a *vestibulum*, mediating the entrance to the central courtyard. In the case of House XVII, the *vestibulum* gives also access one of the stables. In House XVIII, one definitive entrance is to the east, and is the main one according to Osinga. As previously mentioned, it is uncertain if the south gate was already present during the Late Byzantine to Early Umayyad phases, even if I am quite convinced it was, it was possibly not present in the same form as preserved today. Moreover, one further entrance was present on the western side of the unit (Room 17b), but it was blocked at a certain time. The only definitive aspect is that no direct access to the central courtyard from the outside was possible, and in terms of privacy patterns this is a feature we do not systematically find in Umm el-Jimal. Of course, this aspect already underlines the importance – functional and/or social – of the complex. The separation from the outside is also in terms of vertical privacy: both parts of the complex had a minimum of two storeys on each side, blocking the view even from the roofs of the surrounding buildings. Above the *vestibulum* of House XVII there was even a tower, certainly built in a later phase than the original construction, and according to Bulter could had been at least five storeys high.[70] Similarly, the eastern side of House XVIII had at least three floors, as clearly shown by the *bifora* that is still standing (Figure 41c).

Further considering the upper storeys allows the possibility to underline a further difference between House XVII and XVIII. In fact, the inner circulation, and more specifically the access to the upper levels appears to be completely dissimilar. In the first unit, no stairs are visible on the outer faces of the buildings on the courtyard.[71] It is not even clear how the upper floors of the tower could have been entered. The reason of this absence is not clear: one can suppose that either there were some internal stairs that are not preserved today or an intentional architectural choice, perhaps dictated by the physical impossibility to build the typical external stairs in the small and narrow courtyard with the stables. If it was the case, these upper storeys were possibly accessible from the other part of the complex, namely House XVIII. Here, in fact, external staircases are inserted into the faces of the buildings on the courtyard, following the common regional technique, and also showing a more elaborated version of it, like the "V-shaped" stairs on the eastern wall, which finds a direct parallel with House III in the south-western quarter (Figure 42).

Finally, I would like to make one last remark regarding the richness of the building, and perhaps even of its owner(s). The importance of the complex, quite clearly stated by its size, seems to also be reflected by the high quality of decoration, like the preserved *bifora* or the traces of painted stucco on the walls of one *vestibulum*. The two entrances to the house likewise show high architectural care and quality of the structures, starting with the decorated double basalt door and the likely mosaic floor in Room 5 (the Eastern Gate). The paved courtyard is another anomaly in the site, especially considering the large surface it covers.[72] Taking into account all that has been said for the Houses XVII-XVIII complex, I would

[70] Osinga suggests that the construction of the tower followed an increased necessity of security, following the "decommissioning of the Roman army in the later 6th century" (Osinga 2017: 137).

[71] I have not found any reference in the descriptions by Butler, de Vries or even more recently Osinga. In the plans available, no external stairways are indicated and not even an internal one. The only alternative solution is that the upper storeys of House XVII and XVIII were communicating, so it would be possible to enter an upper storey from the several external stairs facing House XVIII's courtyard. Nevertheless, I find hard to believe that all the buildings of House XVII were dependent on the other house to access the upper floors, especially the tower, that would have needed staircases from the third to the fifth floor, since no building nearby is as high.

[72] Nevertheless, we have to keep in mind that the absence of widespread excavations in residential units surely influences our knowledge of the site. We therefore cannot be entirely sure that this building was the only one

Figure 42 - Example of "V-Shaped" stairs in House III in the south-western quarter (photo by the author)

argue – in agreement with Osinga[73] – that its real function wasn't entirely residential, or better yet exclusively as a private dwelling. As for Building XII in Mampsis, it appears to me, that a possible interpretation of the structures as caravanserai is not illogical, considering the clear functional separation of the two areas in the complex and the uncommon size. Of course, one cannot exclude the possibility that it was indeed a private house, belonging to a very large or particularly important extended family or landowner. If it was the case, it is of course very difficult to determine the relative ownership of the different structures (Osinga 2017: 140). Though other buildings, confidently identified as private residences, may not share the same large size as this complex, they do have some of the "urban" features listed above, like a *vestibulum*, a proper inner court or a more refined decoration, which could possibly hint towards some socio-economic stratification in the settlement. For instance, House III in the western quarter, the aforementioned *Praetorium* and "Nabatean Temple" all bear these features. Nevertheless, it's impossible to find a clear pattern in the distribution of such complexes. The emerging trend thus far is that they are in a more or less central location within the quarter along the principal pathways – like the "Nabatean temple", Houses XVII and XVIII – or within the settlement in general – like the "*praetorium*". A possible explanation is likely the development that was undertaken by the buildings themselves in light of the more general history of the settlement. This would help explain the central location of the "*praetorium*", of which its proximity to the Commodus gates, the wider "blank" area of the city, the West Church and the Cathedral clearly emphasizes its prominence.

Lastly, another interesting example is House XIII, also first surveyed by Butler (1930: 202-204). Its typology closely recalls the "Courtyard House", though it was adapted to the Hauranian architectural style, with the typical alternation of smaller and larger rooms and inner circulation system on two (or

with a paved courtyard. What is surely impressive is that such a large surface was entirely paved in a context where a paved courtyard apparently was not a common feature.

[73] Osinga (2017: 140) suggests different hypothesis: military outpost, administrative building, caravanserai, a "desert castle"-like structure or indeed the wealthy private home.

more) storeys. The functional zoning of the dwelling is debated and Butler and de Vries (1998a: 110-111) offer opposite interpretations in regard to the most original feature of House XIII. Namely, the wall separating one of the long rooms from another hall behind it is constituted by screen blocks. De Vries suggests that both the halls were stables, and this wall simply divided different types of animals, while allowing a better air circulation thanks to the holes.[74] It is also interesting to note that House XIII both appears to be in the same block as House XII and was also directly connected to it, indeed through the room with the screen-wall. It is difficult to determine with the documentation available if this conjunction was a later modification or the two complexes were already built as one single unit. What one can quite confidently say, nonetheless, is that the families living here were interested in keeping a direct spatial connection, underlining the belonging to the same larger kinship group, but still maintaining a certain residential independence. This reflects the direct ownership of the single extended family with each nuclear family living in one room or one storey of the different buildings of the complex. This phenomenon can also be found in later periods in other sites, as will be shown in the Syrian case studies (Mseikeh and Sharah, see Chapter 4.2).

4.1.2: Umm es-Surab

The village of Umm es-Surab (Figure 43) is located a few kilometres northwest of Umm el-Jimal, with which it shares several features. Despite some clear differences, probably due to the different functions the settlements had within the regional context, it appears that the spatial organisation for both case studies followed the same guidelines and that the occupational dynamics also proceeded quite similarly, especially as far as the fate of their earlier phases is concerned. As was the case for Umm el-Jimal, Umm es-Surab's origin and first stage of development are almost completely unknown and are not able to be investigated, since the Byzantine structures nearly decimated the older architecture in its entirety. No support from historical sources can be provided to help fill this gap, since its ancient toponymy is still unknown. Nonetheless, it can be quite confidently affirmed that the village was an already well-established entity in the Nabatean period, at least in light of two elements: the epigraphic documentation and the presence of architectural elements dressed in the Nabatean and Roman tradition, which were later reemployed in other structures together with few courses of earlier buildings that survived under the Byzantine complexes. According to one of the epigraphic finds documented by Littmann (Inscription nr. 2 in Littmann 1930: 2-6) during the Princeton Expedition, a temple was built during the second year of the reign of Rabbel II, namely AD 71-72 (Figure 44).[75] In addition to this first and important document, a Nabatean funerary stele was also found in the site that was used as a landing of a staircase in a Byzantine house (Inscription nr. 3 in Littmann 1930: 6-7). Similarly, Butler identifies some reused Roman structural remains, which he connected to a large and monumental building of the Nabatean/Roman period, probably the one mentioned in the first inscription (Butler 1930: 94-95).

---

[74] This hypothesis was completely rejected by Butler, even if he did not offer any precise alternative, suggesting its use as a small industrial building. I agree with de Vries (1998a: 110) in saying that this stable-barn arrangement was "merely a more elaborate version" of the several examples found in the site.

[75] Ultimately, the interpretation of the inscription presents some difficulty, as stated by Littmann himself (Littmann 1930: 2-6). The reference in the Nabataean text to a "square monument" is nonetheless considered more than probable by the scholar, that confidently identify this monument with the "ruins of Nabataean temples" found by Butler in the site, reused in the Byzantine structures. The date to the second year of the reign of Rabbel II is also sure, even if Littmann collocates it in AD 76 instead of AD 72 (as Butler).

Figure 43 - Aerial photo of Umm es-Surab, from West (APAAME_20060911_RHB-0246. Photographer: Robert Bewley, courtesy of APAAME)

Inscr. 2. Scale 1 : 10.

1 דנה ארבענא די עבד מחלמו ועדיו וחורו על עלְ[ת . . . . בשנת]

2 תרתין לרב[אל מלכא מלך נבטו די אחי ושיזב עמה]

1 *This is the cella(?) which was made by Muḥlim and ʿAdī and Ḥūr over the al[tar of the god . . . . in the year]*

2 *two of Rabb[ʾēl, the king, the king of the Nabataeans, who roused and delivered his people].*

Figure 44 - Inscription nr. 2 from Umm es-Surab, with transcription and translation (Littmann 1930: 2-3)

Despite the meagreness of evidence for early periods, it is plausible that a settlement already existed prior to the Byzantine period, probably in connection with the intensive colonisation of this region following the expansion promoted by Aretas IV and especially after the foundation of Bosra as a new capital of the Nabatean Kingdom during the reign of Rabbel II (Bowersock 1994: 59-75; Ball 2000: 62-63). The development of the settlement before the Byzantine period remains nevertheless a huge void in our knowledge. The problem is that only few pilot studies including limited excavations by the Jordanian Department of Antiquities have been conducted at the site and almost all of the information available originates from architectural surveys. This presents two main problems: on the one hand, the later structures entirely destroyed the earlier ones, making it impossible to reconstruct the extension and the features of the Nabatean and Roman phases of the settlement; on the other hand, even if a precise relative chronology was established for some of the complexes, there's almost a complete lack of material that can be dated absolutely, or at least provide a solid chronological reference based on well-established ceramic typologies. The only definitive date available until now is the foundation of the church of SS. Sergius and Bacchus (Figures 45 and 46), in AD 489, which itself certainly reuses an earlier building. This religious complex is also the building that attracted the most attention from scholars studying Umm es-Surab, starting with Butler and the Princeton Expedition (Butler 1930: 95-99). Additional architectonical surveys were conducted by King (King 1983b) and more recently by

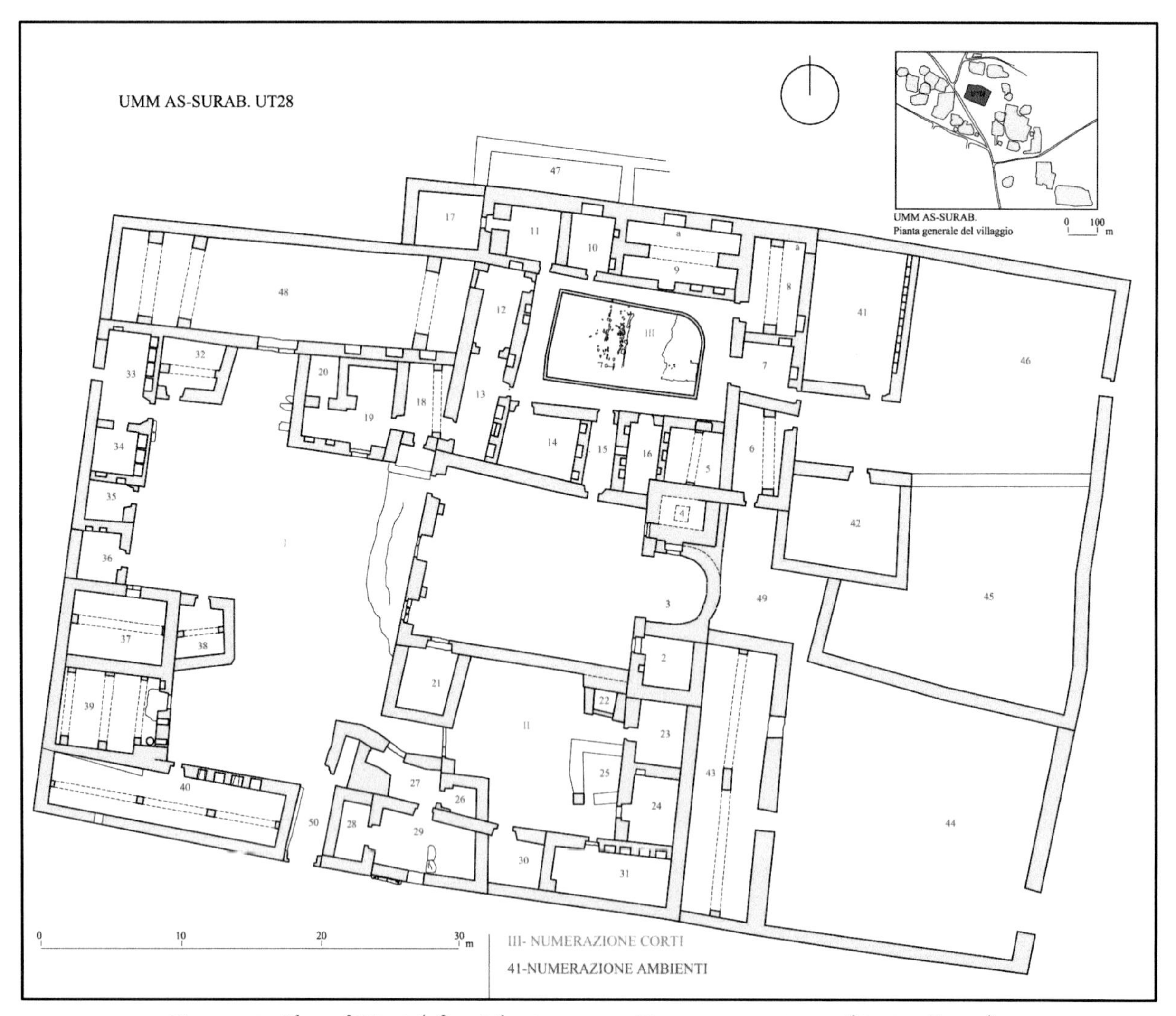

Figure 45 - Plan of TU 28 (after Gilento 2014: 7, Figure 12; courtesy of Dr P. Gilento)

Parenti and the University of Siena[76] (Figure 45). This last team ultimately managed to obtain a detailed relative chronology for the development of the building, though an absolute date for each phase could not be established (Figure 46c). Some of the results are nonetheless extremely relevant, not only for the history of Umm es-Surab, but also for the development of some features of religious building in the region, most notably the first construction and development of the minarets.[77]

The first phase is related to a building from the Late Roman period of which its exact extension or function is not possible to determine. Some remains of this earlier structure are concentrated in the northern and western part of the later complex and suggest that the area occupied by the "Ecclesiastic residence", according to the definition by Butler, was already built in this first stage. The cistern appears to be connected to this building in its initial stage, probably supplying the water required, and afterwards being left unoccupied. Even if the function of the early structure is impossible to reconstruct, one feature is surely of interest: a series of large niches are clearly visible in the roughly built double-parament wall, even though it was not possible to estimate their original height from the ground. As far as the dating of this first phase is concerned, some probes made by the Department of the Antiquities along one of these walls uncovered pottery no older than the Late Roman period, from the 2nd to the 4th century AD. Few other faint traces possibly belonging to this same phase could be found under the proper church, but the later modifications are too great to allow a clearer understanding of the structure at this stage. What is quite sure is that any building present in this area was cleared away by the construction of the church and the connected structures.

The second phase identified by the architectural survey bore witness to the construction of the first version of the church and of another building in the south-western corner of the later complex. The dating of this phase is provided by the inscription engraved on the lintel of the central entrance of the church façade. It is one of the few absolute dates available for the entire site, reporting the foundation of the church on the 12th September 489 and the dedication to the saints Sergius and Bacchus, extremely beloved in this region (Inscription 51 in Littmann, Magie, and Stuart 1930: 57ff).[78] Despite the presence of two different architectural techniques (Figure 47b),[79] it appears that the church reached its

[76] The surveys were conducted by the team of Laboratorio di Archeologia dell'Architettura of the Università di Siena within the *Building archaeology in Jordan Project*, under the direction of Prof. Roberto Parenti.

[77] For the following description, I mainly refer to Gilento 2014. At present, a new project in Umm es-Surab has started (ACTECH Project_France) under the direction of P.-M. Blanc (CNRS) and P. Gilento (University of Paris 1 Panthéon Sorbonne), which includes either excavations and more extensive architectural surveys. Preliminary results are under publication process while the present book is reviewed, and only a limited amount of data was accessible at this stage, thanks to personal communication with P. Gilento. Consequently, the following description remains open to significant changes as far as the chronology is concerned: as it appears from the first communications, some of the development took place in a later period than supposed until now, and more specifically from the 11th to the 13th/14th century.

[78] One discussion is also on the exact meaning of the term *mnemeion* in this context: the most accepted interpretation is *memorial* or *cenotaph*, letting us imagine the presence of relics in the church.

[79] One is extremely refined and elaborated with perfectly squared and dressed blocks laid in regular courses, requiring surely specialised workers is visible in the lowest part of the façade. The other is characterised by roughly dressed blocks and many chip stones levelling the irregular courses and can be seen in the upper part of the façade and in the inner face of the façade. According to Gilento (2014: 10-12), this difference can be explained in light of technical and aesthetic reasons: first of all, the rougher technique is employed only in parts that do not have static and architecturally fundamental functions, as opposite to the use of the better technique in the base of the façade and in all the supporting structures like doorways and pillars; moreover, it is clear that the view of upper part of the façade was cut off from the portico structure in front of the three entrances to the church, rendering a better-looking dressing of the blocks useless. Similarly, the inner faces were all probably covered by layers of plaster and possibly wall-mosaics, hiding the roughly worked basalt.

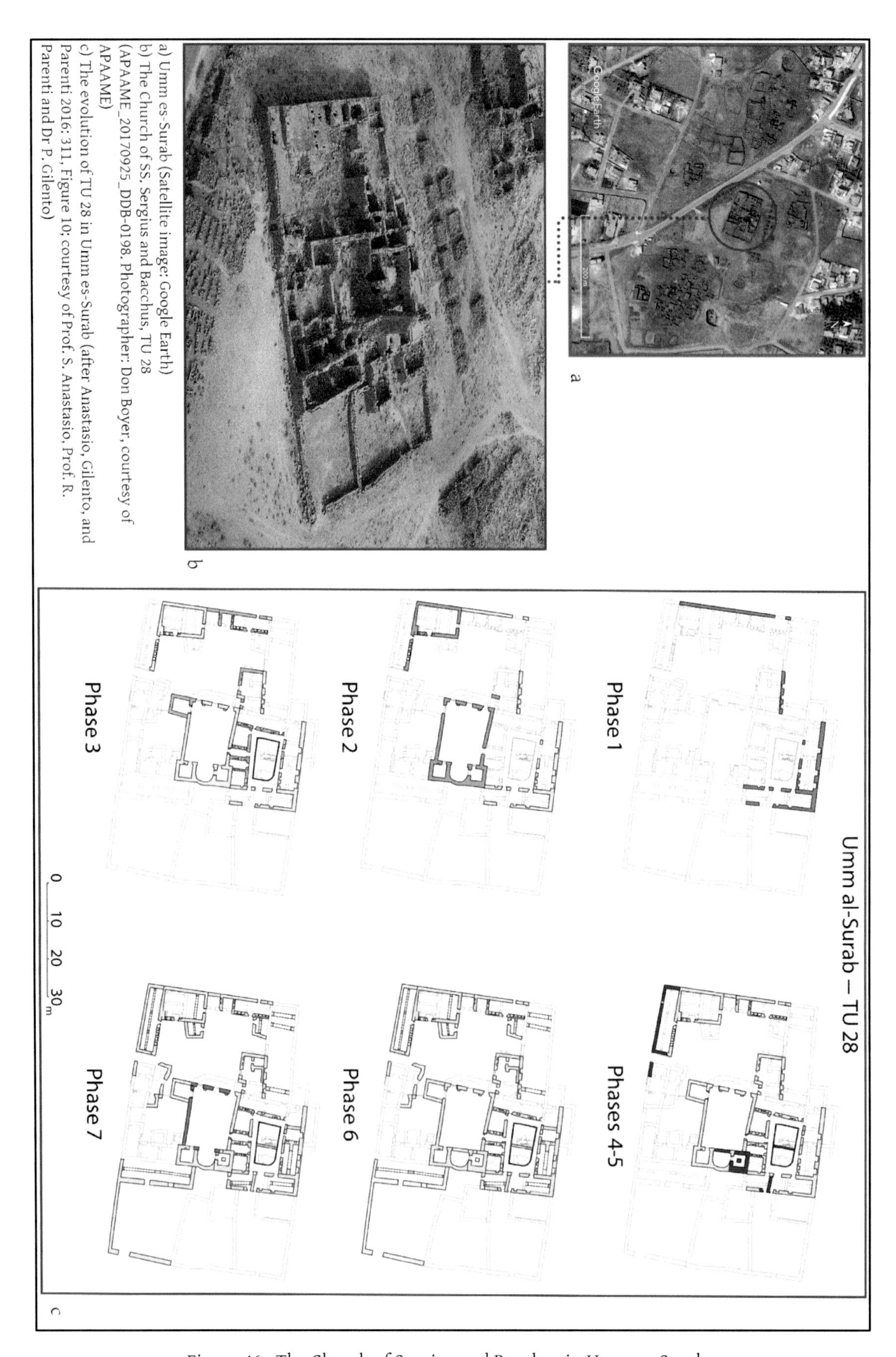

a) Umm es-Surab (Satellite image: Google Earth)
b) The Church of SS. Sergius and Bacchus, TU 28 (APAAME_20170925_DDB-0198, Photographer: Don Boyer, courtesy of APAAME)
c) The evolution of TU 28 in Umm es-Surab (after Anastasio, Gilento, and Parenti 2016: 311, Figure 10; courtesy of Prof. S. Anastasio, Prof. R. Parenti and Dr P. Gilento)

Figure 46 - The Church of Sergius and Bacchus in Umm es-Surab

a) The church of SS. Sergius and Bacchus, TU 28 (APAAME_20170925_DDB-0198. Photographer: Don Boyer, courtesy of APAAME)
b) The facade of SS. Sergius and Bacchus church (elaboration of the author on King 1983b: Plate 29, Figure a)
c) The minaret from the Northern Court, looking South (photo by the author)
d) The apsis area of SS. Sergius and Bacchus (elaboration of the author on King 1983b: Plate 30, Figure a)
e) The entrance of the minaret from the former apsis (photo by the author)

Figure 47 - Details of the church of Sergius and Bacchus

more or less definitive form already in this phase, maintaining it through the following phases until the construction of the tower/minaret and the obstruction of the apsis (Phases 4-5). The relationship – functional and possibly physical – between this building and the other construction of this second phase, occupying the south-western corner of the of the later complex, is not clear. The excellent technique employed here is the same as in other parts of the church, an element that suggest the contemporaneity of the two building activities.

The third phase is also dated to the Byzantine period, and bears witness to the construction of the northern complex, identified by Butler as the dormitory of the monastery or the "Ecclesiastical residence", with only marginal modification of the original church, creating a direct passage between the two buildings. The church's façade itself was probably partially restored during this phase, since a different technique had been employed to rebuild the southern corner (Figure 47b) and a large portion of the inner face. These interventions in any case respected the original plan of the church and may have followed a simple structural slough or a collapse due to the consequence of an earthquake.[80] Whatever the cause was, the church continued to be in use before and after this rebuilding activity, taking place possibly in the Late Byzantine period, likely between the 6th and 8th centuries AD.[81]

The fourth phase marks a decisive moment in the development of the complex: in particular, the church underwent important changes, probably showing a change in the religious identity of the community using it. Specifically, the north-eastern corner of the building was entirely rebuilt following a collapse or an intentional demolition, and subsequently the tower was erected, overlaying the other walls left from the original structure (Figure 47c). The entrance of the *prothesis,* opening to the northern aisle, was blocked and a new door was opened in the southern side of the tower (Figure 47e). Contextually to this, the wall blocking the apsis was probably built, quite clearly signalling a functional conversion of the church into mosque (Figure 47d).[82] Dating this phase, as well as the following phase that sees the reconstruction of the cistern of the Ecclesiastical residence and of some perimetric walls, is a particularly difficult task in the absence of excavation. The survey made by King identified at least two occupational periods in the Islamic era, notably the Umayyad/Abbasid – 7th-10th centuries – and the Mamluk – between 12th-16th centuries, which can be associated with building activity during these phases. Nevertheless, the only chronological indicators thus far originate from the analysis of the typology of the tower and the structural evidence. Albeit the issue is still discussed and no definitive answers had been presented until now, especially in relation with the development of the minaret as an architectural element in general, Gilento considers it more plausible to date the "tower" in Umm es-

[80] Gilento (2014: 18) offers a list of possible earthquakes that could have caused this collapse between the 6th and the 8th centuries. As extremely plausible, the Italian researchers identifies the one of the 9th July 551 or of September 633-634, since these two events had been particularly destructive.

[81] As Gilento reports (2014: 18-19), many churches in the region continued to be in use at least until the Abbasid period.

[82] The wall was removed by the Department during the first preservation works. Nonetheless, the pictures made by Bulter and more recently by King show it clearly. A direct parallel is also visible today in the near village of Samah, where the S. George church apparently saw a similar development to the umm es-Surab case, but still preserves the wall blocking the apsis. It is also another example of tower/minaret built in a later phase of the building and following possibly the conversion of the church into mosque. This complex in Samah was also studied by the University of Siena (for a general description: Gilento 2013: 52-55) and was already surveyed by Bulter (1930: 83-87). In both cases, the interpretation of the tower as minaret is also suggested to the Italian scholar by the contemporary heavy modification of the church under a more ritual perspective: the destruction of the *prothesis* – and therefore the lack of its important liturgical role – and the partial demolition of the apsis are clear proofs of a change in the ritual function of the building. Therefore, the construction of the tower as bell-tower – implying the use of the complexes still as churches – does not appear probable, even if – as mentioned before – reconstructions of churches are attested at least until the Abbasid period.

Surab to the Early Islamic period.[83] Like the chronology established in many buildings of Umm el-Jimal, the evidence from the church of Umm es-Surab suggests an abandonment in the 16th century, lasting until the Druze occupation of the site at the end of the 19th century. This later phase did not see substantial changes in the structures, with building activities normally limited to the restoration of collapsed roofs and the contextual construction of arches easily recognisable from the earlier ones.

As previously mentioned, the Church and the monastic complex are the only buildings in the site that have attracted attention of a plurality of scholars and teams. The rest of the settlement is clearly underrepresented in the archaeological documentation available for Umm es-Surab. The only other detailed report on the development of a building, even if not in absolute chronological terms, is in relation to a residential building on the southern fringe of the settlement, TU24, studied by the University of Siena (Figure 48). It is important to remark on some elements of the process leading to its final structure, which can ultimately offer important information not only in regard to this particular complex, but in general for the development of the entire settlement. The same process can also be described for other residential units in the Hauran region, with some interesting chronological differences. Gilento (2013: 134-140; 2015: 329-360) offers a description of the general development of this farmhouse, based on the structural analysis made by the Italian team (Figure 48c). TU 24 was likely built in an area where an earlier building stood, of which only few differently oriented courses are left under the later constructions. Like in Umm el-Jimal and in the church in Umm es-Surab, the construction of the later complex evidently indicates a radical change if not total destruction of earlier structures, whose function is not possible to determine.

The first buildings of the new residential units do not seem to share initially any physical relation. Gilento confirms them belonging to the same phase in light of the similar ground level and especially for the typology, ultimately a sort of tower-house. According Gilento's hypothesis, the two houses could perfectly match the "simple house" type, even if organised with more storeys and possibly constituting two independent tower-houses. The presence of a small enclosure in front of the building is also stated despite the absence of any clear trace.[84] In a second building phase, one of the two towers was abutted by a long building that constituted the southern aisle of the later courtyard house. Ultimately, this newly obtained building presents the "L-shape" plan described in the introduction, which is quite common in the Hauran region. The later development of the complex in general, clearly shows how all of these organisational forms or typologies are intrinsically transient and temporary, allowing a constant and progressive reform of the inner spaces. In fact, in the third phase identified by Gilento, the proper courtyard began to be developed. An enclosure wall was built at the eastern end of the "L-building", and it is plausible that a new structure connected to the second original tower (CF4) was also

[83] I will not analyse in detail this issue, since it is not a direct concern of the present research. A general report on the discussion on the origin of the minarets in this region and more specifically on the dating of the two examples of Umm es-Surab and Samah is offered by Gilento (2014: 26-29). To summarise, the two positions substantially follow on the one side the hypothesis offered by Creswell (Creswell 1958) and on the opposite side the late dating of Bloom (Bloom 1991: 55-58). The first one considers the pre-Islamic towers found in the Hauran the direct ancestor of the Early Islamic minarets – mentioning also the two examples from Umm es-Surab. These towers would have offered the first Umayyad architects a direct model to develop the form of the quadrangular minaret – the form that will be common in Bilad as-Sham until the late introduction – in the Mamluk period – of the octagonal minarets. On the contrary, Bloom considers this typology as originated in the 9th century in Iraq, under the Abbasid impulse of building the new capital. The minaret form would have reached only later the Syrian area.

[84] I would like to underline the fact that given the temporary and rough construction of these kinds of structures, for instance as in the more recent dwellings of the region, the absence of any trace of this small enclosure in front of "simple houses" is not surprising. They normally do not require any foundation trench or particular structural device, since their main goal is simply to avoid the dispersion of small herds of sheep or goats.

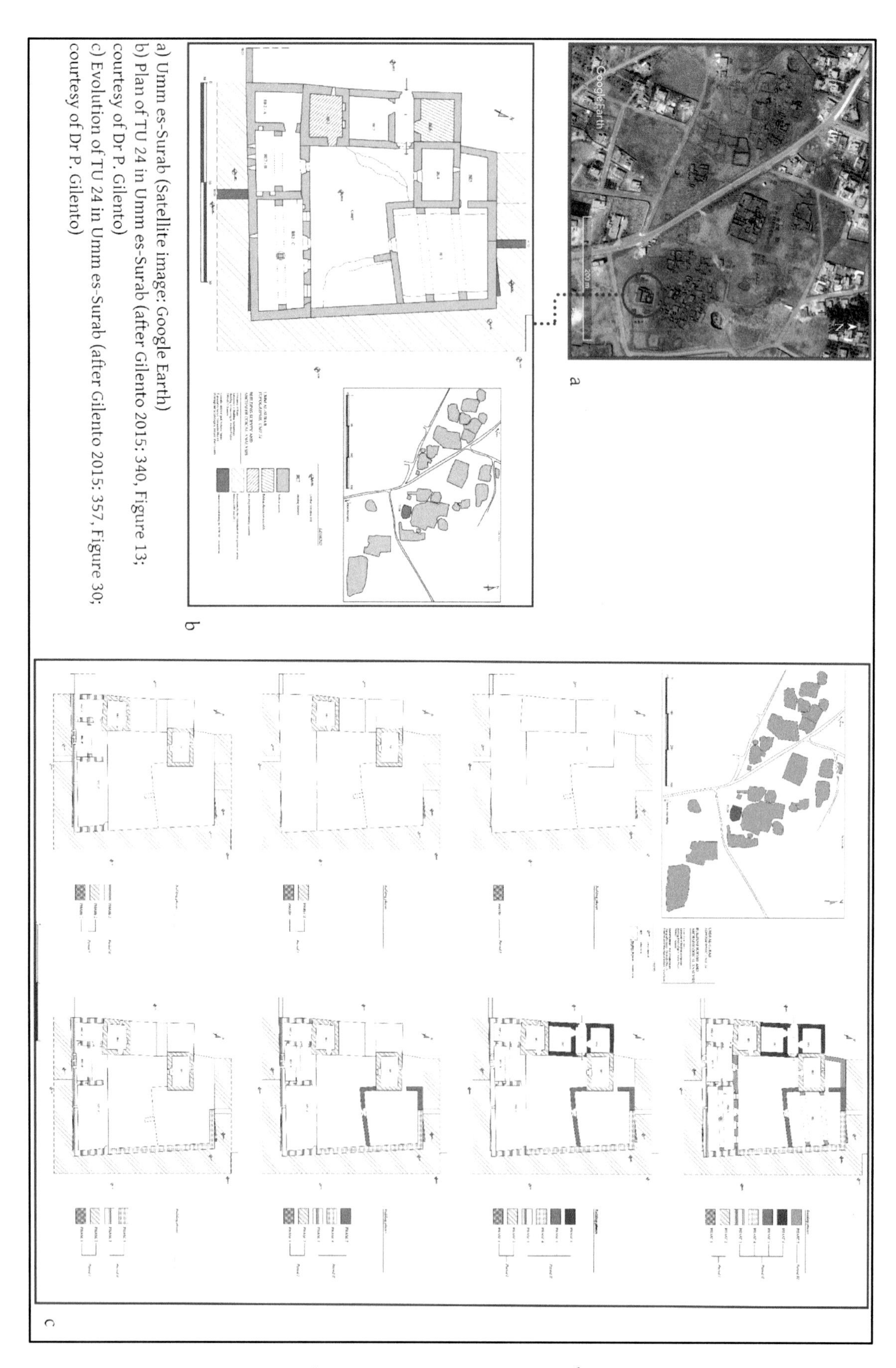

a) Umm es-Surab (Satellite image: Google Earth)
b) Plan of TU 24 in Umm es-Surab (after Gilento 2015: 340, Figure 13; courtesy of Dr P. Gilento)
c) Evolution of TU 24 in Umm es-Surab (after Gilento 2015: 357, Figure 30; courtesy of Dr P. Gilento)

Figure 48 - TU 24 in Umm es-Surab

constructed and later demolished by the construction of CF3. It is interesting that the technique employed for these two last phases is really similar, leading Gilento to conclude a chronological proximity between the two building activities. The "U-shape" plan is obtained in the following phase, that sees the rebuilding of the north-eastern room CF3 (Figure 49). The function of this space is problematic; seeing the large dimension of the room, it was suggested its use as stable, but the absence of mangers does not support this interpretation. Gilento offered an alternative, and even more interesting, interpretation, especially concerning the purposes of the present research: a new residential space in the house to host a new nuclear family. The fifth phase finally bears witness to the completion of the process of progressive accumulation of structures: at this moment the complex is organised around a central L-shaped courtyard, enclosed by rooms. The access to the house is mediated by a *vestibulum* similar to the one seen in many large residential units in Umm el-Jimal. The total absence of decorations or wealth symbols, as suggested by Gilento, may be indicative that the family living in this complex had any particular socio-economic or political rank within the local society. The structure is afterwards maintained in its actual form, and only some marginal restoring and reroofing by the Druze can be identified (Phase 6, Figure 50).

Figure 49 - Example of re-roofing of the Druze period in TU 24 in Umm es-Surab (photo by the author)

Figure 50 - The North-Eastern room in TU 24 of Umm es-Surab (photo by the author)

The major problem in relation to this complex is reconstructing an absolute chronology for its development, especially since no epigraphic or excavation finds can help. The analysis and comparison of the architectural techniques employed led Gilento to believe that Phase II of the complex, namely the "L-Shape" house, could have taken place in the Early Byzantine period between the 4th and 6th centuries AD[85]. As previously stated, Phase 3 is technically similar to the previous one, and should also fall within the same chronologic horizon. Tentatively, the complex reached at least the "U-shaped" plan some time during the Late Byzantine period. The big question is when the proper courtyard system was finally reached: the parallel examples of Umm el-Jimal suggests within the Early Islamic period.

In summary, both complexes analysed in detail offer the same chronological framework, which, also after the comparison with Umm el-Jimal's chronology, can also tentatively be extended to other part of the settlement. Generally speaking, a progressive accumulation of buildings around a central empty courtyard can be suggested for many other complexes in the site. The initially sparse occupation of the settlement, with few isolated buildings, gradually intensified until it grew into the dense settlement we can see today, at least on a quarter/block and house level, following the same kind of process documented for Mseikeh and Sharah (see Chapter 4.2). Umm el-Jimal is the first immediate example, but the same densification is also visible in the Negev settlement. Chronologically, the beginning of such a process seems to take place in the Early Byzantine period – around the 4th and 5th centuries – and its conclusion tentatively in the Late Byzantine/Early Islamic horizon, by the 6th – 8th centuries. One can confidently suppose that the settlement reached its final and today visible form in this period, undergoing only a few modifications in the Mamluk period before it was abandoned. Currently we have no clear indication regarding when this final stage occurred.[86]

The settlement spread over *c.* 10 ha, clearly presents an open form, without any wall enclosing the urban space (Figure 51). The village area is nevertheless somehow defined by a series of enclosures or field delimitations surrounding it in its entirety, as is clearly visible in the aerial picture from 1953 (Figure 4). This "ring" circumscribes the surface where clearly five distinguished quarters and two isolated structures were built. The relative position of these quarters is also interesting, since they are dispersed around the central monastery of SS. Sergius and Bacchus. The religious complex seems to play the role of the ideological centre of the settlement: its prominence and centrality is not only underlined by its relative position in the village, but also by the "empty" areas surrounding it. Like in Umm el-Jimal, these supposed "empty" or blank spaces appear to function as gathering places, as suggested by at least two elements. First of all, the presence of small enclosures, probably used temporarily or seasonally, potentially providing a space for local markets. The second, and probably more important element, is the presence of the village's large public cisterns. This pattern reflects exactly the same situation described in Umm el-Jimal around the Commodus Gate: large water facilities, enclosures to temporarily

[85] In particular, Gilento finds a really close similarity of the technique of this phase with the ones used for the third building phase of the Barracks in Umm el-Jimal and the second phase of the church in Umm es-Surab, both dated to the Early Byzantine.

[86] As mentioned in note 77, recent excavations and new architectural surveys seem to point to a later development of the site: yet, the process described by the archaeological data remains the same, with a progressive shift from a more scattered built environment to a more clustered one. If previously this development was supposed to be in the Late Byzantine period, it seems now to date from the 11th to the 13th/14th century. This change in the chronology of the site does not invalidate the description offered in the following pages. As in the case of Mseikeh (Chapter 4.2.2), it seems possible to me that the later development heavily impacted earlier structures, from where the building materials were taken to build or modify the new ones: this re-use of structures and materials could have determined the impression of a more scattered Late Byzantine settlement.

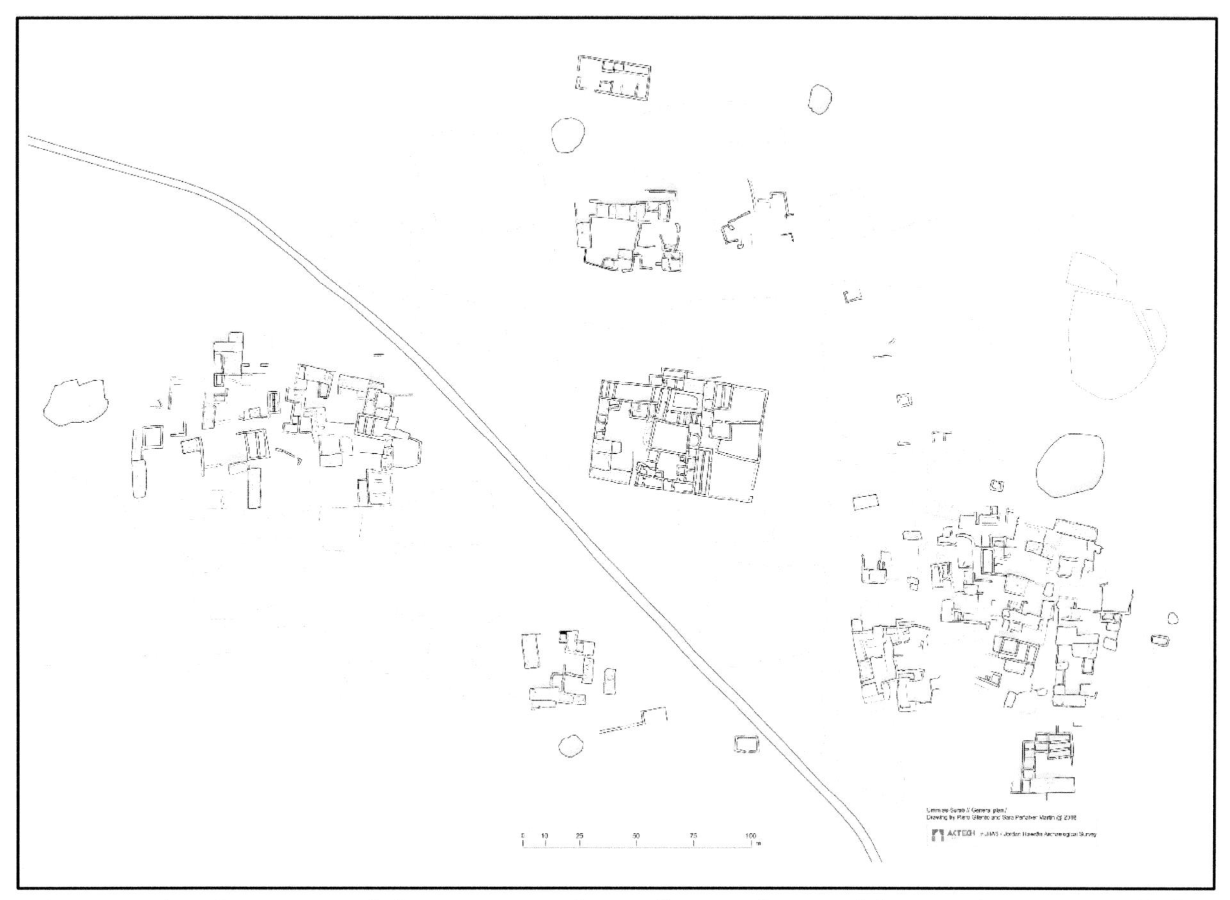

Figure 51 - Plan of Umm es-Surab (2018, courtesy of Dr P. Gilento and Dr S. Peñalver Martin, ACTECH Project)

keep animals, and the presence of public and isolated buildings. Likewise, a church plays a central role in both the sites, spatially and ideologically. In my opinion, the relative location of such areas in the settlement depends to a large degree on the different typology of settlement. Whatever the exact status of Umm el-Jimal had, it clearly shows some semi-urban elements that are completely absent in Umm es-Surab, such as the city wall. It is therefore not surprising that in a settlement where the distinction between the inner urban part and the hinterland is clearly marked, these gathering spaces are in correspondence with the major and more formally signalised gate.

A rural settlement like Umm es-Surab did not likely have a city wall, since the limits of the village are simply marked by the presence of a corollary of enclosures, while leaving several accesses to the core of the site itself. Apart from this difference, the spatial characteristics of Umm el-Jimal are closely reflected in Umm es-Surab: there is no formal distinction between functional zones, nor a centralised administration managing the development of the settlement, which again appears to follow a spontaneous and unplanned expansion following a growth in population. Similarly, the inner circulation does not appear to follow any rule: not even proper paths can be identified, but simply broader unoccupied spaces between the quarters are left allowing individuals to reach the different areas within the settlement. Only some radial passages can be pointed out through the enclosures buffering around the entire settlement and connecting it to the regional street network. Umm es-Surab is not far from some of the principal regional roads, like the one coming from Gerasa and crossing the *Via Nova Traiana* about five kilometres north-east from the site. Likewise, the distribution of the quarters does not reflect any sort of economic or social stratification. Overall, three distinct quarters can be confidently identified, distributed from west to south in respect to the central church.[87] Under all these aspects, they all seem to be homogeneous: not one shows any features suggesting a difference in importance or that it a prominent role in the settlement, which can be better grasped when considering their structure in greater detail.

The quarters are conceived as autonomous units like the ones in Umm el-Jimal. The homogeneity in this matter is striking and similarly involves the blocks and even the single houses. The buildings are always gathered together, forming completely independent dense clusters, both functionally and socially, with clearly marked limits, offering an almost continuous blind face to the outside. Only few narrower throughways between the buildings are spared and allow the inner circulation (Plate 11). The blocks are similarly structured, and sometimes the paths running between them are so narrow that it would make little sense to distinguish them as different units. What is important to underline is the complete absence of buffer-zones between the private and public areas within the quarter. All the blocks open directly to the inside, showing no mediation, and no particular devices to increase the separation from the outside. It is also interesting to note that the absence of spatial features such as *cul-de-sacs* or courtyards determines the fact that each unit of a block has its own separate access from the street.

Generally speaking, apart from the central monastery, almost all of the buildings in the village do not show any kind of public or exclusively private vocation, and are normally deeply embedded within a quarter. As in Umm el-Jimal, this compactness and relative isolation in the settlement's frame reinforces the impression of a direct reciprocity between quarters and discrete social units, even more extremely in the case of Umm es-Surab. Only two buildings escape this picture apart from the SS. Sergius and Bacchus complex: one almost squared structure north of the monastery, of which its function is not clear yet though it does not seem entirely residential, and a farmhouse close to the south-eastern quarter that nonetheless does not show any distinctive feature in comparison with the other dwellings of the site.

Similarly, the way the blocks are distributed throughout the quarter, and also the houses within the blocks, do not show any particular pattern. It can be confidently deduced that the development of the complexes proceeded gradually and spontaneously according to the availability of building areas and probably to the proximity of closer familiar groups. The relative position of the block in the quarter may also determine specific choices. As for the western quarter in Umm el-Jimal or in the most external blocks in Shivta (Chapter 4.3.1), there's a tendency of having rows of dwellings on the liminal part of the quarter, determining a more regular appearance to the block they form. This regularity is not determined by a formal plan of the spaces, though it may have met the requirement of having a continuous blind face to the outside of the cluster. In fact, since each dwelling needs to have direct access from the street but no openings to the outside of the quarter, the only solution is to align the units in a more regular arrangement. This pattern is particularly visible in the western quarter (Figure 52), which nonetheless shows some other peculiarities, especially as far as the distribution of house

[87] It has to be pointed out that we do not have an exact idea of how many were present in the village, since the modern disturbance had clearly compromised the easternmost part of the settlement: some indications can be obtained by the aerial picture taken in the early 20th century by the Royal Air Force, which shows a possible large cloud of building in that area. The modern village does not entirely overlap with this part of the ancient site, but the ruins are not entirely clear, and do not allow for any possibly analysis. Although we can suppose that the features are similar to the other better-preserved quarters of Umm es-Surab, as suggested by its isolation and lack of planning.

types are concerned. In fact, if the rest of the settlement seems to show a more homogeneous presence of complex houses and a low variability in the house typology. At the same time there are some single houses present and at least an extremely dense complex, with one of the few tower-houses of the site still standing, which closely resembles House XVII and House XVIII in Umm el-Jimal. Nonetheless, as previously mentioned, the presence of more regular clusters in the "periphery" of a quarter, or even of a settlement, is only a trend, and there are many other examples at this site and elsewhere showing how the organisation does not follow a standardized rule. Similarly, it is important to stress once again that it does not reflect any effort towards planning the quarters, but represents a practical solution to meet a specific need. Functionally, they are no different than the other "irregular" clusters, since they both are characterised by polyvalent areas.

It is extremely rare to find any kind of specialisation within the units, as each one has both residential and productive facilities. The only exceptions are the churches. Interestingly, in Umm es-Surab there is at least one, but probably two, example of a religious building exclusively accessible from a private courtyard (Figures 52b and 52d).[88] This case is not exceptional, since sometimes these churches are actually part of "urban monasteries" or monasteries that are deeply embedded in the settlement's layout. In these cases, the "residential" units attached to the religious buildings normally have specific features like a cloister, closely recalling the "Ecclesiastical residence" of the Ss. Sergius and Bacchus.[89] In other instances, though, there is no clear indication of a facility designated to a community of monks, but really only evidence for a private dwelling closely connected to the religious complex. The western quarter of Umm es-Surab seems to fall into this last case.[90] These examples may presume a close connection between some religious facilities and discrete social groups in the settlement with a clear distinction between public religious buildings and complexes prerogative only of some inhabitants or discrete social groups.

As for the quarters and the blocks, the same homogeneity can be found in the residential typology across the entire settlement. The vast majority of the dwellings are "complex" farmhouses, spontaneously developed through the gradual addition of structures. Only few examples of "single" houses can be found and even more rarely some "complex" dwellings present particular features, like tower-units - surely at least three storeys high if not more - similar to the ones noted in Mampsis in Building n.2. It is difficult to determine the development of such complexes without excavations or even the removal of collapse. One can hypothesize that their development finds some parallels at the site like that of TU24, with earlier isolated "tower" structures, progressively embedded into as courtyard system; this development also has parallels in Shivta. Likewise, these houses could have followed a development similar to those in Umm el-Jimal, where a unit in the "Courtyard" House is elevated in order to have more domestic space. Surely though, the occurrence of such type of houses with units higher than three storeys is rare; few examples can be found in the region.[91] In fact, most of the houses have one or two storeys, and only in some structures: this feature is evident in Umm es-Surab, but seems to be typical for domestic architecture in the Hauran.

---

[88] Butler (1930: 95-99) and King (1983a: 82) mention three churches in the settlement, but the recent French expedition identified a forth one. The largest is of course SS. Sergius and Bacchus; other two are in the Western quarter, surely one but probably both accessible only through the courtyard of a private house; and one in the norther part of the settlement.

[89] There are parallels in Umm el-Jimal, in the south-western corner of the settlement and possibly also in the eastern quarter, and in Shivta, around the northern Church.

[90] This situation recalls the Nilus's house in Mampsis, even if here no direct entrance from the house to the religious building was guaranteed. Moreover, the house and the church constituted an isolated

[91] Like Umm al-Quttayn: see Gilento 2015: 335-337.

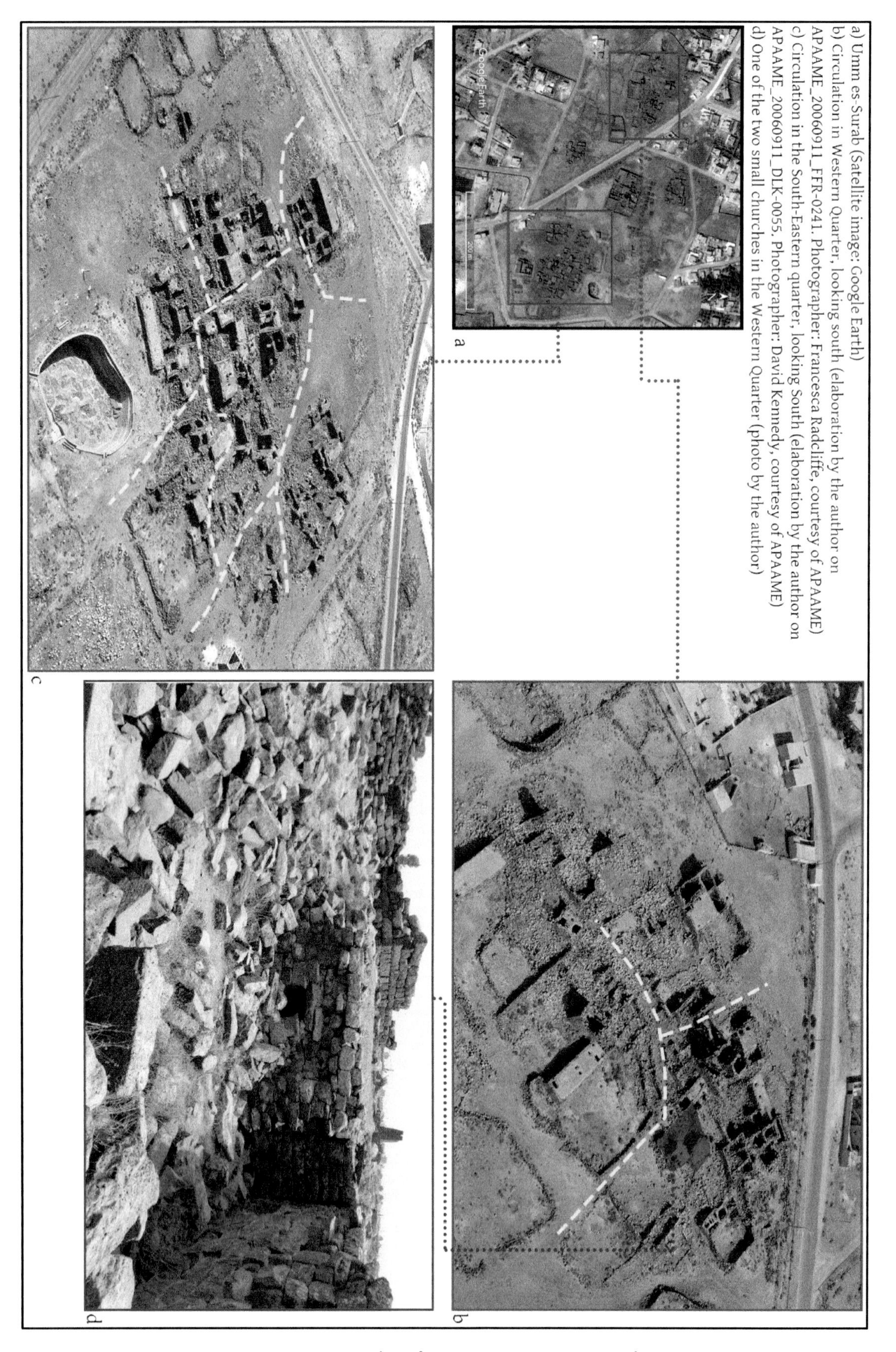

a) Umm es-Surab (Satellite image: Google Earth)
b) Circulation in Western Quarter, looking south (elaboration by the author on APAAME_20060911_FFR-0241. Photographer: Francesca Radcliffe, courtesy of APAAME)
c) Circulation in the South-Eastern quarter, looking South (elaboration by the author on APAAME_20060911_DLK-0055. Photographer: David Kennedy, courtesy of APAAME)
d) One of the two small churches in the Western Quarter (photo by the author)

Figure 52 - Examples of quarters in Umm es-Surab

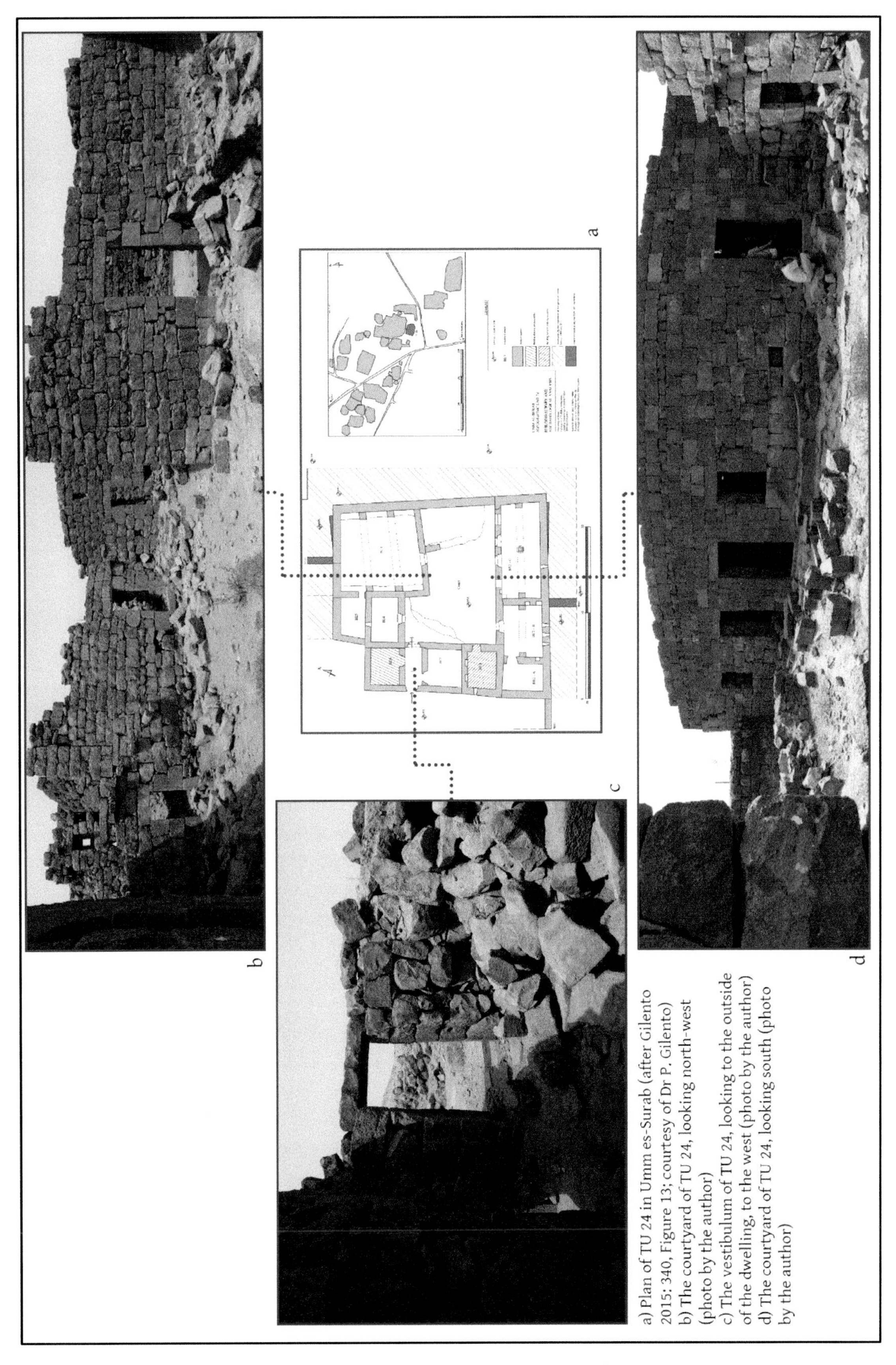

a) Plan of TU 24 in Umm es-Surab (after Gilento 2015: 340, Figure 13; courtesy of Dr P. Gilento)
b) The courtyard of TU 24, looking north-west (photo by the author)
c) The vestibulum of TU 24, looking to the outside of the dwelling, to the west (photo by the author)
d) The courtyard of TU 24, looking south (photo by the author)

Figure 53 – Views of TU 24's courtyard

It is also rare to find completely isolated houses, since almost all of the dwellings in Umm es-Surab are grouped in clusters, sharing at least a dividing wall between the courtyards or having structures abutting against the neighbouring buildings. The only exception is TU24. Being the only residential unit clearly separated from the rest of the quarter, though still close enough to be still considered part of it, it shows some features that are not present elsewhere in the site, but are commonly found in the region. The general appearance is exactly the same as the other houses in the settlement, except the entrance that opens to the central courtyard (Figure 53). This can be easily explained when considering the isolation of the building. While in the other houses the access from the pathways to the inner court is direct, in TU24 there is a vestibule flanked by two further small rooms. This situation closely recalls the more "urban" residences like some "Courtyard" Houses in Shivta and Mampsis or the larger and richer complexes in Umm el-Jimal, such as Houses XVII and XVIII. I would suggest that the presence of these features is a solution to meet the need for privacy: the absence of a quarter around the building providing a "screen" for the dwellers' privacy led to the adoption of another solution, exclusively on the house level, guaranteeing the same privacy standards. Therefore, the vestibule avoids a direct access from the public to private sphere. What is more difficult to determine is the reason why this house is the only one isolated. It is not too far from the quarter where is can be supposed that it is inhabited by a family of "outsiders" and at the same time, its architectural features do not suggest any particular socio-economic status, while it also shares all the characteristics found in the other residences of the settlement.

4.2: The Central Hauran (Syria)

The Central Hauran (Figure 54) shows a strong cultural continuity with the southern part of the region located in modern Jordan. Nevertheless, some differences can be identified on the regional settlement pattern, especially for the region of the Leja. This diversity is mainly determined by the absence of larger urban centres, or better yet, officially recognised *poleis*. In the case of the Southern Hauran, Borsa, not only a *polis* but a provincial capital, is clearly the political and administrative centre of the entire region, interconnected with a series of smaller rural villages and a limited number of semi-urban entities like Umm el-Jimal. In contrast, this latter semi-urban dimension seems to be extremely frequent in the Leja, and can be related to the vacuum left from the absence of *poleis*, combined with the economic and military strategic importance of the region. Importantly, this is an area for which an official use of the term *metrokomia* is attested in the Roman period and probably also in the earlier phases of the Byzantine administration, which is the closest definition to the modern concept of "town" we can probably find in the ancient sources (Sartre 1999). Even if it is not included in the list of the official *metrokomai*, the first case study in this region, Sharah, perfectly exemplifies the semi-urban dimension mentioned, and that unsurprisingly shares many aspects with Umm el-Jimal.

4.2.1: Sharah

The site is located in the northern fringe of the Leja and is often referred to as the Syrian Umm el-Jimal for its exceptional conservation, and also due to the fact that the modern settlement does not overlap the ancient one (Clauss-Balty 2010: 199-214). Sharah is one of the major centres in the region, with an extension (*c.* 20 ha) closer to that of Umm el-Jimal than all the other case studies. Along with its

Figure 54 - The Leja region in the Syrian Hauran (Satellite image: Google Earth)

dimensions, the presence of public structures also suggests a certain local prominence.[92] Butler was the first scholar to offer a report on the site, and he immediately affirms the impression of being in a settlement of a well-established importance from the Roman into the Islamic period (Butler 1930: 438-440). Despite the condition of the site, which at the time occupied by a small community of Bedouins, he recognised the presence of major monumental buildings: Roman Baths[93] and fragments of a not

[92] If morphologically and the presence of large public complexes, might suggest a similarly higher rank in the settlement's hierarchy, possibly being a *metrokomia*, there are some elements discouraging me to state such hypothesis. First of all, the lack of any epigraphical or historical sources, second, and decisively, the presence of a confirmed *metrokomia* – *Phaena*-Mismiyeh – only 5 km to the north.

[93] It is the only building described in detail by Butler in this site. It was found along the paved street in the northern part of the town. Its conservation is described as excellent, despite the fact that from the outside the building was barely visible underneath the neighbouring collapsed structures. The interpretation is mainly based on the identification of some water pipes. Seeing the high quality architectural workmanship, Butler (1930: 438-440)

located temple, some also with Nabatean features, towers of the Byzantine period, and two mosques from the Islamic occupation phase of the site. He also mentions a large arched complex in the western part of the town – still roofed – that the locals referred to as "the khan", whose architecture features Butler considered as belonging to the early phases of the settlement.

There are no clear indications about the chronology of Sharah, but it seems to follow the general evolution of the sites in the Hauran area. The only absolute dating is still offered by Butler, who reported the presence of a reused block with inscription in a mosque, dated to AD 161-169. This has a direct parallel with some excavation data from the French expedition under the direction of J.-M- Dentzer on the eastern rampart and stronghold of the town (Bruant 2010); in fact, it appears that the first phase dates to the mid-first to the mid-second century AD. The walls appear to have been destroyed in the Byzantine period, and more specifically in the 5th-6th centuries, possibly by a seismic event. The remains of the fortress connected to this part of the rampart had been by then reoccupied and converted to a large residential complex, as the presence of fiddles suggests (Bruant 2010: 220). Unfortunately, they did not determine the date of this development. Nonetheless, this defensive structure does not necessarily seem to belong to the first phase of the entire settlement.

Similar to other sites, some structures in the settlement had a military function, at least in light of their organisation, not corresponding to either domestic or cultic architecture. Even though it begs to question if a squared plan and the presence of small and modular rooms organised around a courtyard should automatically imply the presence of forts or military oriented buildings,[94] it is quite clear that their conversion in many cases into domestic complexes only occurs in a later phase. Determining the period in which these later modifications occur is difficult in the absence of an extensive excavation: the general trend for the case studies in the present analysis is that Late Byzantine or Middle Islamic phases alter older structures. The major problem presented by having exclusively architectural surveys is that often the Middle Islamic phases reuse Byzantine materials, which in turn could have already reemployed earlier structures. This makes it extremely difficult to obtain an absolute dating, only allowing the relative development of the complex to be reconstructed.

Sharah has been often compared to Umm el-Jimal in archaeological reports, and several features – aside the unusually good conservation – are indeed shared by the two sites. The size is also comparable, since Sharah reaches an extension of *c.* 20 ha (Figure 55). Nonetheless, some elements on the general structure of the settlement clearly mark some differences between them. Different from the Jordanian site, the morphology of the terrain had a deeper impact on Sharah. If in the other case, the town had available a vast plain area, the Syrian centre had to face the hillier context of the Leja region. Inside the settlement, in fact, several peaks can be found.

The first major feature of the site that is surely striking is the extensive defensive system surrounding the entire settlement. When compared with the walls of Umm el-Jimal, several major differences can be pointed out: first of all, the buildings of the settlement appear to be embedded less frequently in the

---

suggests that the buildings could belong to the 2nd – 3rd century AD. The location is not clearly indicated in the recent plan of the French expedition, nor the presence of such building is mentioned in the report.

[94] Similar buildings may have also served as trade-post or caravanserai or even storerooms. The case of Mampsis is just one possible example. It is clear that considered the context and in light of the widespread brigandage in the area, a military purpose of these buildings is the easiest guest, but the concentration of at least four hypothetical forts looks suspicious to me. Surely, the determination of a precise chronology for each building could help in clarifying the issue.

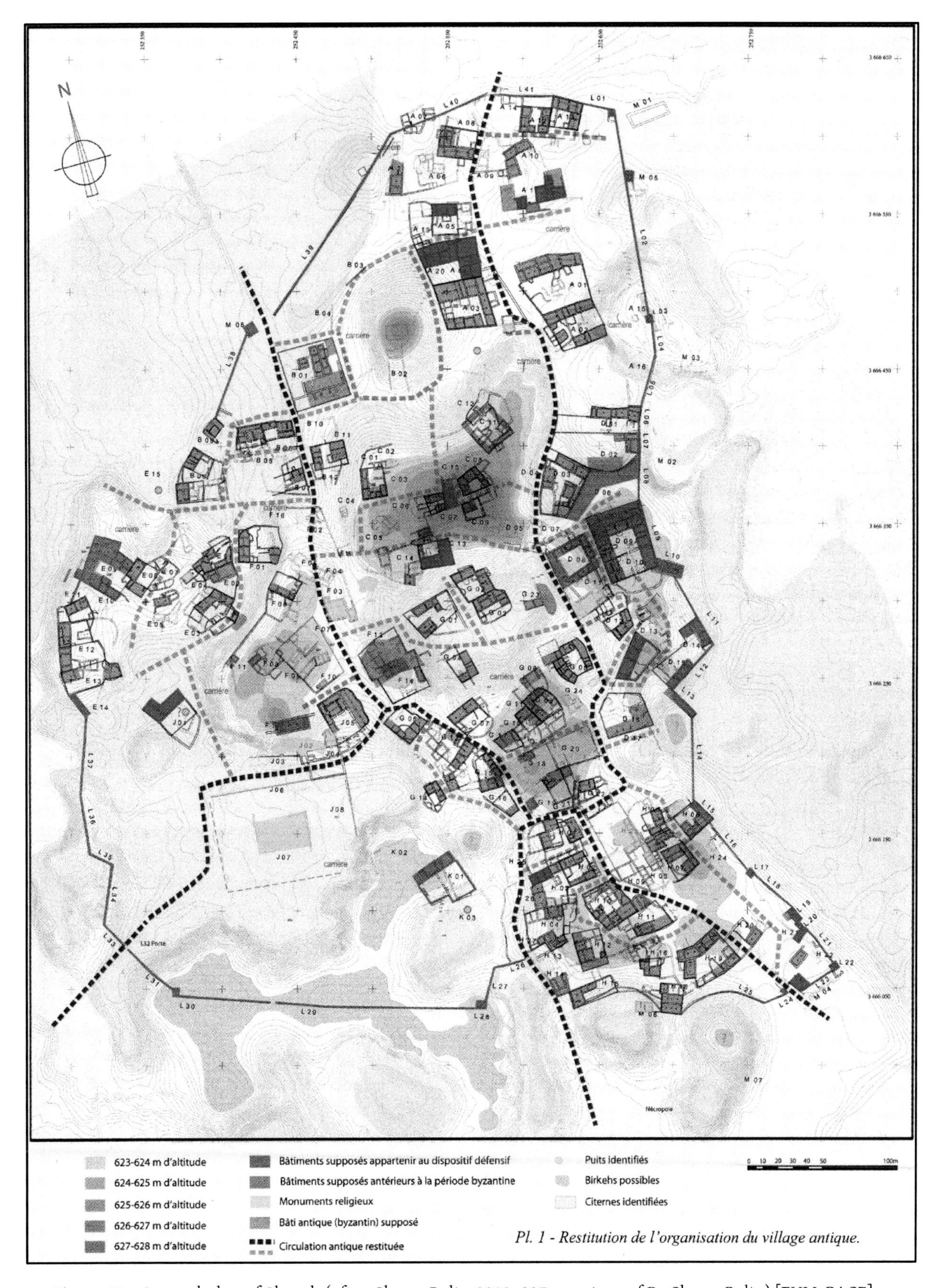

Figure 55 - General plan of Sharah (after Clauss-Balty 2010: 207; courtesy of Dr Clauss-Balty) [FULL PAGE]

fortifications. On the contrary, these seem to closely follow the topography of the site, and hence several towers were built in the higher spots along the walls. As mentioned above, the chronology is still tentative, but according to the trial trenches, it appears that already in the early phases of the settlement – between the mid-1st and mid-2nd century AD – a defensive infrastructure was present. Nonetheless, like in Umm el-Jimal, the walls are too thin to have provided a formidable barrier from a siege or organised attacks. As suggested by the authors of the archaeological mission, the walls, with a thickness of 0.90 – 1.20 m, simply seem to have offered a sufficient deterrent for brigands (Clauss-Balty 2010: 200).[95] A similar function may had been played by the "enigmatic" buildings in the northern part of the settlement (A11, A05, A04-20), tentatively interpreted by Clauss-Balty as defensive structures on the route from the *metrokomia* of *Phaena* and crossing the Leja.

Apart from this, the settlement shows all the typical features previously described for the other sites in the southern Hauran. The most evident features are the lack of any plan or geometrical organisation of the space, and consequently the absence of any clearly built street network.[96] The settlement seems to be vaguely organised on two axes, running one parallel to one other. The westernmost might have been the most important one, at least looking at the kind of buildings that are flanking it, notably the religious monumental group formed by the *mithraeum*, the cave and the Roman sanctuary, and also the only isolated church and mosque. More or less in the centre of the settlement, this western axis splits in two, with one street leading south-eastwards and the other south-westwards. This second ramification connects with the other major axis of the site, running also in the North-South direction. These two axes identify at least five principal entrances in the city-walls, to which one more to the west can be possibly added, since a wide interruption in the fortification was found. Of these accesses, only one gate is probably monumentalised, namely the south-eastern.

The network of secondary paths determines the different quarters and blocks and appears to be randomly and spontaneously developed within the walled area like in the other case studies. It is quite difficult to clearly identify quarters: all the structures are quite scattered in the walled areas. One does not have a compact urban texture, as is the case with Shivta (see Chapter 4.3.1), nor distinguished but compact quarters as in Umm el-Jimal. Instead, in Sharah there is simply a major concentration of buildings in some areas, but the absence of a proper street network and a clear demarcation between groups of structures makes it difficult to identify proper quarters. Nonetheless, when evaluating the plan provided by the French expedition, it appears that the south-western corner of the site is almost empty (Figure 56): here the only structure is a large Roman sanctuary, closely located to a *mithraeum*, which is in turn associated with a cave. It is certain that no domestic structure was located here. A higher concentration of buildings is surely present in the opposite corner of the settlement, to the east and the southeast. It is likely that analysing the morphology of the terrain more closely will play an important role in clarifying this situation: in fact, at least four hillocks are rising and are partially occupied by the structures (Figure 56).

[95] The problematic context of the Leja is well known already in Roman sources. In particular, the widespread brigandage – helped by the geography of the region, offering many and easy hiding – was one of the first issues that the Roman administration in Syria had to deal with (as written by Josephus or Strabo; for a precise reference to the sources, see Clauss-Balty 2010: note 6). The impression of a particular concern towards security in Sharah – and in other sites in the region – is suggested also by the structure of the single houses (see below).

[96] It is interesting to note that Butler, in his report on Sharah mentioned paved streets and even some colonnades flanking some part of them. This aspect, together with the other features of the town mentioned above, suggest him that the site should have been of a certain importance in the Roman and Byzantine period, and even in the Islamic period.

Figure 56 - Sharah: The South-Western area, with the large Roman religious complex (in yellow) and the South-Eastern concentration of dwellings (in blue) (Elaboration by the author on Google Earth satellite image)

Generally speaking, the same undifferentiated functional distribution of the areas is advisable: no specialised quarters – perhaps with the exception of the sanctuary – are clearly distinguishable in the settlement's frame, and most all the functions are coexisting. It is nonetheless interesting to note that the only pattern one can find in the distribution in Sharah, where no other socio-economical differentiation is marked, is that the complexes having the largest stables. In general, they are one of the key features of the residential units in the settlement and are notably concentrated in the most peripheral dwellings in the settlement. Here, the possible functional connection with the entirely rural hinterland of Sharah is at its highest (Clauss-Balty 2010: 204). The contrast between the marginal houses and the ones more occupying the inner part of the town is indeed quite striking already from a statistical point of view: more than half of the houses do not have a stable or have only one. Slightly different is the number of the mangers, which can nonetheless reflect a different specialisation in the breeding species, normally bovines and sheep/goat. In contrast, complexes with three or four stables are extremely rare, being not more than the 11% of the total. Their concentration in the most liminal part of the settlement may suggest a different economic orientation towards breeding in the settlement, or at least a different role in the economy of the single household. On the one side, a subsistence-oriented investment in animal-breeding seems to be suggested, with a production aiming to the simple supply of animal products for a family: in this respect, goats or sheep are probably the most common choice to meet the needs of most households and could also be kept more easily in the most central areas of the settlement. On the other hand, larger areas that were also closer to the cultivated areas may see a higher percentage of bovines or other beasts of burden from which labour could be better used in the rural hinterland, for instance for ploughing.[97] As mentioned above, a deep-rooted link between the settlement and its hinterland can also be indirectly confirmed by the distribution of the most densely occupied areas: the most notable one, the south-eastern, is in fact the closest to the large plain extending to the south of Sharah.

The intermediate level in Sharah presents a totally different organisation from the other case studies: no clearly distinguished districts or quarters can be identified like in the sites in the southern Hauran. However, at the same time there is not the same density that characterises the Negev settlements. The buildings here are loosely grouped into different areas of the settlement, rarely forming proper blocks. Each dwelling tends to be independent and physically autonomous, but at the same time built close to the others. This organisation appears to be particularly evident in the Byzantine period, where an even larger part of the houses appears to have been separated (see infra), being unified with neighbouring farmhouses during the Middle Islamic period.[98] Obviously, the settlement could have looked even more scattered and sparse than it is today. The absence of proper quarters to a certain extent implies the absence of "buffer zones" between the private and the public areas on a higher level than the house: *cul-de-sacs* and courtyards are not represented anywhere in the settlement, not even where clusters of dwellings are formed, since each unit appears to have a direct access from the paths.

Morphologically, the south-eastern district, the one with the highest concentration of buildings, does not differ from the rest of the settlement or present other features than the general ones described above. Still, one can hardly speak of an exclusive quarter, since clear limits of the district are hard to identify. What is different is the density of buildings, clustered around a religious complex occupying a

[97] This is a mere hypothesis, which could only find confirmation with a precise study of the stables' architecture, identifying differences related to the different animal species, and of course with archaeozoological studies, which can also provide information on the economic orientation of the breeding of each species.

[98] In the reports of the French excavations, the period mentioned is a generic "medieval time". This definition is extremely vague, embracing a multiplicity of different phases. We can suppose that the most probable chronological horizon would be the Middle Islamic one, according to the more general context.

more even surface between three small peaks. Going northwards, this concentration – that nonetheless does not correspond to compactness – decreases and a progressively looser pattern can be identified. A possible explanation is the proximity of a large plain south of the town that may have encouraged the settlement of a larger portion of the population. Nonetheless, the absence of clear marks separating these structures from one another and simultaneously the sort of loose continuum of buildings connecting all the parts of Sharah suggests that no strong grouping was present also on a social level.

One other sector of the town presents some peculiar features: a group of structures occupying the middle section of the eastern border of the settlement. According to Clauss-Balty, this complex should probably have functioned as a fortress in the earlier phases of the occupation of Sharah. This interpretation is plausible, seeing its topographical collocation – on a higher ground on the larger north-south axis leading to the street to Phaena – and the plan of each unit. Nonetheless, even more striking in the light of this probable functional specialisation is the fact that this district is in no way separated or marked from the surrounding structures, similar to what will be shown in the eastern quarter in Mampsis (see Chapter 4.3.2). The "fortress" here can be ultimately seen as a grouping of larger squared structures, separate from one another, loosely occupying a liminal section of the settlement and deeply embedded within the rest of the town.

Lastly, the south-western part of Sharah is the only area apparently free from private dwellings. However, one can hardly speak of a proper functional zoning or of a clearly separated quarter, since it is quite evident that this part of the site was not considered "off-limits" by the inhabitants. Nonetheless, the concentration of several major religious complexes reinforces the idea of a cultic specialisation in this quarter, a feature that is not shared by any other district in the settlement. The other districts' functional zoning at this time remains blurry.

The major similarities between Sharah and the other sites in the Hauran area can be found on the house level, both for the architectural technique and organisation of the spaces. Also for this settlement, simple and complex houses are almost exclusively the types of dwellings represented, a development which seems to follow the same stages described for the UT 14 in Umm es-Surab or the *praetorium* in Umm el-Jimal. According to the reports from the French expedition, almost half of the houses present a "I" Plan, like in the dwelling we described as simple.[99] The other dwellings have "L" (16 units, corresponding to the 29%) or "U" Plans (9 units, 16%). Only two houses are constituted by two aisles facing on the other.[100] The progressive accumulation of structures, forming a central courtyard or even the progressive merging or former isolated dwellings is a phenomenon documented throughout the entire site.

Once again, the most problematic issue is the precise chronology of the additions or modifications to the site. Two different phenomena can be tentatively recognised: on the one hand, the creation of proper "injunctive" houses with a courtyard surrounded on three or four sides by structures and on the other hand, the fusion of two separate units into one single larger complex. Both processes are not exclusive to Sharah, but these trends in development seem to have been realised at a later date,

[99] Precisely 23 on 55 houses (42%) identified as ancient (Clauss-Balty 2010: 202).
[100] It is interesting to see the change in the percentage of these type of plan of the houses identified as "modern" (Clauss-Balty 2010: 202): we see a drastic decrease of the single houses ("I" Plan: 10 houses of 74; 13%) and "L" types (10 houses, 13%) and a clear increase in the more complex plan ("U" plan: 15 units, 20%; and the completely closed plan: 6 houses, 8%). This evolution seems to confirm the progressive nature of the development of the domestic architecture in the region. The ultimate stage, namely the more closed plans, that in the other case studies are clearly reached in an earlier phase, is only a later achievement in Sharah. A characteristic that the site shares with the other settlement of the Leja we consider here, Mseikeh (see Figure 59).

especially the second one. In fact, the French researchers consider these phenomena belonging to the "medieval" phase.[101] Whatever the dating of such label exactly defines, it appears that the creation of larger houses, resulting from the fusion of two earlier separated dwellings, dates later than the chronological horizon considered for the present project. It is nonetheless interesting to underline the similarities that exist between these complexes and other buildings confidently dating to the Late Byzantine or Early Islamic periods, like the large complex formed by Houses XVII and XVIII in Umm el-Jimal or Building XII in Mampsis. The absence of clearly defined quarters or blocks, and also consequently of the devices creating "buffer zones" on these levels, do indicate a careful development of the privacy and in general security "systems" on the house level in Sharah. This compensation is particularly evident and determines a situation close to an example from Umm es-Surab (UT 24), not surprisingly also a completely isolated house in the site. It is evident that these structures, not benefiting from a "collective" defence, must provide for their own security, a need that appears to be particularly felt in the Leja region.[102] The inverse correlation between the complexity of the quarter/block and the "defensive" character of the single house had been already expressed elsewhere, though in Sharah this aspect is particularly evident.

4.2.2: Mseikeh

The other Syrian case study is also located in southern part of the Leja and despite macroscopic differences with Sharah, it does share two elements with the previous case study. First of all, it appears that the chronological development of the site follows more or less the same phases, with an important development in the Byzantine period, and in later periods a decisive increase in size and the adoption of determined spatial solutions like complex houses with a court completely surrounded by structures. Most notably the excavation's report (Guérin 2008) suggests a hiatus in the Early Islamic period, and a re-occupation only in the 12th century until the 15th, which could also possibly offer an important insight into the vague chronology of Sharah. Secondly, the organisation of Mseikeh also seems to be heavily influenced by the topographical conformation of the area. In fact, it occupies a basaltic plateau at 600 meters a.s.l., dominating the surrounding cultivated areas. The settlement, *c.* 4.5 ha large, is spread out over the entire surface of this plateau, which slopes offer a natural defence (Figure 57a). These two elements distinguish the two Syrian sites from the settlement considered in the southern Hauran in Jordan. More specifically, some elements of Mseikeh notably differ from other village or entirely rural settlements like Umm es-Surab and Shivta. It is interesting to note that this Syrian site could represent a sort of midway between the other two, sharing on the one hand clearly distinguished quarters like in the Jordanian case studies and on the other hand a generally more compact character of the settlement similar to the case studies from the Negev.

Once again, chronology is the most problematic issue. The absence of the ancient toponomy describing the village worsens the general picture, precluding hints from historical sources. The studies conducted so far allow some similar settlement dynamics already described for other case studies in this thesis to

[101] "L'époque byzantine correspond à une phase de développement importante, avec l'agrandissement de maisons existantes et la construction de nouvelles. Mais les périodes médiévale et moderne ont vraisemblablement été tout aussi importantes, marquées par un processus comparable d'occupation, amplifié par la transformation en habitations de bâtiments qui avaient à l'origine une autre fonction" (Clauss-Balty 2010: 203)

[102] "Parmi les cinquante-cinq habitations romano-byzantines recensées, certaines comportent des sortes de greniers qui servaient de cachettes et qui confortent l'hypothèse qu'une certaine insécurité régnant dans la région entre la fin du IIe et le VIIe siècle, et sans doute au-delà, cette insécurité étant probablement liée aux populations instables de l'intérieur du Leja" (Clauss-Balty 2010: 206).

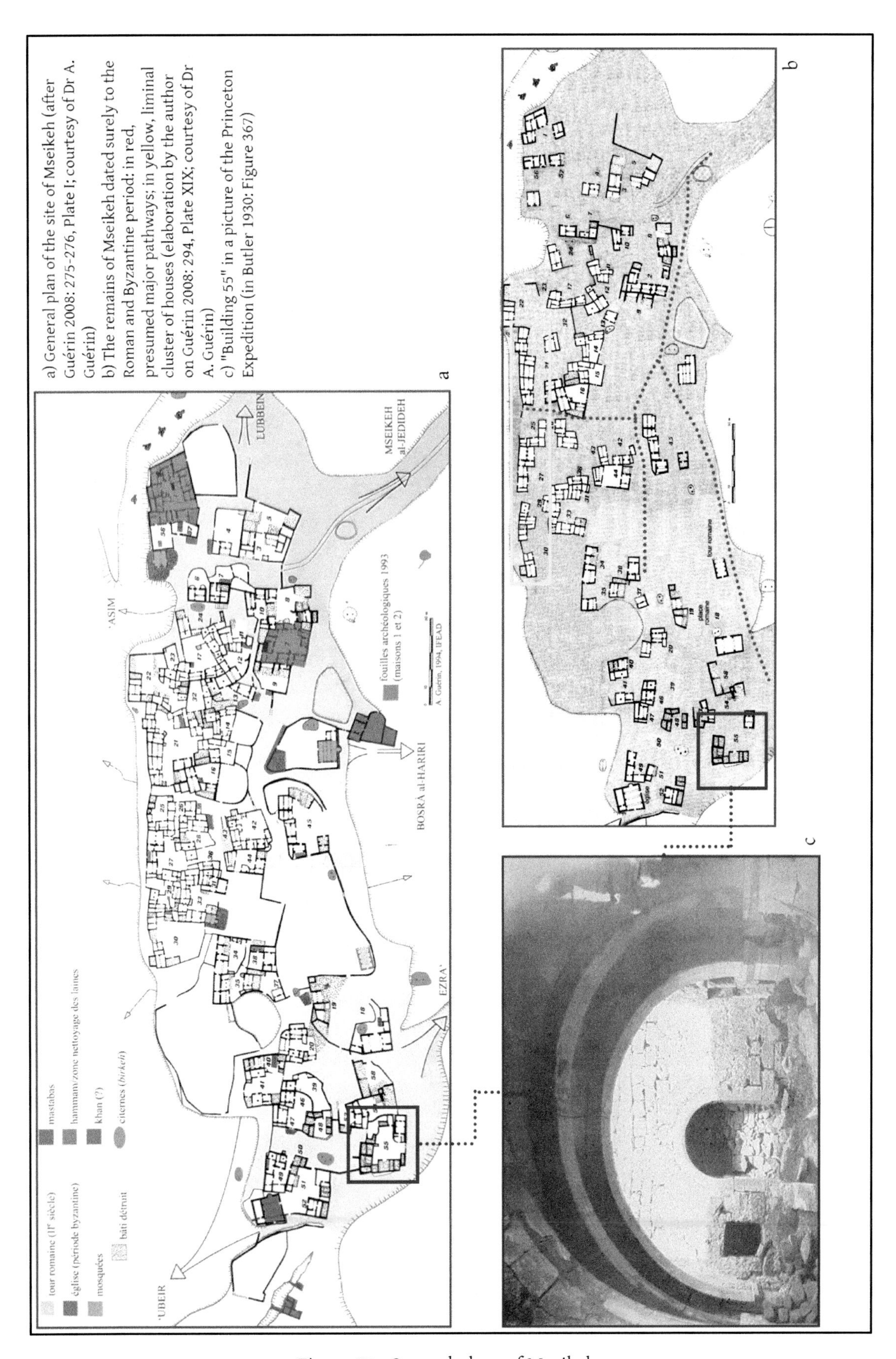

Figure 57 - General plans of Mseikeh

be identified. Like in Sharah, it appears that the first occupation of the site took place during the Roman period, with the construction of a defensive structure in the central area of the upcoming village and, according to Butler, a pagan temple. The extent of the site at this earlier phase is unclear, but after the survey conducted by the French expedition, it seems that the occupation was concentrated around the supposed tower (Guérin 2008: 259). The defensive character of the first occupation is suggested by the form and the orientation of the structures connected to the tower, namely Building nr. VI and two houses (nr. 18 and 19). They in fact were organised around an empty area and all the openings are oriented toward this central space. This phase probably belongs to the first half of the 2nd century AD, according to three altars dated to AD 133 and 136 and an inscription (now lost) of the tower also from the same chronological horizon.[103] The general impression Guérin gained from this Roman phase is that this initial fortified group of structures represented the original nucleus of the settlement (Guérin 2008: 259).

Another building, which interpretation is extremely problematic and that had been described in detail by Butler (1930: 424-426), is also of particular interest (Figure 57c): it is the last complex on the southwestern fringe of the settlement (nr. 55 in the French survey) and is the place where two of the three altars were discovered. The plan of the complex is unclear, and so is its dating. It is composed by a long hall and adjoining smaller structure with two rooms and a portico. From this portico and the southern end of the first hall, two of the above-mentioned altars were found, suggesting an early date for the structure, since as Butler suggested, they would not have been moved far from their original collocation.[104] Nonetheless, seeing also the extreme quality of the masonry and the presence of high quality painted decorations in the internal long hall,[105] Butler supposed some sort of public destination for it: he in fact decided to name it the "*Praetorium* of Mseikeh". Despite the questionable interpretation of the complex as *praetorium,* similarly to what seen in Umm el-Jimal, it is clear that this building had a certain importance within the settlement.

A major expansion of the settlement occurred during the Byzantine period, when a monastic complex was built on the western fringe of the village together with at least 20 of the 58 houses recorded in the site. A sort of dichotomy in the settlement organisation had been detected by the French expedition (Guérin 2008: 259-260). On the one side, it appears to have an extremely scattered pattern of houses without any identifiable courtyard,[106] spread across the entire surface of the site; on the other side, there are at least two more compacted groups of structures, namely the monastic complex and the tower's *ensemble.* It is clear that the following phases heavily modified the Byzantine structures, which therefore further affects our interpretation. Nonetheless, according to the excavations in two houses,

[103] Bulter (1930: 424) reports the finding of this inscription, but offers a later date for it, that would have been dated between AD 247 and 249.

[104] At the beginning of its report, Butler actually investigated the chance of its dating as an Early Christian church. Afterwards, seeing the extreme quality and the peculiarity of the complex he made this interpretation, though without suggesting a clear building date (Butler 1930: 425f).

[105] Butler affirmed that "the stonework is of the best quality throughout, much better than that of most of the churches of Southern Syria" (Butler 1930: 425).

[106] After the plans offered in the publication, I understand that in this case the presence of courtyards in general is not denied in itself. What is not identified by the authors of the report is indeed the typology of dwelling we have described as "Courtyard House", so with a central and regular court surrounded on all its sides by structures. The consequence is that apparently in this phase, we have a predominance of "simple houses" or at least rarer "Complex Houses". Moreover, the overlaying of the later structures might have easily erased any trace of enclosures connected to the units, offering us a slightly difference appearance of the settlement in general. I will consider in detail this matter below.

it appears that the occupation of some of the Byzantine buildings continued into the Umayyad period, according to the ceramic remains.

A second major expansion took place after the "Islamic conquest". According to the ceramics found in the site, a reoccupation of the site roughly begins in the 11th century under the Fatimids followed by a major development and densification in the Seljuk or Ayyubid period, between the 12th and the 14th century (Guérin 2008: 260). The relationship between the Later Middle-Islamic site and its earlier phases is still unclear, but it seems that there was a shift of the settlement towards the East, leading to the gradually abandonment or at least a less dense re-occupation of the western part of the village. The rate of construction of new houses and modification of the older ones is impressive: in total 58 dwellings are dated to this Islamic phase. Even more interesting to note is the augmentation of single modular cells constituting the dwellings, actually the number of rooms: from 91 cells formatting 20 houses in the pre-Islamic period to 450 cells in 58 dwellings in the following phases (Guérin 2008: 261). This striking fact may find a direct correlation with the densification of rooms documented in the residential structures in some other case studies. At the same time, one can probably identify the same process in the domestic architecture clearly documented in Sharah though less systematically shown in the other case studies: the fusion of two residential units or more into one single complex and simultaneously the creation of completely enclosed courtyards. In this respect, namely the fact of having houses gain direct access to others as if they constitute a single unit, the remark by Guérin is particularly interesting: "cette facilité de passage entre les habitations suppose une organisation familiale étendue" (Guérin 2008: 261). This is of course a direct parallel to what is testified in other settlements in earlier phases. The construction of the three mosques, the hammam and the khan that can be found in the site should belong to the same chronological horizon. Mosque II, in light of its connection with a *mastaba*, is considered by Guérin to be the main mosque of the site. The other two are tightly embedded within the urban frame. All these structures reuse older buildings, which is a common pattern in the development of the site in this phase.

The general conformation of the site is difficult to clearly reconstruct for the Byzantine period, considering the heavy intervention of the following developments. However, when considering the plans reporting the Roman and Byzantine remains, one can nonetheless obtain some interesting information (Figure 57b). First of all, it is evident that the Byzantine settlement extended from the original nucleus – the fortified group in the western part – to occupy more or less the entire surface of the plateau.[107] The manner in which the limits of the village were marked are still unclear, but a useful remark was offered by Butler, who affirmed that the entire settlement was "enclosed" by a wall; its traces had also been partly recorded by the most recent French expedition. Nonetheless, the borders of the village were somehow clearly marked: the separation of the site from its hinterland is underlined primarily by its morphological aspect, with steep slopes delimiting the urban space. In fact, even if it's not possible to have any chronological indication about the construction of the aforementioned enclosure, the access to Mseikeh had been always somehow ruled by well-marked "entrances" that were in turn determined by the topography of the site.

The viability of the site appears to be affected by these slopes as well, since it's clearly linked with the entrances that open in three directions: east, south-east and south-west. Five roads leading to the

[107] According to Guérin (2008: 249f), in all the quarters of the Islamic period with the only exception of the "Quartier Est", also materials from the Roman and Byzantine periods had been found, with a particular concentration – also of architectural remains – in the "Quartier de la tour", namely the original nucleus of the settlement.

village had been identified and are the continuation of the two major axes running inside the settlement. This is certainly true for the Islamic phases, but one can confidently extend this description to earlier phases as well. In fact, looking at the areas where the buildings concentrate during the pre-Islamic and the Islamic periods, and consequently pointing out the "empty" spaces, one can observe that there is not much difference between them. The spontaneously developed path-system was respected in its general organisation, and to a certain extent was determined by the topography of the site.

Two major paths – or better empty tracks that could have been pathways in the Byzantine period – can be understood as the result of spare ground leftover from the building activity (Figure 57b). The first runs along the entire southern edge of the village, starting from the western entrance to the plateau, grazing against the *praetorium*, the original fortified group and the so-called "Roman place", and finally heading to the south-eastern access to the site. Comparing the two plans of the pre-Islamic and Islamic phases, one can see that this track was always left unbuilt, suggesting its continuative use as circulation axis. Similarly, another track runs parallel to the previous one, but toward the centre of the settlement. Despite the fact that the better-defined medieval quarters determine a clearer track for the path, it can be supposed that this case wasn't dissimilar from the earlier one. Interestingly, these two East-West axes were not the only lines maintained in the later development of the site. One smaller and narrower path already present in the Roman and Byzantine settlement cuts the site in two northward-southward, starting from the exact centre of the village, where it intersects perpendicularly the second East-West axis,[108] and leads to a door in the northern enclosure, decorated with an engraved lintel.[109] The relationship between these hypothetical path frame and the Byzantine quarters is not clear. Surely, the sparser and more scattered occupation of the plateau during this phase bound the single complexes to these three axes less tightly. Something different is noticeable during the Islamic period, when it is clear that each quarter had a direct access from at least one of them. Similarly, no clear connection between buildings and communal spaces is clearly identifiable, apart from the so-called "Roman place" close to the tower, which is the only surely detectable open space of the earlier settlement. Another "blank" space is possibly opened to the east of this complex – as also seen in the Islamic phase – but it is impossible to suggest any function, since no kind of infrastructure datable to the Byzantine period had been found.

The hydraulic infrastructure is another point in our knowledge lacking for the site, especially for its earlier phases. It seems that no communal major infrastructure for the supply and management of water was present in Mseikeh and like the other case studies, private individuals appear to have faced this issue with the typical domestic rain fed cistern in their domestic complexes. It is nonetheless interesting to note the absence of larger cisterns or reservoirs, destined to more communal or public purposes, which especially coincide with the "blank" areas, even in the Islamic period.

One can say also very little on the inner distribution of the complexes or on their stratification, functionally and socio-economically. It is quite clear that, despite the small dimension of the village, some important complexes were present. There could also be a tendency of having them in peripheral positions: the monastery is most notably the first building encountered when entering the village from

[108] This intersection is marked in the Islamic phase by the presence of a double door, giving access to the quarters to the North. It does not seem that this kind of organisation of the space was also present in the Roman and Byzantine period, when the occupation of this area – though denser than in the western part of the site – appears to be insufficient to justify such structure.

[109] Littmann did not propose any dating for this lintel (nr. 795/11 in Littmann, Magie, and Stuart 1930: 420), but according to Guérin (2008: 249) the kind of decoration – most notably the "couronne de laurier" is a typical Roman motif and lastly used in the Byzantine period.

the west; the *praetorium* is situated at a short distance and on the periphery of the western quarter; lastly, on the southern fringe of the settlement, in a more central position on the East-West axis, is the "fortified group" of the tower.

While the analysis of the general form of the settlement already presented several difficulties, the single quarters and blocks present further complications because of the heavily renovated organisation that took place in the Islamic phases. It is impossible to determine how much of the earlier spatial distribution and organisational principles was maintained, but there is no reason to exclude the possibility that some aspects of the later developments could have been applicable to the Byzantine period. Moreover, some features may find direct parallels with other case studies considered here.

The scattered and sparse character of the settlement in general is also visible on the quarter level, where one can hardly find the same density of buildings for instance present in the closer South Hauran's sites. Under this aspect, the similarities with Sharah are evident: the tendency in both case studies from the Leja is that they have more frequently isolated complex or at least clusters of a maximum of two or three dwellings, spontaneously constituting an irregular block. There is, though, a quite striking difference between two areas of Mseikeh, namely the western (and evidently older) district and the possibly later Byzantine expansion of the village. Even if it's not possible to speak of two organically distinguishable quarters, the different density of structures in the two areas is clearly a feature differentiating them. While clusters of houses and in general the presence of larger dwellings is quite frequent in the eastern part of the site, the western one seems to be dominated by isolated structures and simple houses, with rare – if identifiable – clusters.

Looking at the later Islamic development of the site, a sort of imprint of the situation just described appears throughout the new quarters, which are entirely enclosed and accessible only through specific points marked by well-built doors. These entrances give access to intermediate courtyards – direct parallels for the "privacy devices" mentioned in other case studies. From these semi-public or semi-private open spaces, the single residential units are then accessible in most of the cases directly and without any "vestibule". This way of organising the quarters – maybe except for the officially established entrance doors – is not a new pattern, but is already visible in earlier phases of the other case studies, especially the ones presenting more accentuated urban features, like Mampsis and especially Shivta in the Negev. Although no certainty is available that the later spatial organisation of Mseikeh closely resembled the Byzantine one, it is interesting that the intermediate "black spaces" are maintained between the two phases (see Guérin 2008: 249-259 and 289-293). I would suggest that the same spatial process for the single houses is ultimately to be described on a quarter level, with the progressive creation of enclosed spaces and a densification of the occupation. Still a marked difference between the eastern and the western part of the settlement was maintained, since the western quarter ("Quartier de l'église" in the French report) kept its more sparse and scattered aspect if compared with the highly dense western quarters – in particular the "Quartier de la mosquée".

One other aspect quite confidently recognisable already in the Byzantine phase is the northern border of the settlement, which closely resembles the "liminal cluster quarters" seen in Umm el-Jimal. The structures – in this particular case only cells whose relative relation is uncertain – constitute a continuous and almost entirely closed front to the outside (Figure 57b). It is not certain at present if this "barrier" was part of an enclosure surrounding the entire settlement or was limited to this single part of the site. This part of the village, though, was clearly felt as peripheral and exposed, and therefore needed some sort of solution to mark the separation from the outskirts of the quarter (or the entire

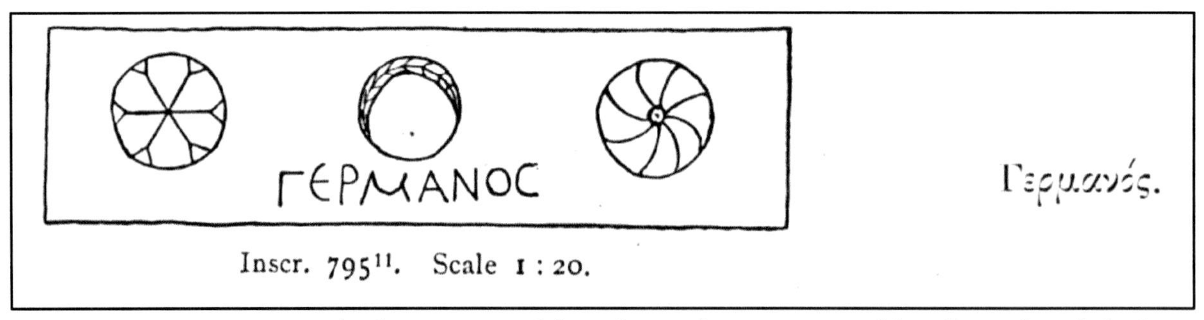

Figure 58 - Inscription nr. 795 (with transcription) from the Northern entrance of the village of Mseikeh, close to Building nr. 25 in the French survey (Littmann, Magie, and Stuart 1930: 420)

settlement). The maintenance in the Islamic period of the Roman or Byzantine door with the inscription (Figure 58) as the entrance to the North-South axis of the settlement, supports such a view.[110]

The excavation of two houses by the French expedition offers important information about the development of the residential architecture and its spatial organisation from the Roman-Byzantine period to the Islamic period. Despite a different and later chronology, it is possible to see the same formation process described in particular in Umm es-Surab and in Sharah.[111] This process ultimately consisted of the same densification of the building spaces documented for the settlement in general and for the quarters. It is nonetheless interesting to see that not only the original forms but also the final products do not differ from the other case studies previously mentioned.

In "Maison 1" (Figure 59b), for instance, the excavators identified at least three earlier structures, initially separated and then progressively unified into a single residential complex starting from the 12th century (Guérin 2008: 234-237 and 279, pl. IV). The typology of these original houses is difficult to determine because of the later modifications. For example, the "maison antique n.56" – on the north-western corner of the Islamic complex – is actually only a section of a Roman-Byzantine dwelling of which only passage rooms are preserved. The "maison antique n.1" is even more interesting, where the opposite corner of the later house, which clearly represents an example of "L-building" like the one documented in Sharah, is composed by three large halls flanked by smaller cells on two storeys. Finally, in the southern part of the Islamic courtyard, there was the third Roman-Byzantine structure, again with a large hall and one smaller room; unfortunately, the remains do not allow for a clear typology of the structure. No particular installations – apart from a large niche in the "maison antique n.56" – had been found, and therefore suggesting any functional destination for any of these earlier structures is not possible. All these dwellings seem to respect the typical Byzantine typology of the region, as seen in Sharah, with a changing and to a certain extent random combination of larger halls and smaller side

[110] I am not sure that the door had the same function of "entrance" of the settlement or of the quarter also in the earlier phase. The inscription reporting only a name (*Germanios*) and the modest dimension of the opening itself could also identify the entrance of a private dwelling. Nonetheless, this do not invalidate the hypothesis offered above: this would be in fact – according to the plans of the French expedition – one of the few openings towards north of the Roman and Byzantine period.

[111] This remark results to be particularly important especially for one of the case studies considered in last part of the chapter, Tall Hisban. As we will see, in fact, a village developed in the Mamluk period reused and overlapped earlier structures, tentatively dated to the Byzantine period. One of the major questions was indeed to determine the degree of modification in the spatial arrangements of this newly built village, and possibly to understand how the domestic architecture change in function of a different social structure. The fact that houses completed in the Middle Islamic in some sites do not present substantial differences from the complexes terminated surely before this period, suggest that the organisational frame adopted did not undergo much modification.

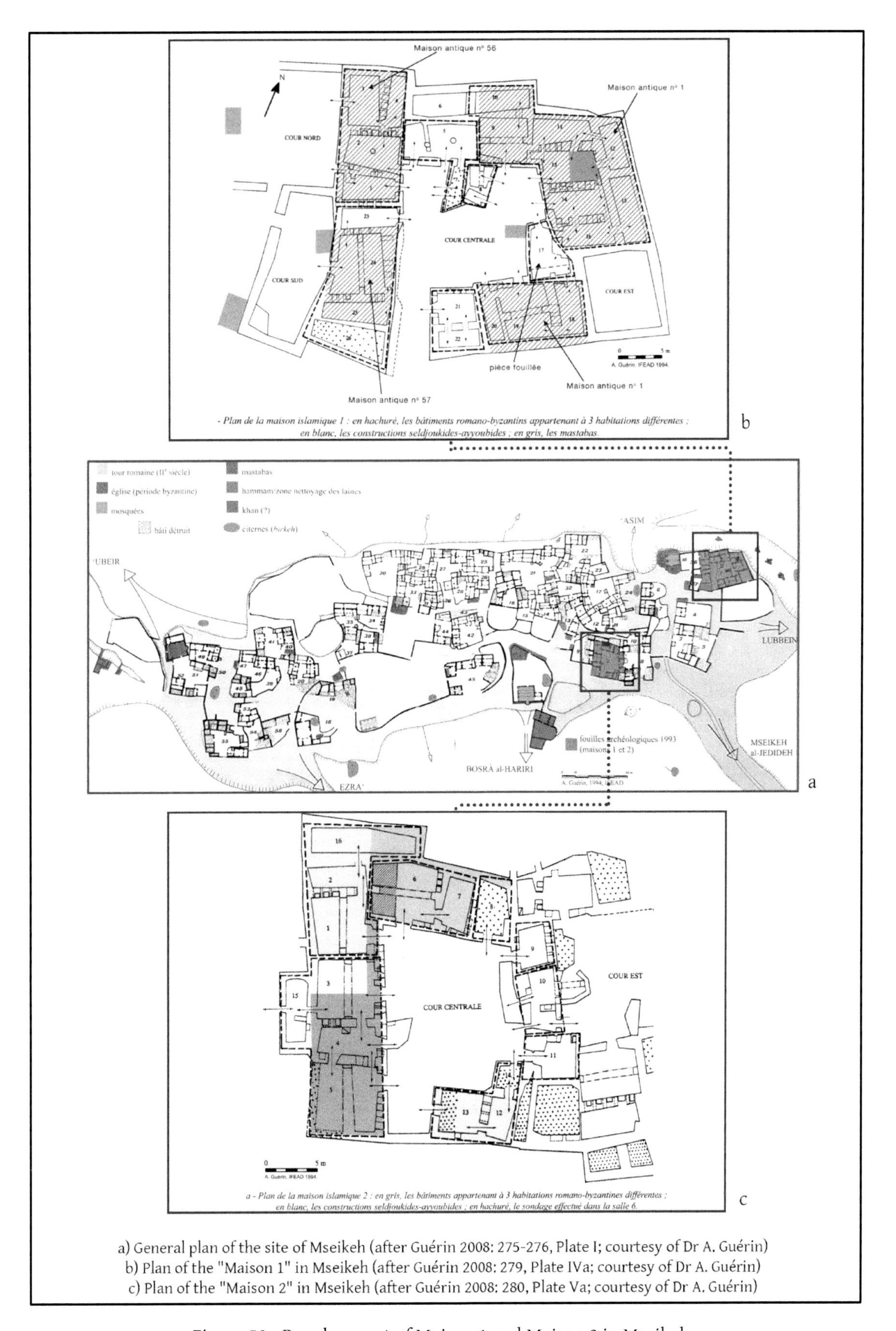

a) General plan of the site of Mseikeh (after Guérin 2008: 275-276, Plate I; courtesy of Dr A. Guérin)
b) Plan of the "Maison 1" in Mseikeh (after Guérin 2008: 279, Plate IVa; courtesy of Dr A. Guérin)
c) Plan of the "Maison 2" in Mseikeh (after Guérin 2008: 280, Plate Va; courtesy of Dr A. Guérin)

Figure 59 - Development of Maison 1 and Maison 2 in Mseikeh

rooms on two storeys, organised tendentially into L or U forms. These separated complexes were then unified in a later phase around a common central courtyard, in Mseikeh and Sharah starting from the 12th century, and in Umm es-Surab – and plausibly Umm el-Jimal – in the Late Byzantine period.

The same indications and an even more detailed chronology come from the "maison 2" (Guérin 2008: 237-243 and 280 pl. V; Figure 59c). According to a probe made in the western wing of the complex, the foundation of the first structures can be dated to the third to the beginning of the 4th century. Four successive phases dating to the Byzantine period bear witness to a progressive abutting of new cells against the earliest one. At first, the occupation was limited to the north-western corner of the later Islamic house, which was quite confidently a simple house (cells 1, 2 and 16 on the plan by Guérin) with a fold, or an unpaved roofed room, hypothetically reserved for animals. The second and third Byzantine phases, tentatively belonging to the 4th – 5th centuries, bore witness to a clear development of the house eastwards, with the addition of cells 6 and 7 and subsequently 8 and 9, possibly independent in origin. The fourth and last Byzantine phase experienced the definitive development of an L shaped dwelling, with building of cells 3, 4 and 5 and the erection of a second façade. No material culture from the Early Islamic period was found anywhere in the complex, and a long hiatus seems to characterise its chronology, from the 7th to the 12th century. In fact, it is only in the 12th century that occupation resumed with the development, as in the previous house, of a central court surrounded by roofed rooms.

As these examples suggest, the process in Mseikeh is very similar to the one described in Umm es-Surab, and perfectly fits the theory behind the spatial conception of the so-called "injunctive" houses: a progressive and spontaneous addition of cells, possibly dictated by an increase in the number of dwellers or by additional needs arising, like new reception spaces or productive installations. What differentiates the examples from Mseikeh is the late development of an entirely enclosed central court, never before the 12th century, as opposed to Byzantine dating of such structures in the other case study. It is clear that in both cases, the chronology is complicated. On the one side, the possibility of having already some examples of "complex" houses in Mseikeh before the Islamic phases cannot be excluded, for which traces might had been completely erased by the following intensive and highly destructive occupation. On the other side, a potential later dating for some structures in the other sites must also be taken into account, since the constant reuse of building materials makes it indeed extremely difficult to determine the chronology of the cells with certainty, especially if no aid from excavation is available. What is extremely interesting to me, nonetheless, is the fact that the spatial logic ruling the development of the houses, and also the functional distribution of the spaces, appears to be quite similar, without undergoing such a radical modification between the pre-Islamic and the Islamic period.

4.3: The Negev

The case studies of Shivta and Mampsis[112] are both located in the Central Negev (Figure 60), more specifically in the Highlands preceding the real desert located in the southern part of the Late Roman and Byzantine Province of *Palaestina III*. Settling here was only possible through an intensive effort made toward water catchment and management; therefore, it is not surprising to find extensive hydric systems around and within the settlements, allowing agricultural exploitation of the surrounding land

[112] These two sites – together with two other towns (Elusa and Avdat), four fortresses (Kazra, Nekarot, Makhmal, and Grafon) and two caravanserais (Moa and Saharonim) – had been included by UNESCO in the World Heritage List in 2005, with the official designation of "The Incense Route – Desert Cities in the Negev" (see the official UNESCO website: http://whc.unesco.org/en/list/1107).

Figure 60 - The Negev Highlands region (Satellite image: Google Earth)

and ensuring the survival of the inhabitants. With different degrees and typologies of intervention, ranging from the centralised effort to the more spontaneous enterprise of the dwellers themselves, the landscape had to be constantly modified in order to allow for a more intense agricultural use and possibly support a denser population. Intervention in this direction are normally related to the first occupation of the region by the Nabatean: archaeologically, the evidence of Nabatean efforts in shaping the countryside is rare, but it is possible that later intensive developments have erased earlier traces, especially during the Byzantine period (see Chapter 7.1). Furthermore, many settlements show a clear commercial vocation, thanks to their proximity to the caravan route leading from the Red Sea to the Mediterranean. While this is the case for the pre-Byzantine phases, it is more uncertain for later periods, when a likely change in the commercial routes took place, with an augmented importance of the local and regional trade, and simultaneously some clear functional changes occurred in the structures of some of the settlements. Nevertheless, the Byzantine period bore witness to a gradual increase in the importance of the phenomenon of the pilgrimage, which surely contributed to maintaining a certain degree of prosperity in the region. It is furthermore difficult to find large settlements in this region: the only exception may be Elusa, the capital of the district and possibly of the Province after Petra. The vast majority of sites are small settlements, namely villages or isolated farmsteads. The more frequent occurrence of farmsteads seems to be a phenomenon further north and in the coastal strip, while there are few in the Negev region. Despite this, many of the sites present morphological features closer to urban entities than completely rural ones. Because of this it is not surprising to find sites like Mampsis, Shivta, Avdad, Nessana and Rehovot often mentioned in archaeological publications as "towns", even if no clear marks of "urbanity" are to be seen, at least in terms of functions of the settlement (Hirschfeld 2003: 408; Avni 2014: 260-267). This particularity of the Negev settlements may have different reasons, probably operating simultaneously. For instance, a longer Hellenistic tradition may have led to a more frequent adoption of determined architectural and spatial features also in a rural context. However, the influx of a different social framework in the settlement is not to be underestimated in regard to this phenomenon.

4.3.1: Shivta (Sobata/Soubaita)

Located in the centre of a heavily cultivated area just on the fringe of the Negev desert, Shivta had been a relatively prosperous settlement, though never assuming any important administrative status (Baumgarten 2004). Despite its clear agricultural specialisation, some organisational features of the village do find precise parallels with more urban contexts and ultimately is the case study of the present analysis that could be more closely compared with the so-called "Islamic city". Few exact indications are available for its chronological development; once again, reconstructing the earlier phases is particularly problematic because of the later heavy modification of the settlement. The first nucleus of the future Shivta is probably linked to the Nabatean colonisation of the Negev Highland (Negev 1980: 21f; Segal 1985: 323; Röhl 2010: 77f), under the reign of Obodas III (30-9 BC) or Aretas IV (9 BC- AD 40). According to some of the excavations and survey conducted in the site, the part occupying the lowest part of the slope – in particular, the area east and south-east to the Double Pool – appears to be the first one built, since the only Nabatean pottery for Shivta was found here (Segal 1985: 323f). It is also the only area where "architectural elements of unmistakably Nabatean nature" had been identified (Negev 1980: 21f), especially concerning the technique used for arching, the construction of the staircase-tower and stables. All these elements do find direct parallels with the architecture in Mampsis, more accurately dated to the Middle and Late Nabatean period.

The settlement likely underwent an important development in the Byzantine period. The exact dynamics are still not entirely clear, and there is a slight difference between the early hypothesis posed by Segal (1985: 322-325) and the hypotheses from more recent excavations. Segal suggests that the settlement had been gradually expanded toward the top of the hill, which was ultimately occupied by the larger religious building, the North Church (see also Bar 2005: 56f). The ceramic finds show a peak in the settlement between the 6th and 7th centuries, when the maximal extension was reached. The growth process also contextually implied the modification of earlier Nabatean complexes and a general densification of the urban space, and appears to start already in the 4th century, as testified by the "Stable House" (Segal 1985: 94) and the South Church (Röhl 2010: 80)[113].

[113] The particular plan of the South Church appears to have been determined by the fact that the religious complex was built on an earlier structure – possibly a Nabataean temple as suggested by Segal – when the settlement in that areas was already developed (Segal 1985: 322-323).

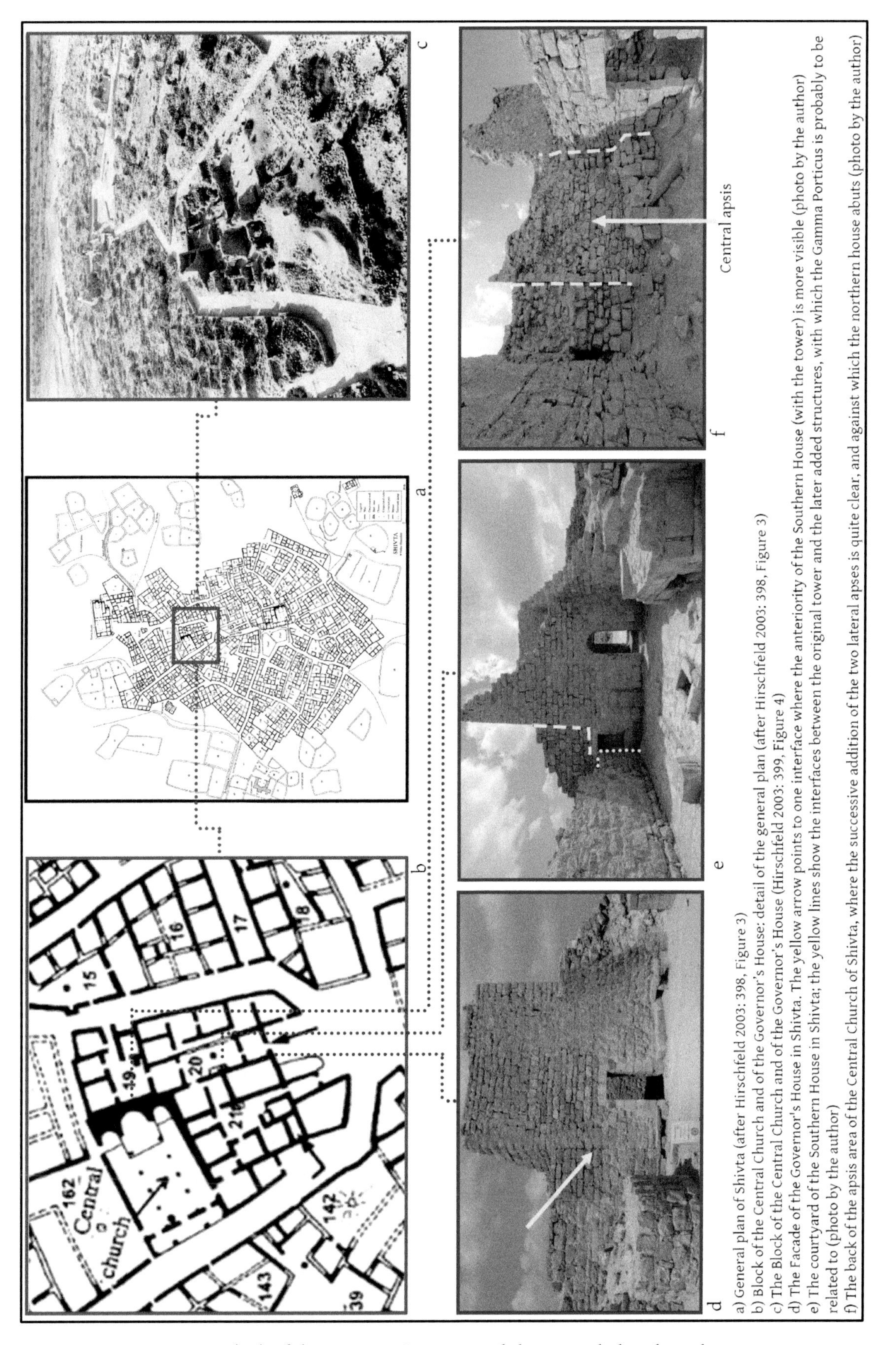

Figure 61 - Block of the Governor's House and the Central Church in Shivta

Some of the larger complexes appear to have been built already by the 5th-6th century in the central sector, like in the case of the Pool House (Segal 1983: 37). In this regard, a particularly interesting example is the block formed by the Governor's House and the Central Church (Figure 61). The development of this area appears to be extremely problematic and no indications for an absolute dating of the entire complex or the single structures composing it are available. The hypothesis offered by Segal does not convince me entirely, even if some elements of the general framework seem to be correct[114]: it is quite clear, for instance, that the four complexes in the block had been heavily modified at a certain moment in the Byzantine period, especially with the building of the church, encroached in an already settled area. The relationship between the different "residential units" and especially their development is less convincing. Segal clearly considers their construction contemporary, speaking exclusively of later modifications to connect the earlier isolated units. After a quick autoptic overview of the structures I grow more sceptical of this hypothesis, either regarding the development of the single units and more importantly of the entire block (Segal 1983: 151-167).[115] In my opinion, the so-called "South House" was the original structure of the block, which was probably isolated in an early phase. Only in a subsequent moment, the "West House" and the "North House" were added, also implying some heavy modifications to the organisation of the "South House" (Figures 61d and 61e).[116] It is likely that this last building is also later than the first phases of the later Central Church, even though the construction of this religious complex and the progressive edification of the three apses generated a great deal of modifications (Figure 61f). Understanding the relative chronology clearly would not have been possible without a more detailed analysis. The most important aspect, though, is that the surly progressive accumulation of structures - partially modifying earlier buildings not only architecturally but also possibly functionally - is a phenomenon that took place in the Byzantine period and characterises the historical development of the settlement.[117] In this circumstance, there

[114] I think a more detailed analysis of the architectural remains is more than necessary. A renewed survey of the standing structure at least could offer important information on the precise relative chronology of the complex, but also on the function of the entire block. Ultimately, if excavations were also conducted and the possibility to link such relative chronology with a more absolute temporal framework, the study of this area could shed light on the development of the entire settlement.

[115] It is not the purpose of the present research to offer a detailed report of the structures, and it would not be scientifically correct, since I have not conducted a systematic study of the architectural remains. I can simply mention here some impressions I gained while visiting the site. It also has to be said that the recent restorations make it quite difficult to see the original appearance of the walls, and this aspect is important to keep into account if a proper architectural survey will be planned.

[116] The first impression is that especially the actual conformation of the courtyard underwent several modifications. The tower was surely the earliest building built, and was connected in phase with part of the enclosing walls of the courtyard (Figures 61d and 61e). The staircase that we see today is probably contemporary, since the entire western enclosure seems to be built together with the tower. I am not entirely sure that the "Gamma-portico" and consequently the paving of the courtyard already belonged to the original construction. The later creation of the covering structure can be suggested by the fact that the pilaster abuts against the wall of the tower and a series of chick stones are employed. Even if this element does not necessarily imply a later construction, but be simply a not particularly refined architectural technique - surely distant from the Nabataean one - the correlation of the Gamma structure with the wall abutting against the nice is in my opinion a clear mark of its later construction. This wall in fact appears to belong to a series of later construction heavily modifying the original aspect of the first complex.

[117] In this regard, Segal excludes the use of the "West House" as Monastery, as suggested by the earliest researchers of the settlement - more notably Woolley and the Colt Expedition in 1936 (Segal 1983: 151-152 and 156). Even if I am too not so convinced by the hypothesis, I cannot confidently refute it, since there are other examples of urban monasteries without any productive installations, such as in Umm es-Surab and Umm el-Jimal. However, I am also not so keen in excluding other functions different from the administrative sphere, especially for the later phases. In particular, I am not convinced by Segal's conclusion from this combined negation of the Monastery use and the

apparently weren't four independent and contemporary dwellings, later unified into one single unit. On the contrary, only one or two were original structures – and again their original residential function is not certain – against which successive buildings were later attached.[118] Excavations in the site (Margalit 1987) offered more precise chronological indications: more specifically the probes in the North Church and in a residential unit (nr. 121, after Hirschfeld's survey) in the western border of the site (Erickson-Gini 2013).

For both complexes an initial occupation in the 4th century is confirmed, followed by a later and substantial reconstruction in the 6th century, especially in the case of the North Church, possibly following a seismic event. In particular, the reconstruction of the North Church could also offer a clear explanation for the changes seen in the other churches, notably in the apsis-area, that show in all the three examples in Shivta heavy modifications in a later phase. The shift from mono-apsidal basilica with two-squared *pastophoria* to three semi-circular apses[119] is considered by Margalit as a sign of a "fully developed cult of the Saints and the Martyrs in the region" (Margalit 1987: 117-118), a development that could have supported the further and definitive development of the site in the Late Byzantine period.

The two sets of data, namely the progressive development towards north proposed by Segal and the archaeological data of the recent excavations, are not necessarily contradictory. The new archaeological dating simply shows that the area of the settlement was already entirely occupied by the 4th century, undergoing some radical changes in the 6th century, probably with a major concentration of the built environment. This is also suggested by Hirschfeld, who sees structural elements in the tower-houses, like the Governor's House and the Pool House (Figure 62), that guided the development of the site: they would have worked more specifically as centres of aggregation for later structures (Hirschfeld 2003: 407-408). Likewise, the same intensive investment can be seen in the agricultural hinterland of the settlement, which also bears witness to a peak in activity during the Byzantine period.[120] It is quite clear, in fact, that the prosperity of Shivta in this period mainly due to surpluses

presence of an "urban" Church and administrative building, seen as a mark of a marked consolidation of the secular-urban leadership of the local community between 6th and 7th centuries (Segal 1983: 166). At the same time, I am definitely sceptical about the interpretation of the complex South House-West House in its earlier use – such as the Stable House to this extent – as a structure economically oriented towards horse-breeding, as in Mampsis. The structure of the stables is in fact are extremely different between the two sites, and the typology in Shivta is surely closer to the complexes found for instance in the Hauran. Breeding of animal is surely a possibility, but the economic orientation of the site towards the commerce of horses in the light of these two complexes appears to me like a step too far. Furthermore, I would like to underline – as also Hirschfeld noticed (2003: 407f) – the similarities between the Governor's House complex – especially in its last configuration – and other structures in the site, such as the "*Vicarius* House" and the Pool House. Another interesting aspect is the proximity with structures with stables, in the latter case the Stable House (also in this case, a more detailed architectural analysis would clarify many aspects not only of the problematic chronology of the block, but also of its function). Röhl (2010: 112-113) interprets all the structures of this block as primarily residential units, in the case of the South House with some official functions. I agree more with Hirschfeld that considers these differences as marks of a socio-economic stratification in the local community.

[118] To this regard, I would suggest that the absence of a cistern under the "West House" (Segal 1983: 163) could be a confirmation of its later construction as an addition to the "South House", and of the fact that it never was an independent unit.

[119] This modification, as said, is dated by Margalit to the early 6th century, that was ultimately the presumed date of the first construction dating of the North Church by the earlier studies and some recent ones, based on some epigraphical finds dating from AD 505 to 679 (Röhl 2010: 78).

[120] Particularly interesting is a study on *Columbarium* towers around Shivta, that were connected not only with the production of meat, but also of fertilizer for the fields (Hirschfeld and Tepper 2006).

Figure 62 - The facade of the Pool House in Shivta, where the central structure resembles the Governor's House's Tower (photo by the author)

from agricultural production, probably sold in regional markets, as suggested also by productive and commercial buildings identified in the site (Hirschfeld 2003: 408).

As far as the transition to the Islamic period, radical changes in the settlement do not seem to occur, and an abrupt and complete abandonment is not likely. A small mosque was built abutting against the Baptistery of to the South Church (Figure 63)[121], which nonetheless appears to be continually used. The other Late Byzantine structures were also maintained, even if with less care than in the past, as documented for House 121 (Erickson-Gini 2013). Nevertheless, some changes took place in the site and in the hinterland and as stated by Erikson-Gini: "significant economic and political changes took place, transforming the character of the site in its latest phase of occupation during the eight-ninth centuries CE" (Erickson-Gini 2013). Still, the abandonment of the site is difficult to date precisely, but it is clearly not linked with a single violent episode (Hirschfeld 2003: 396).[122] In fact, in two areas of the site, namely the northern and the central, the careful sealing of the main entrances to private houses was documented (Figure 64): 21 doors had been blocked and are most notably concentrated in clusters of courtyard houses. This was interpreted as the sign of a simultaneous abandonment of discrete sections of the settlement by some large family groups (Tepper, Weissbrod, and Bar-Oz 2015). I would like to add one more aspect related to this phenomenon: other examples where similar behaviours can be documented are sometimes associated with the burial of hoards, like in Mampsis, and could testify the manifested intention by the dwellers to return to the house that could ultimately not be realised. One can therefore suppose that a general sense of instability and threat of an unfortunate economic

[121] The precise date of construction is still unclear, but several inscriptions on plaster had been found: they are in Kufic script and are dated to the 8th-9th centuries (Hirschfeld 2003: 396).

[122] Röhl (2010: 80) considers a later occupation of a not specified structure in the 12th century as a "Caravan station" plausible, but no clear data are given. What appears quite clear is that the atrium of the North Church underwent heavy modifications, with the blocking of the *porticus* and the creation of single rooms in the *intercolumniation*, in a manner closely recalling the encroachment processes seen in the *cardi maximi* of the larger *poleis* in the Byzantine East.

Figure 63 - The Mosque of Shivta (photo by the author)

Figure 64 - One of the blocked doors in Shivta (photo by the author)

situation could have pressured some discrete social groups of the local community to plan a temporary, but ultimately definitive, abandonment of their houses (Avni 2014: 323-324).

On the most general level, the first self-evident element of Shivta is its inner compactness (Figure 65a). The entire settlement - occupying *c.* 8 ha - appears like one large and unique quarter, with no elements allowing for the clear division of the inner space into districts.[123] This characteristic even more clearly marks the outer limits of the village, despite the absence of any enclosure or wall. The boundaries are "spontaneously" constituted by the most external blocks, offering a blind face towards the countryside and leaving only few narrow throughways between the houses, ultimately serving as accesses to the settlement. In total, twelve passages open to the outer front of the village. This solution certainly allowed flexibility, since new blocks could be gradually added, without having a fortification to modify or rebuilt. The only "formal" gate to the city is to be found in the northern part of the settlement, next to the North Church (Figure 65b).[124] Moreover, in the immediate surroundings of the village, pens or small enclosures and other large public facilities, like reservoirs or cisterns and also productive structures like wine presses (Figure 65c), creating a sort of outer ring or "buffer zone" with the proper hinterland. The northern part of the settlement had been always considered the more recent, even if recent excavations showed that it was already apparently settled in the 4th century – and it's the area where a principle of urban planning might be more clearly seen (Figure 65d). Of course, one cannot recognise the classical raster of streets, but certainly an effort toward a more rational organisation of the spaces and monumentality is detectible: there is a marked difference with the southern part of the settlement.

In general, the village does not lose it completely unplanned and "spontaneous" layout as a result of the progressive addition of further buildings, without any street system framing the different blocks or defining functionally isolated quarters like in Mampsis. As Hirschfeld correctly points out, the impression is that, though lacking a planning on the entire urban scale, "the street alignment was apparently mostly established in advance, for the streets have many straight sections" (Hirschfeld 2003: 397). This organisation is clearer in the northern part, while in the southern part narrower paths and larger blocks are more frequent. In the former, in fact, two larger and longer paths – still unpaved – run from the north-western corner of the settlement eastwards and south-eastward and seem to have functioned as major axes of the settlement. The only comparable "street" is the north-south axis connecting the "North Square" (Figure 65e) and the "Double Pool Square" (Figure 65f). These are ultimately the two larger public spaces to be identified in Shivta, also having the only two clearly communal water infrastructures of the settlement, supplied by an aqueduct terminating in the southwestern part of the settlement. Both of these "plazas" are seen by Hirschfeld as eligible to house local markets and social gathering space, as attested by the presence of benches along the walls of the structures facing these openings (Hirschfeld 2003: 397). In particular, the extremely cured aspect of the northern plaza is the only paved public space, suggesting that the northern section of the village saw a particular effort in its realisation. Only one additional, larger place – on which the so-called Governor's House opens – may have also functioned as gathering place, even though it appears more likely to have been a simple junction of the two of the major axes of the settlement, namely the North-southern and the North-west – South-Eastern ones (Figure 65g).

[123] A subdivision of the settlement into districts and quarters was proposed by Segal (1985: 318-325). In his analysis, he distinguishes seven different quarters (Northern, West-central, East-central, Early eastern, Early western, Southwest and Southeast), which are organised in three sectors (Northern, Central and Southern). The three districts are not a mere spatial division, but should follow the chronological development of Shivta. Although I may understand the purposes of such subdivisions, I find it extremely arbitrary, especially for the quarters. In fact, no elements allow the quarters to be clearly distinguished, such as the main streets or particular infrastructures, with very few exceptions. I would therefore suggest that in the case of Shivta it is more practical to not differentiate between larger quarters and consider only the subdivision in blocks.

[124] Hirschfeld (2003: 398) identifies in its plan two further gates in the south-eastern part of the site, but are simple doors like the "Eastern Gate" in Umm el-Jimal, being therefore not entirely definable as gates.

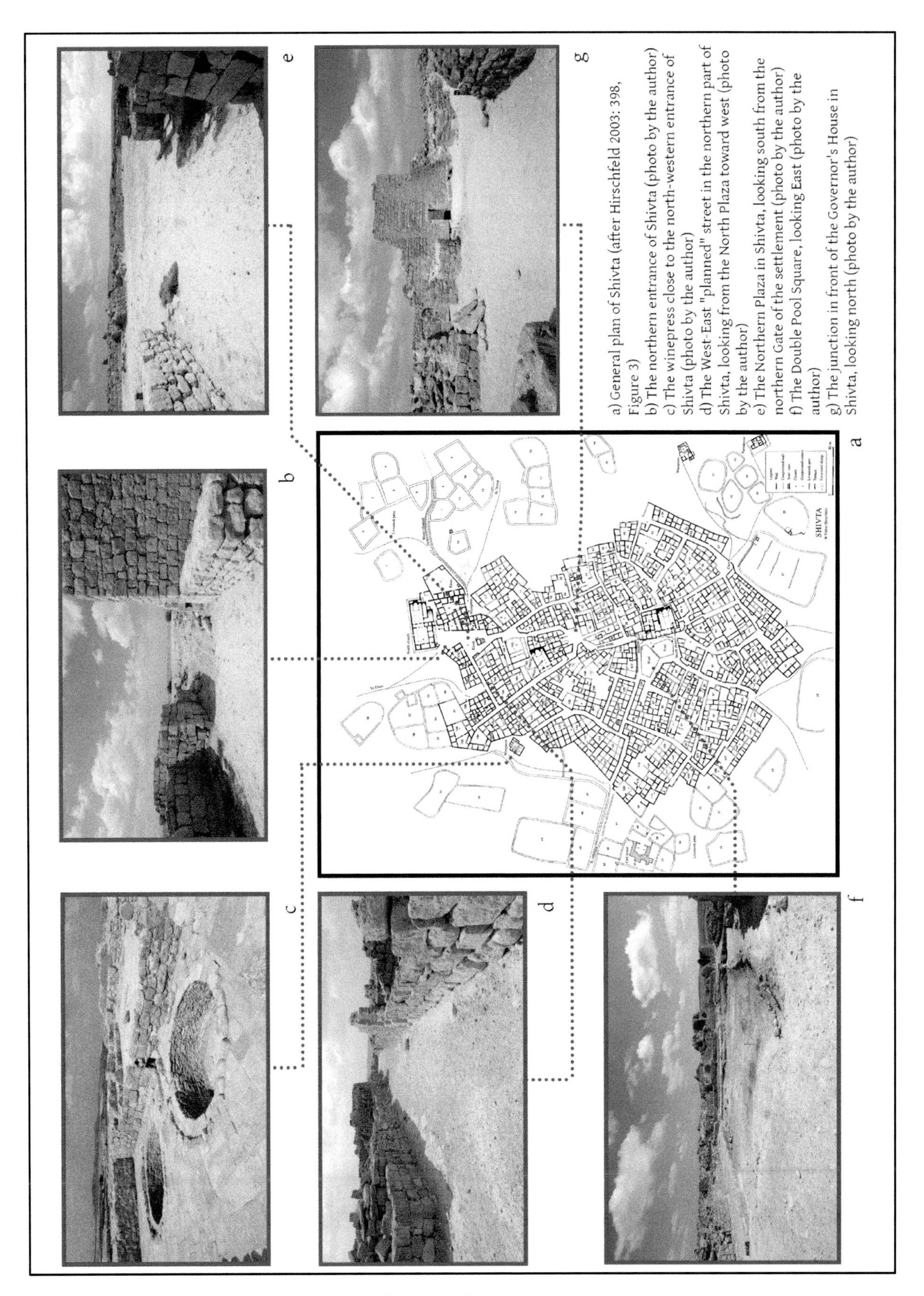

Figure 65 - Shivta: settlement organisation

Figure 66 - Shivta: possible public buildings

A building plan, and more importantly the maintenance also undertaken during the heavy modifications on the settlement, is visible in the water system, especially in the central and southern sectors. A network of channels is widespread throughout the entire village, feeding the major "Double Pool"[125] and also collecting rainwater to the cisterns of the private dwellings and also public structures, most notably the South Church. The impression that is ultimately given is that a public effort in regimenting the water supply was undertaken.

The spatial organisation of the settlement, and partially its compactness, also determines the fact that it is not so easy to identify a socio-economical or functional separation of the spaces, let alone the original nucleus.[126] As seen for other case studies, residential and productive or commercial[127] functions are mixed together, often sharing the same spaces and facilities (see infra). In several circumstances, small rooms interpreted as shops open either onto the pathways or to the inner courtyard of complexes, like in the case of Pool House. It is certain that data concerning this aspect are limited: the major problem that is also shared by the other case studies is the lack of extensive excavations and the absence of clear installations. Only one exception can be found: the so-called "kiosk" in the North Plaza (Figures 66b and 66c) that is interpreted by Hirschfeld as a communal mill or press that could have also functioned as an administrative structure related with tax collection or inspection (Hirschfeld 2003: 401). Nearby, one other complex, neighbouring the "*Vicarius*'s House", has a wine press, though it seems to be a private installation. To locate other similar public facilities, one would need to go to the immediate proximity of two of village's entrances, where two or more wine presses with a nearby storage facility, allowing for the separation of each individual's raw materials, would need to be present in order to attest communal use.

Similarly, no clear administrative centre is identifiable, despite the label that is applied to some structures in the north-western part. This interpretation is mainly justified by the dimensions of some of the complexes present; such structures, however, are also present elsewhere in the settlement, perhaps showing some sort of economic stratification among the residents of Shivta. It is not entirely surprising though that a major concentration of such larger structures is found in the northern part of the village, the only one showing some sort of planning and relative "monumentalising" efforts. Hirschfeld also suggests a potential public nature of a complex in the same block of the "*Vicarius's House*" because of the presence of two rooms with benches all along the walls (Figure 66d).

Despite the clear presence of a certain socio-economic stratification among the 170 dwellings identified in the settlement, no clear zoning can be identified. Only few structures seem to be located in a position that could underline their prominent importance, and most of them are clearly public buildings: the North Church religious complex, facing one of the two "squares"[128]; the Governor House and Central

---

[125] Two coins found in the plaster are dated to the 4th and 6th centuries, also suggesting a date for the structure. The cisterns in general were periodically cleaned and taken care of by some inhabitants, as testified by four ostraca (Hirschfeld 2003: 397).

[126] As mentioned above, Segal (1985: 322-323) affirms that the first Nabataean settlement was occupying the south-eastern part, what the author calls the Early Eastern and Early Western quarters. Like in other case studies, the later development of the settlement determined an almost complete obliteration of the earlier phases.

[127] As suggested for some smaller rooms connected to large residential units. See for instance the Pool House.

[128] Segal suggested that the northern part of the settlement could have been conceived as an autonomous quarter, a sort of settlement within the settlement, formed by the North Church, the monastery and the structures attached to it to the south. He states in fact that "a unit, unique in character, was established which functioned without contact with the other parts of the town" (Segal 1985: 326). I think this hypothesis is a little too extreme: this view could eventually describe an earlier phase, but it is quite clear that once the entire northern sector of the settlement was realised, also this "unit" is tightly integrated with the rest of the town, at least physically. At

Church in the centre of the village, at the crossroad between the major "north-south" axis and one north-west running branch; the South Church and the Pool House, facing the square with the double pool. Although, a potential trend in socio-economic stratification can be detected, especially considering the probable evolution of the settlement: tower-structures always appear to be in the centre of larger blocks, supporting the hypothesis mentioned above that they worked as a core for the development and expansion of the settlement. This is the case for instance of the aforementioned "Governor's House" in the centre of the village, "Pool House" in the south and "Vicarius's House" in the north, but other examples can be found elsewhere. One final structure is interesting for its position, and especially for the fact that it's the only complex directly opening to the outside of the settlement: the last building to the north-east, neighbouring with the North Church (Figure 66e). It was interpreted often as khan, but a more recent hypothesis was proposed that it functioned as a military depot or barracks (Di Segni in Hirschfeld 2003: 402).

The absence of major throughways – with the only partial exception of the north-south broader path – and the general compactness of the settlement make it appear as though it is structured in a unique quarter, divided only in blocks. It is indeed difficult to recognise large functional or physically distinguished sections, as mentioned above. The general tendency is to have large and irregularly shaped blocks, developing from a spontaneous and gradual addition of single units next to one other. It is nevertheless possible to identify at least two different types of blocks, insulae-inspired and cluster blocks, though they likely originate from a completely different spatial and possibly social background.

The "*insulae*" inspired ones are geometrically shaped and they normally have units opening directly on the public pathway (Figure 67b). It is interesting to note that there is a correlation between this kind of block and the proper "Courtyard Houses" described in the introduction of this part: it appears that a more "Hellenistic" influence simultaneously involves both levels of the settlement. The reasons could be either chronological, or in my opinion, more likely dictated by socio-economic factors. The important issue to underline here is that each complex, or unit, opens directly to the communal ground, with no mediation between the private and the public spheres except on the house level. The spatial organisation therefore suggests that the dwellers in these blocks did not necessarily identify as a discrete and compact group within the settlement's community. This impression is suggested by the second type of block found in Shivta, which follows the proper "cluster" model described in the other case studies, with normally three or four units sharing walls. This kind of spatial structure is mainly concentrated in the south and south-eastern part of the settlement. Nevertheless, these clusters distinguish a peculiar feature, a "buffer zone", also present in the other sites. In certain cases, in fact, some blocks do present systems creating a "buffer zone" between the inner part of the quarter and the accesses to the single units and the public paths. This is obtained through smaller pathways that lead to inner courtyards or to the *cul-de-sac* (Figure 67c). Despite the different spatial organisation of these two elements, the goal is clearly the same: divide the exclusively private and the public areas through an extensive "semi-private" zone. It is of course not a systematic circumstance, but from what it is possible to evince from the available documentation, such devices seem to be more developed and frequent here than in the other sites analysed in the present research. The impression from this feature is that there was a deliberate intention to mark the presence of larger social groups, possibly large extended families or even clans (Hirschfeld 2003: 404-405): the easier solution is to create a first inner courtyard on which each complex could open, but that was somehow restricted to the members of that

the same time, we do not have any structural element, like walls or enclosures, underlining a clear functional separation of this area, as for instance in the south-eastern quarter in Mampsis (see Chapter 4.3.2).

a) General plan of Shivta (after Hirschfeld 2003: 398, Figure 3)
b) One example of insula of Courtyard Houses in Shivta (detail from Hirschfeld 2003: 398, Figure 3)
c) One example of cluster of houses with intermediate courtyard, in Shivta (detail from Hirschfeld 2003: 398, Figure 3)

Figure 67 - Shivta: quarters' types

particular extended family. It is ultimately a development of the *cul-de-sac*, which only seems to be used in small blocks or to give access to smaller complexes. Nonetheless, no doors or proper gates had been found to clearly mark the entrances to these devices, as in later periods or properly urban contexts.

The choice between the two different types of blocks, the *insula* or the cluster, could have also been determined by the development of the block itself and from the space available to build new structures. Even though no clear pattern can be found, it is interesting to note the concentration of large single units organised into more regular *insulae* in the north-western part of the site, also with more "planned" paths (see Figure 65d), in respect to the more "clustered" and irregular situation in the southern part of the settlement. In several examples, moreover, larger clusters also allowing inner circulation between the different units inside the blocks, as perfectly demonstrated by the block of the Central Church and the Governor's House.[129] The different spatial organisations just described reflect a different privacy pattern. If the devices clearly were created to interpose a semi-private in defence of

[129] This is more evident on the house level of analysis and has direct parallels in the other case studies of the present analysis. This appears to be a development taking place during a later phase of occupation, where the units that were originally separated from one another and were only connected later.

the totally private areas are present, it can then be understood as a solution to face a higher demand of privacy by the inhabitants of the block or a more evident desire to mark the discreteness of a determined social group. This need, on the contrary, does not seem to be so urgent when such spatial devices are not applied. The development of the two kinds of blocks also seems to follow different paths. If in the "cluster" blocks the process of progressive addition of new buildings to the old ones is quite evident - according to the model of the "conjunctive" houses - the *insulae* have another pattern, where the subsequent modifications do happen within the outlined shape of each unit with at least some union of some complexes in the block. These two last aspects have direct parallels on the house level, and more evidently in relation to the position of the courtyard in the complexes.

As far as the functional zoning in the blocks is concerned, the substantial coexistence of both productive/commercial and residential areas described for the settlement in general is a constant feature on this level of analysis. One can hardly find any kind of specialisation, since domestic structures appears to be connected to "*tabenae*" (Figure 66f) or other non-residential areas, even if the direct access is in some cases to be verified.[130] Only few cases had been considered "specialised" by earlier studies, but this hypothesis does not seem valid anymore. One was the northernmost block, including the North Church, seen as a large monastic complex. This hypothesis had been contested, since the Church does not have any direct access to the other structures, which on the other hand appear as independent from the religious building (Hirschfeld 2003: 401-402). Another example was the central block with the Central Church and the so-called "Governor's House", understood as the administrative and civic core of Shivta. In fact, in my opinion the interpretation of those two buildings as the political centre of the settlement is not sufficiently supported by the evidence available. Moreover, as stated before, the presence of such large complexes with towers is attested in other blocks throughout the settlement and can be simply a sign of a socio-economic stratification.

170 dwellings units have been identified in the site, covering a tendentially small area (average 364 sq. m), and normally organised on two storeys (Hirschfeld 2003: 401-402).[131] The technique employed is

[130]Clarification is needed in this context. For Shivta an extremely detailed analysis is available - structure by structure - in the dissertation by Röhl (2010). She suggests the functional destination of every single unit of the settlement based on structural evidence and excavation reports. A clear distinction of spaces in "economic" and "residential" specialisation is set: therefore exclusively "economically oriented" units are clearly distinguished from the residential ones in a block. There are a few remarks that should be made on this matter: first of all, it is not clear which feature and criteria allowed such stark separation. Rarely, in fact, do we have installations connected to some sort of productive or commercial specialisation, or even assemblages of finds completely oriented towards such destinations. Moreover, sometimes such installations may also be connected to clearly residential areas, confirming the tendency of having mixed residential and productive/commercial units, as seen also in almost all the other case studies and in general in more rural contexts. The further aspect is the graphic basis used for the analysis, namely the map originated by an aerial photograph in 1981 (Brimer 1981: 227-229), instead of the more recent and accurate photograph made by Hirschfeld (2003: 398, Figure 3). The major problem is that it does not allow the limit of each unit to be clearly distinguished: in fact, the openings of the different rooms and areas are often not show, a problem that was also identified by Hirschfeld, and one of the major justifications for his subsequent survey. Therefore, if the functional identification is quite clear for some cases and also confirmed by other scholars, in most cases it is hard to possibly reconstruct the eventual "specialised" structural features of a unit. Of course, even the present analysis is limited as far as the functional interpretation of each single space in the settlement. Though this is not the purpose of the present research, what is most important to show is the coexistence and the sharing of spaces for different functions.

[131] In particular, Hirschfeld noticed that "135 (80%) [houses] fall in a range between 250-550 sq. m. Only 10 dwellings are smaller than 200 sq. m and only 8 are larger than 600 sq. m" (Hirschfeld 2003: 404). The tendency, therefore, is to have generally smaller dwellings than in Mampsis for instance.

normally of a high standard and is uniform across the entire site; it also suggests, as stated by Hirschfeld, that the units were constructed by local groups. As mentioned earlier, though, some kind of social diversification, even if in a tendentially homogeneous society, may be reflected by the variety of house types found in Shivta. In fact, almost all of the examples described in the introduction to the chapter are represented among the dwellings in the settlement, from the simpler to the more complex. The most important aspect of this context is the radically different spatial conception that underlies the various type choices. The distinction that can once again be made is between the "injunctive" and the "conjunctive" models of complexes. The difference had been already explained, but some elements specifically in relation to the context of Shivta can be pointed out.

The first interesting element is that the two spatial conceptions are present next to one another throughout the entire settlement, without any clear differentiation in chronology or spatial distribution. Nevertheless, it appears that "conjunctive" houses became more frequent when earlier structures had been modified in the Byzantine period, with the creation of extremely large complexes as seen with the Governor's House and with the Stable House. In these circumstances, the courtyard system itself seems to undergo some changes, and clearly the theoretical spatial framework defining a traditional "Courtyard house" becomes more flexible. There are therefore more or less irregular units like in the South and West Houses in the "Governor's House" that more closely reflect a "Complex house", with a progressive accumulation of structures occupying the space available. In this case, moreover, a tower-structure is also combined in a courtyard system: this example is more clearly visible and better studied in the case of Building I in Mampsis (see Chapter 4.3.2).

For both cases, it brings to question if its only (or principle) function was residential: Röhl and Segal see a combination of residential and administrative functions. While I agree with the concept that it is a polyvalent complex, where the typical mix of functions (in most cases residential and productive) is even more extreme than in other contexts, for this particular case, I am not sure if the administrative function is represented. Its similarities with the Pool House complex (Figures 68b and 68c) are in fact striking, at least as far as the South House of the "Governor's House Block" is concerned. Both complexes also present a tower-building, dominating the entrance to the complex, and integrated a courtyard system. In addition, the Pool House does not seem only to serve a residential purpose, if at all, but also a commercial function with a series of *tabernae* opening to the Double Pool plaza (see Figure 66f). If a residential wing of the complex was present, it could have possibly been on the second storey, opening to the possibility that the dwellers were also the owners or workers of the *tabernae* in the northern side of the complex, as proposed by Hirschfeld also for the wine press in the complex close to the "*Vicarius's* House" (Hirschfeld 2003: 407). Moreover, it would be interesting to see which exact physical relationship existed between this complex and the neighbouring Stable House (Figures 68d, 68e, 68f), which likely underwent heavy modifications apparently at the same time or shortly after the Pool House was first built. If a direct connection of the two structures existed, it would even more so be a direct parallel with the Governor's complex. Though in general it is interesting to note the association of Tower-Courtyard Houses with Complex Houses and stables within the same block. It was said that Complex Houses were more frequent in presence of modified older structures. Although in the northern expansion a stricter adherence to the "Courtyard House" model seems to have been followed, and less changes in the inner organisation took place.

In general, no real pattern, chronological or topographical, can be identified. Both types of houses were built at the same time, sometimes next to one another. Nonetheless, some concentrations can still be found, such as the more marginal blocks on the southern limits of the settlement that are entirely composed by Complex Houses. These were probably more or less contemporary to the aforementioned

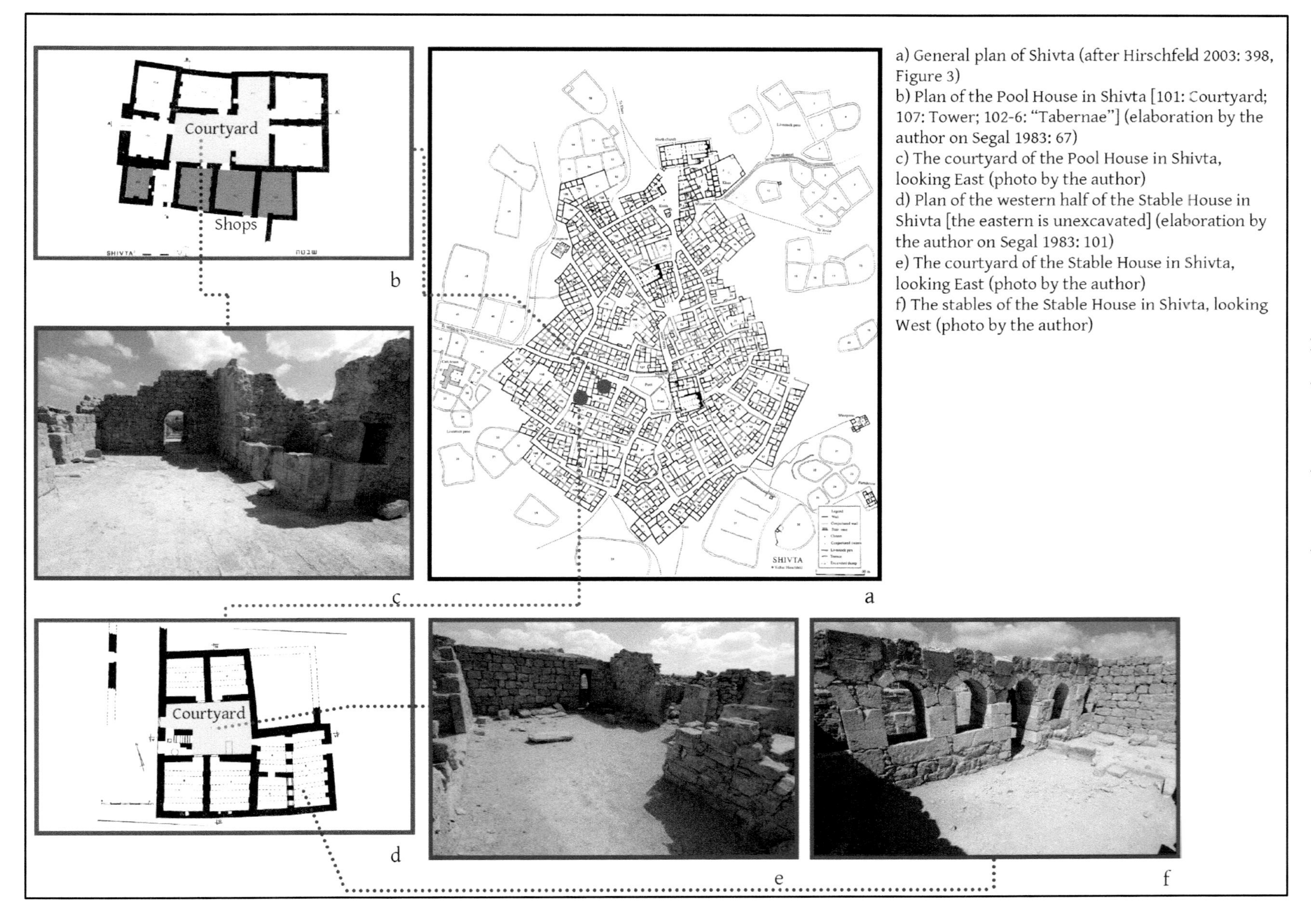

a) General plan of Shivta (after Hirschfeld 2003: 398, Figure 3)
b) Plan of the Pool House in Shivta [101: Courtyard; 107: Tower; 102-6: "Tabernae"] (elaboration by the author on Segal 1983: 67)
c) The courtyard of the Pool House in Shivta, looking East (photo by the author)
d) Plan of the western half of the Stable House in Shivta [the eastern is unexcavated] (elaboration by the author on Segal 1983: 101)
e) The courtyard of the Stable House in Shivta, looking East (photo by the author)
f) The stables of the Stable House in Shivta, looking West (photo by the author)

Figure 68 - Shivta: Pool House and Stable House

northern complexes. Thus, if no clear pattern is detectable, the spatial organisation of the single dwellings simply shows a different approach to the development of private areas. The factors determining these choices are indeed difficult to unravel. The relative position in the settlement and the space available likely play an important role, but in my opinion, they are not sufficient to exclusively influence the choice of the original spatial conception. The influence of local traditions could be one further consideration. Several studies on settlements from the Iron Age were able to show the direct connection between the so-called "Four room" House - also explicitly referring to the Israelite population, with a clear religious and ethnic characterisation (Faust and Bunimovitz 2003) - and the "injunctive" house also found in the Byzantine period. This could possibly lead to the extreme hypothesis that the "conjunctive" houses could have been developed by determined social groups that did not belong to the local, and hence architectural, tradition.[132]

Furthermore, one key factor of reading the spatial structures that in my opinion should not be disregarded is the need for privacy and its manifestation in the architecture. When looking at the distribution of the "injunctive" houses, one may notice the following trend: they are more frequent in smaller blocks that do not present the ""buffer zone"" devices seen in the larger blocks. It follows that some solution had to be created in the house itself, like "*vestibulum*" areas separating the entrance from the inner courtyard. Such elements are nonetheless also present in some of the "conjunctive" houses, even though they are sometimes combined with the other "privacy" devices. One aspect that nonetheless appears to be more strictly followed is the vertical privacy, so not only the visual inaccessibility from the street but also from the other houses or structures. Some features should be considered in relation to this aspect. First of all, the disposition of the roofed structures, especially the one with more than one storey: they are disposed in a way to block the sight from neighbouring buildings toward the open areas. Secondly the position of the courtyard that in the case of the "injunctive" house does not represent a simply spared ground but in fact appears to be a planned space. Thirdly, the inner circulation patterns, especially in relation to the collocation of the stairs, namely if they are on the inside or outside of the building. This differentiation in privacy care may reflect different degree of intimacy between the inhabitants of a block, also implying different social ties between them. On the one side, the "injunctive" houses with a stricter respect of the private spheres - more closely recalling properly urban examples such as the so called "patio" or "peristyle" house - and probably reflecting a lower degree of social proximity of the inhabitants with the neighbours. On the other side, the "conjunctive" houses, with their more "open" structure could relate to tighter family connections among the block dwellers.

Despite the deep differences between these two conceptions, several aspects are actually shared, most important being the relative autonomy of each unit. In fact, each unit is to a certain socio-economic degree independent, able to answer the daily needs of the family living there. The underground cisterns, which were either connected to the central hydraulic system or were rain-fed, are present in every single complex in the settlement. Use of the public cisterns most likely would have been limited only to a reduced number of cases, since in normal environmental circumstances the private water supply should have been sufficient to meet the requirements of a family. Moreover, each type of unit had a high degree of polyvalence of its spaces, especially in the courtyards.

[132] The present "state of the art" does not allow us to put this hypothesis to the test. It is a possibility to be taken into account, even if it is extremely difficult to prove even in light of a better knowledge of the architectural structures. Nevertheless, the possible role of nomads in the development of Shivta is suggested by Röhl in the conclusions of her dissertation (2010: 151-161), and more specifically for the liminal Complex houses I referred to earlier (2010: 132-133).

4.3.2: Mamshit (Mampsis/Kurnub)

Mampsis, modern-day Mamshit, which in the past was also known by the Arabic name *Kurnub,* had been thoroughly investigated in the 80s (resulting in two main publications: Negev 1988a; Negev 1988b). Despite the limited extension of the site of *c.* 2.4 ha, which at least by the Late Roman - Byzantine Period appears to have been roughly half the size of Shivta, this settlement shows an accentuated urban character. For this reason, its description as a village is extremely reductive: albeit this definition is more plausible after its extension, one will derive an extremely different picture when focusing more on the function and the structures found.

Mampsis is likely not an entirely urban centre, but its semi-urban dimension and especially its diversity from the most rural contexts has to be underlined in some way. To this end, the lack of an agricultural orientation is demonstrated by the absence of structures in the hinterland of Mampsis that, for the Negev, are normally connected to intensive land exploitation like the ones surrounding Shivta. Unfortunately, no clues are available on the official status of the site during its history, but its importance in the regional context seems to be confirmed, at least for the Byzantine period, by the few historical sources that are available. For instance, in the Nessana Papyri, and more specifically Papyrus nr. 39 dated to the mid-6th century, Mampsis is mentioned several times and appears as fourth on the list believed to be by Kraemer as an "account of allotment by villages" (quoted in Negev 1988b: 5-6). This is particularly impressing especially considering that the only city in the region, Elusa, is the second to last entry in the list (with a little bit more than the half of Mampsis' quantity of *solidi*). For this reason, the small Negev settlement attracted the attention of the scholars. After also considering the other settlements present in the list, Negev argues that the list is not an account of tax payments, but of economic support for defensive purposes that was guaranteed by the central authority to the sites in the Negev. The first sites listed, in fact, are ultimately important military centres, included in the *limes* system – Chermoula and Birsamis – or at least fortified towns – Nessana, Mampsis and Oboda. Perhaps more intriguing is a second list with smaller amounts of money recorded in the Papyrus, and Negev interpreted these sums as the annual fees paid by semi-pastoral groups around the settlements, including Mampsis (Negev 1988b: 6). The importance of Mampsis in this period is furthermore confirmed by its presence on the so-called "Madaba map", the large mosaic floor of the St. George Church in Madaba (Jordan), also dated to the 6th century.

Understanding how this picture is applicable to the earlier phases of the settlement is nevertheless a challenge, since the material evidence for this period is extremely scattered and unclear (Figure 69). That being said, precise dating of the founding of Mampsis is not available at present. According to archaeological records (Negev 1988a: 1-8), no construction appears to predate the reigns of Obodas II or even Aretas IV from the second half of the 1st century BC – early 1st century AD. The origin of the site can be confidently connected, however, with the development of the trade route under the Nabatean control of the area from the Hellenistic – Early Roman period. As for other settlements in the area, like Oboda and Nessana, their commercial vocations in fact appear to be quite strong. A major distinction, however, should be made for Mampsis: its development appears to be stimulated only through private initiatives, whereas other towns clearly attest a direct intervention by the royal court for their initial development (Negev 1988a: 3-4). Although certain caution is nonetheless necessary in this interpretation, since the later phases of the settlement had extensive earlier structures that led to an almost complete destruction of the Middle Nabatean structures. What is indeed evident is that the later Byzantine settlement occupies an already built-up area, reusing part of the structures or collecting building materials from other parts of the earlier town. Nonetheless, one can confidently affirm that

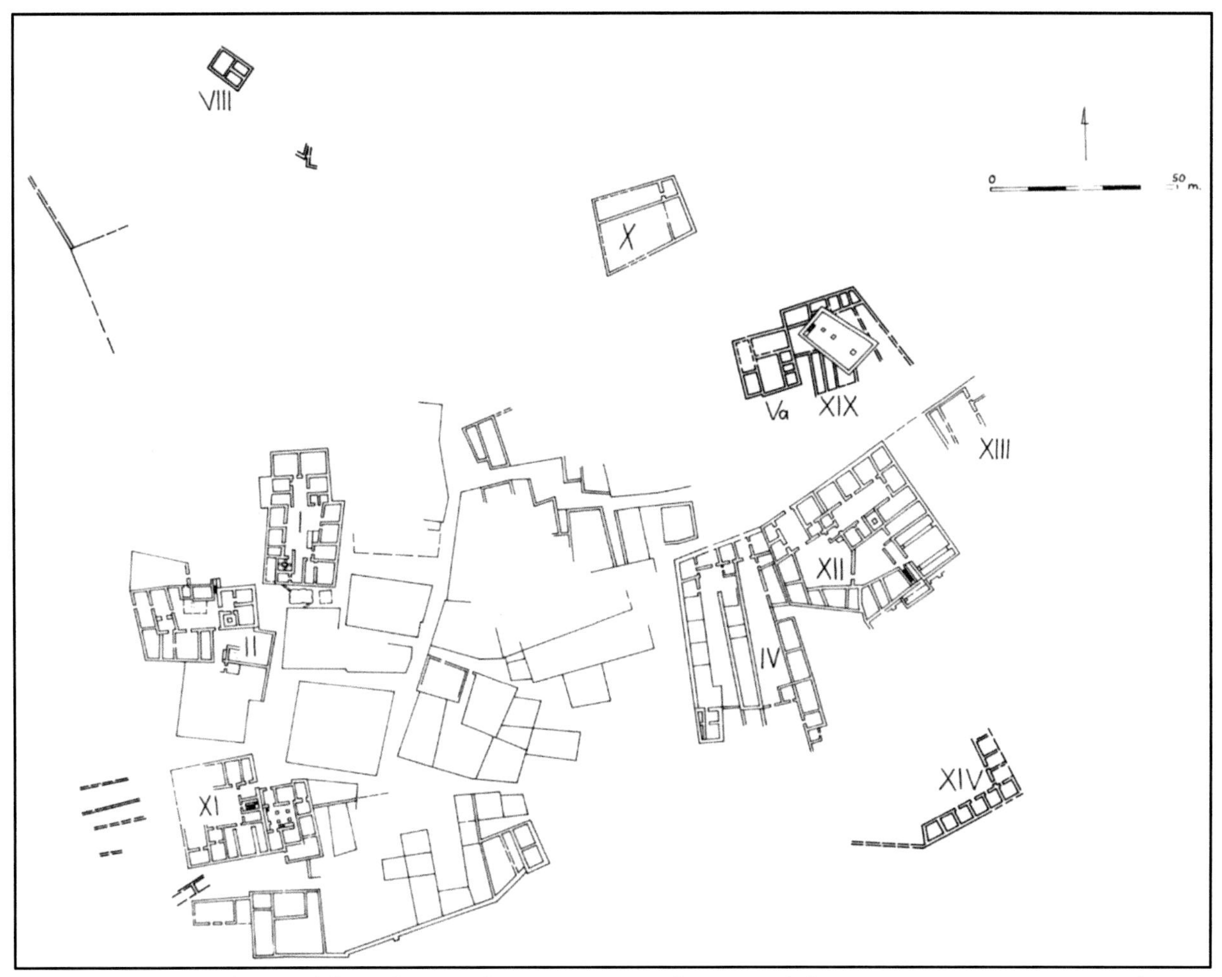

Figure 69 - Plan of the Middle and Late Nabatean remains of Mampsis (after Negev 1988a: 30, Figure 2; Plan reproduced with the permission of the Institute of Archaeology, The Hebrew University of Jerusalem)

Mampsis was originally a Middle Nabatean settlement, expanding upon a larger surface than that of the Late Roman – Byzantine period (Negev 1988a: 26-49).[133]

Some of the major complexes of the site, which most likely played a central role also for the local administration, especially so-called "Building I" and "Building II", were first built in this period. However, only few traces of these earlier phases remain, since the town underwent an important development during the Late Nabatean – Early Roman period, between the 1st and 2nd centuries AD. In this period, most of the Middle Nabatean structures experienced radical interventions, such as an expansion of the residential areas and a ground-level destruction of the entire structure, like the so-called "Caravanserai" (Building VIII). According to Negev, Mampsis appears to have benefitted especially from trade and horse breeding; he justifies his thesis by the presence of at least two buildings with very large stables. What is certain is that the settlement benefitted from a particularly positive economic situation: the quality of both architecture and decoration, in the best Nabatean tradition, is incredibly high and unexpected in such a small settlement. Other major buildings were founded, sometimes replacing earlier structures. This is the case for the market, overlaying previous defensive

[133] Traces – as mentioned – are few and scattered. Middle Nabataean rests can be found in buildings VIII (the Caravanserai), I (The Palace), IX, IV, XIV, XIX and Va and further under the city walls and the West church.

towers, and for the bath and the connected public reservoir, both causing the destruction of at least two large dwellings.

The transition to the Late Roman and Byzantine periods also appears to be quite traumatic for the inhabitants of the site. Some hoards appear to be hidden in several complexes at some point during the 3rd century and were never recovered. The single houses experienced quite an important modification in their living spaces, with the widespread subdivision of the larger rooms into smaller units. The large stables also appear to have changed in use, being converted in other productive or residential areas. This phase is dated to the creation of the city wall and the enclosure separating the Building XII, the market and the East Church from the rest of the settlement. The East Church is one of the two churches built in Mampsis. The other is the so-called "West Church", which was built on the western wing of an earlier dwelling.

Some of these complexes allow us to follow the aforementioned progressive evolution of the settlement, and it is worth considering them a little bit more in detail. The evidence for remains from the Middle Nabatean rests is extremely limited (see Figure 69). One of the few exceptions come from the so-called "Palace" (Building I; Negev 1988a: 40-43 and 50-77): the re-planning of the Late Nabatean period in the early 2nd century included the construction of the entire northern wing, and more importantly the creation of the staircase-tower and the upper storey (or storeys as it seems), but largely reusing earlier structures (Figure 70b; see infra). The impression, though, is that the Middle Nabatean structures belonged to two different dwellings, which were then incorporated into a single unit in this phase. The final form of the building is ultimately a "traditional" Negev courtyard house: a complex with a central courtyard entirely surrounded by structures and accessible only from one entrance from the outside, passing through a *vestibulum*. What is more striking though – and actually constitutes the exceptionality of Mampsis's architecture – is the quality of the structures already at this stage, and more importantly the degree of elaboration of some architectural solutions like the staircase-tower and the Gamma-Porticus in the courtyard. These devices appear to have been constructed by workers that already perfectly understood these typologies of structure, possibly having learned them elsewhere and more specifically in a properly urban context.

The organisational change of the Late Nabatean period clearly met the need to create a larger living space, after a general demographic increase and perhaps also a general increase in size of individual families. What is interesting – and ultimately makes this complex an "injunctive" house – is that the outer limits are respected, and the new unit is planned from the outside towards the inside, leaving few space for the "indiscriminate" addition of new rooms in later phases. It is clearly difficult to say if this was determined by a cultural factor (for instance a more Hellenistic tradition) or simply by the lack of enough room in the quarter. It is also interesting to see the changes that the complex underwent in the following and last phases identified by Negev, approximately dated to the Late Roman or more likely to the Byzantine period; also in this circumstance, the dwellers had to face an increase in the population. At the same time, however, an augmented need for privacy arose. The most likely explanation in my opinion is a change from a nuclear or limited "extended family" – perhaps the parents with the first son's family – to a larger extended agnatic group. The result is the subdivision of larger rooms into smaller spaces, not in communication one to the other. It is also likely that a functional change of some areas of the house occurred, judging from the modification of the room interpreted originally as the archive or *tabularium* of the Palace – in the light of the several in-wall cupboards (Figure 70c).

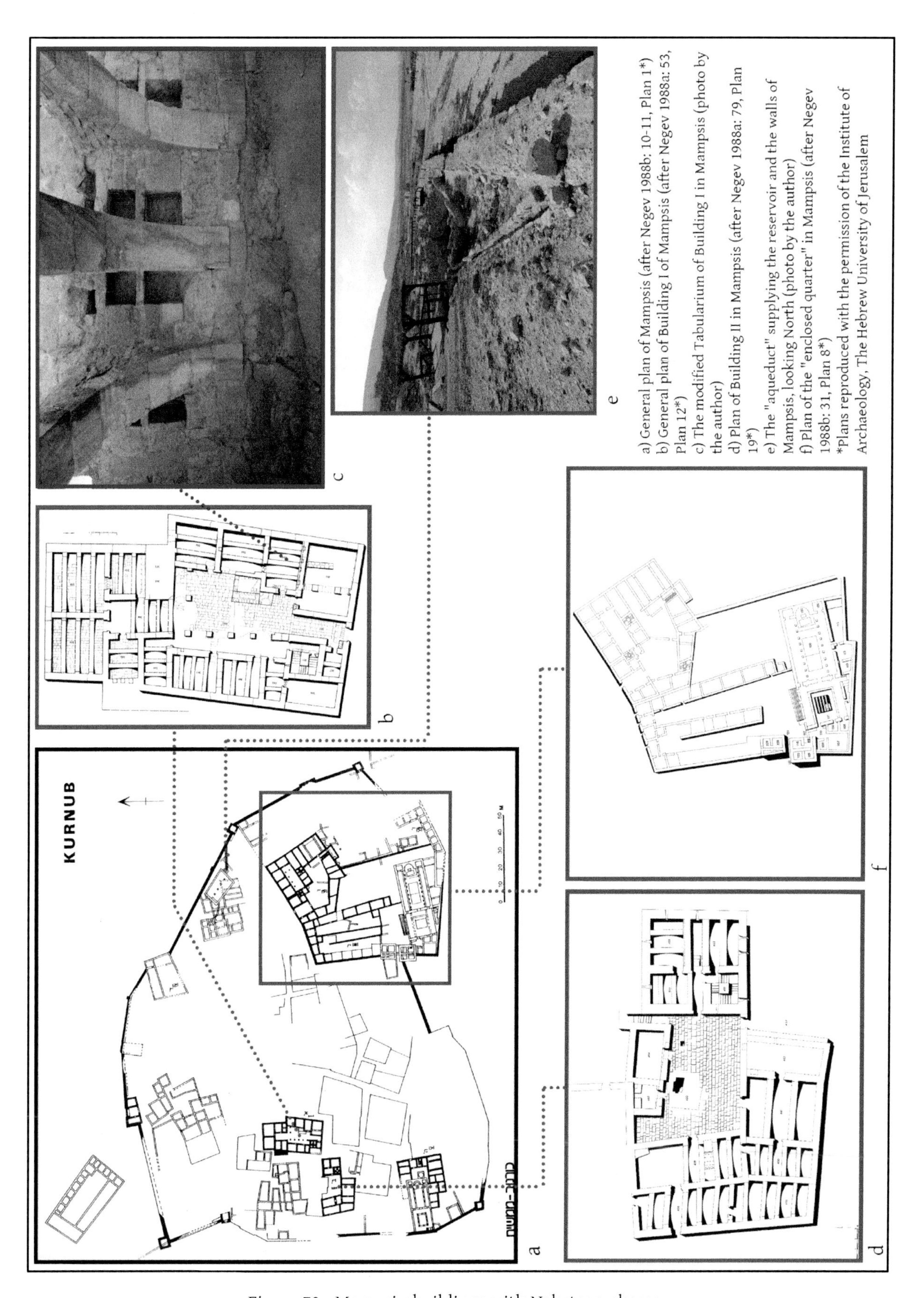

Figure 70 - Mampsis: buildings with Nabatean phases

Likewise, "Building II" (Figure 70d) also held an administrative function in its earlier phase, and bore witness to a similar progressive densification of the spaces through its history (Negev 1988a: 77-88). In this case, Negev identified two distinguished phases. In the first, the tower structure, the southern rooms, the courtyard and the cisterns were built.[134] Still within this original building activity, though probably a little bit later than the original structures, a series of storerooms on the western side of the later complex were added. This phase is dated by Negev to the Middle Nabatean period, mainly because of the extremely large blocks employed in the construction of the tower that was rare after the Herodian period. Functionally, it was suggested that this building was the "office" of the inhabitants of the "Palace"-Building I, but could also meet defensive purposes, since at that time the settlement had no wall. Nevertheless, the presence of the southern rooms and the large dimension of the complex in general are clues that convinced Negev to exclude an interpretation of Building II as a small garrison post, in favour of a more important administrative function. In a second phase, not precisely dated, the complex was furnished with a kitchen and a series of communicating rooms and with the stairs leading to the platform of the tower. Only with these interventions, the proper courtyard system was completed, and the difference in the evolution of the complex in comparison with Building I described above is quite clear. Even if the reasons are still questionable, the conjunctive pattern in the organisation of the structures is evident, with a progressive addition of new units without any respect of pre-established outer borders. Its relation to the large complex extending to the south can also be clearly understood even though it hasn't been investigated in detail yet. Building II continued to be in use in the Late Roman – Byzantine period, even if further modification is not evident.

A complete destruction of the Middle Nabatean phases can be seen for Building XIX and Va, two dwellings with courtyard that were entirely overbuilt by Building V (Negev 1988a: 167-181) and Building VII (Negev 1988a: 181-191) in the Late Nabatean period. Building V in its earliest phase was a bathhouse that underwent some expansion probably in the 4th century, but never changed its function. Its physical and functional relation with Building VII – the only public reservoir in the settlement – is very likely. Their construction appears to have taken place at the same time, and both the Baths and the reservoir underwent some modifications in the 4th century, namely the addition of a second smaller basin connected to the main cistern and the short aqueduct supplying the complex. The *terminus post-quem* for their creation is the edification of the city-walls (Figure 70e; Negev 1988b: 9-29). Surely, this building activity was one of the most destructive in the site, at least as far as written sources have indicated. In order to enclose the settlement with this defensive structure, in fact, several Middle Nabatean buildings were deconstructed or heavily modified, like Building XIX and, more interestingly, a possible earlier city wall found underneath Building IX. According to some coins and materials found in probes along the fortifications, it appears that the wall was built between the end of the 2nd and the 4th centuries, so between the latest phase of the Late Nabatean and the Late Roman. It is likely by the time of its construction, all of the site's Late Nabatean buildings were already completed; similarly, the walls were already present when the churches were built, as it is clearly visible near the West Church.

One other complex reusing (or overlaying) earlier Middle Nabatean is Building IV, the so-called "Market" (Figure 70f; Negev 1988a: 163-167). The foundation of a tower underneath the southwestern end is clearly visible, such as the presence of another building under the Eastern street. The Market itself can be dated later than Building XII, since it abuts against the western wall of the outer courtyard (see infra). Negev suggests the Late Nabatean period for its construction, with a continuation of use into

[134] This construction constitutes in my opinion a direct and earlier model to the South House of the Governor's complex and possibly also of the Pool House in Shivta.

the Byzantine phase, when it constitutes an enclosed quarter isolated by the rest of the settlement together with the East Church and Building XII.

Building XII (Figure 71) is particularly interesting for several reasons, mainly because of its dimension - with a surface of *c.* 1600 sq. m- and its function, probably connected in origin with horse breeding (Negev 1988a: 111-147). Another aspect that is particularly challenging is its chronology. No clear Middle Nabatean remains can be found here. Negev identifies four phases of development of the complex, starting with the Late Nabatean period, although I have some doubts specifically on the first nucleus of the construction. The complex is presented by Negev as a single large unit already in its original form, with a marked separation of functional zones, namely the residential area and the productive one. Thus, the north-eastern corner of the building would have been the residential unit, organised around an independent courtyard and according the same architectural typology seen in Building I, with the same structural devices such as the staircase-towers and the Gamma-Porticus. The other outer courtyard would have embraced this residential unit and organised the productive/commercially oriented area of the complex: the principal features are the large stables, following a sort of three-nave "basilica" plan -also found in Building XI - where the central nave, however, wasn't roofed. This element suggests Negev that in Building XII a superstructure was probably built, and allowed the costumes to see and choose the horses before the purchase.

I am not certain that these two areas - the residential and the productive/commercial - had been thought of as a unit already in the first phase, even when considering the dating of the building to the Late Nabatean period. When reviewing certain aspects of the building's construction, I am simply sceptical in considering this period as represented by a single homogenous phase. In my opinion, in fact, the entire western part of the complex - around the Courtyard 400 - did not belong to the original building activity. When taking that into consideration, it seems to me that the original nucleus was only the eastern part of the building, especially the southern wing of the "residential unit"[135] and possibly also the stables. Later, but still in the Late Nabatean period, the rest of the residential unit was built, likely with an expansion of the Gamma-Porticus, as well as Courtyard 400 as a later expansion of the complex. The heavy modifications made in the later part of this period make it extremely difficult to reconstruct the precise relative chronology. Furthermore, the difference in the preparation of the building materials is quite clear, especially between the southern wing of the residential wing - even if modified later, especially the staircase-tower, that is in fact typologically different from the others - and the rest of the north-eastern unit. Similarly, the later construction of the northern wing of the western courtyard is clearly shown by the outer wall. What is nonetheless certain - and it is true especially for the eastern part of the complex, that results in the form one can see today - is that during the third decade of the third century the occupation is interrupted, as the deposition of a hoard testifies. The fact that it wasn't recollected, but "forgotten" in its hiding-place, suggests that the new dwellers reoccupying the building in the 4th century weren't the original ones. Negev dates the subdivision of the outer courtyard in two to the 4th century, indicating probably the presence of different nuclei of an extended family (Figure 71c).[136] This subdivision is also implied by the opening of a new entrance to the complex: the staircase-tower 437. The third phase - taking place in the first half of the 4th century - also modified the organisation of the complex, following a process already described in other complexes in the site, namely walling doors connecting neighbouring rooms, especially in the north-eastern unit. The stables underwent some changes too, but still a semi-public or commercial function

[135] Some wall paintings had been found in room 411 and are stylistically dated to the 2nd – early 3rd century (Negev 1988a: 147-162).

[136] The belonging to the same ethnicity as the earlier dwellers is suggested to Negev by the careful removal of bones in the Tomb 108 close to the site (Negev 1988a: 146).

is suggested. Finally, a fourth phase – not necessarily distinguishable from the earlier one – is present, which Negev associates to the spread of Christianity and is documented by the engraving of crosses and palm branches on lintels and other architectural structures. Within this phase, Negev suggests the use of the complex as a monastery.

A similar and even more radical change in the Byzantine period can be described for the so-called "Nilus House" (Building XI, Figure 72; Negev 1988a: 88-109; Negev 1988b: 52-63). In its original form, dated to the Late Nabatean period, it should have been a house with a central courtyard surrounded on all sides by structures and accessible from the street through a *vestibulum*. It is difficult to say if it followed the injunctive or the conjunctive model, but the former option seems more probable. One of the most important features of the complex in this phase is the large stable occupying the entire southern wing, which could house up to 16 horses. Together with the one in Building XII, this is the only large stable complex found in the settlement. The residential part of the house appears to be on the upper storey, accessible by a staircase tower in the eastern wing, as also seen in Building I. The technique employed follows the Nabatean technique used also in the other structure, suggesting the contemporaneity of their construction.[137]

Attached to the eastern side of this complex, Building XIa is relatively dated to the Late Nabatean period, but to a later phase than the construction of Building XI (Negev 1988a: 109f). Abandoned in the Byzantine period, which according to Negev was due to its poor construction, it seems to have connected Building I with another complex to the east, even though their functional relationship is not clear. It was constituted by a central courtyard surrounded by rooms and accessible through a *vestibulum*, and the structures appear to have been used for residential purposes.

The Byzantine period – probably during the 4th century – represents a radical change for the entirety of Building XI: the entire western wing was entirely demolished by the construction of the West Church. Negev affirms the possibility that the owner of the partially destroyed house could have been the client of the church, namely the same Nilus mentioned in the dedicatory inscription. A connection between the two structures seems to have been present though a series of rooms in the southwestern corner of the former dwelling. At the same time, the other three wings also underwent a complete reorganisation, with the creation of four independent living quarters, similarly to what is seen in Building I. Larger rooms were divided and communication doors were blocked. Most notably, the stables were converted also into residential spaces and divided into different rooms, marking a clear and radical functional change. The same process will take place also in Building XII, signifying probably a general shift in the economic orientation of the settlement. A further phase had been identified, seeing a systematic re-roofing in almost all the rooms, probably following a seismic event in the 5th century.

Despite the radical changes that involved the dwellings' spatial organisation in the course of time, though, a more pronounced "stability" can be seen on a more general level. The spatial distribution of the quarters did not undergo considerable changes, and the construction of the city-walls and the enclosure around the eastern Quarter did not alter the settlement organisation substantially. Nonetheless, the construction of these two walls – the first delimiting the settlement, and the second separating one quarter – represents a change not of secondary importance, and not just from a purely physical point of view. In fact, on the one hand it testifies the need of a marked division between the urban space and the countryside, and on the other hand it attests a functional and social differentiation

[137] In the case of Building I, we have seen that the staircase tower was a later addition in the original Middle Nabataean period. The re-organisation and the superelevation of the building are therefore to be seen as contemporary to the construction of Building XI.

a) General plan of Mampsis (after Negev 1988b: 10-11, Plan 1; Plan reproduced with the permission of the Institute of Archaeology, The Hebrew University of Jerusalem)
b) Plan of Building XII (after Negev 1988a: 113, Plan 27; Plan reproduced with the permission of the Institute of Archaeology, The Hebrew University of Jerusalem)
c) The outer courtyard (photo by the author)
d) The stables (photo by the author)
e) The vestibulum (photo by the author)
f) The residential part (photo by the author)

Figure 71 - Mampsis: Building XII

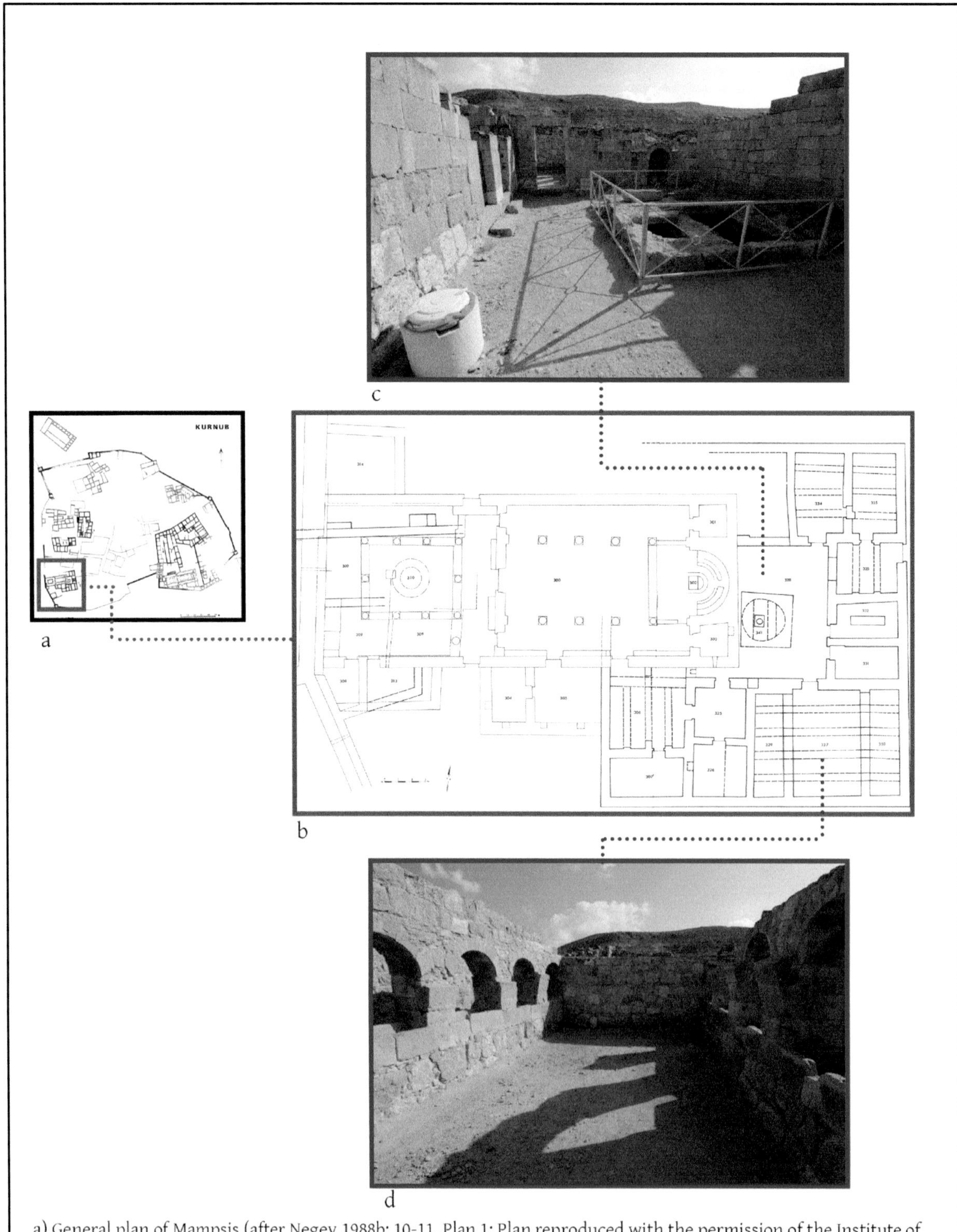

a) General plan of Mampsis (after Negev 1988b: 10-11, Plan 1; Plan reproduced with the permission of the Institute of Archaeology, The Hebrew University of Jerusalem)
b) General plan of Building XI (after Negev 1988a: 89, Plan 21; Plan reproduced with the permission of the Institute of Archaeology, The Hebrew University of Jerusalem)
c) The courtyard of the so-called Nilus's House, looking North (photo by the author)
d) The stable of the so-called Nilus's House, looking North (photo by the author)

Figure 72 - Mampsis: Building XI

within the settlement itself. It's not clear what ultimately led to the creation of the city-wall; Negev sees a probable explanation in the fact that with Diocletian's reform Mampsis became a crucial hotspot in the regional defence system. Therefore, the construction of the walls could have been the expression of a direct intervention of a central authority (Negev 1988b: 27). The inner wall is slightly more intriguing, for which Negev does not offer any interpretation. I believe that its construction reflects a clear intention to establish a social or functional or even administrative differentiation between quarters in the settlement.

In summary, even if a precise absolute chronological framework is not available because of the lack of inscriptions and written sources, the architectural analysis and the excavations allowed the tracing of the general phases of development of the site. Most notably the changes made between the Middle Nabatean and Late Nabatean periods - with the almost complete destruction of the earlier structures and later again between the Late Roman and Byzantine periods - with a clear densification of the living spaces and a general reorganisation on the domestic levels that nevertheless did not directly involve the settlement's pattern. The abandonment of the settlement is another huge question mark in the story of Mampsis. There are no clues about what may have ultimately led to the abandonment of the site and, more importantly, when it took place. Islamic material culture was thought to be absent in the archaeological strata and since no church appears to be converted into mosque, it was often suggested that the site was abandoned by the end of the Byzantine period. A re-analysis of the ceramic material, however, showed that the settlement continued to be occupied throughout the 7th century (Avni 2014: 263). Negev already confidently excluded that Mampsis' abandonment was due to the "Arab Conquest", despite traces of a destructive event in several parts of the site, most notably in the latest phases of the North Gate and of the West Church (Negev 1988b: 8). He was nonetheless convinced that the abandonment of the site preceded the end of the 6th century and was determined by the increasing unrest of nomadic tribes near the site and the concomitant lack of support from the central authority.[138]

One final remark should be made before considering the detailed analysis of the settlement. Despite the lack of a widespread documentation for the Byzantine-Early Islamic transition phase - that quite clearly represents a strong continuity in the structures from the 6th century - the site shows an urban structure and more importantly a development that is crucial for the present research, for at least three reasons: first, many of the same phenomena attested here especially between the Late Nabatean-Early Roman and the Late Roman-Byzantine are also found in other sites, and were often wrongly related with the "Islamic conquest"; second, it allows the visualisation of the differences between the Middle Nabatean and the Late Nabatean spatial patterns - that had been mostly erased in the other case studies - and the Late Byzantine's; third and lastly, it shows how even a small settlement - much smaller than many other settlements considered here - may actually present several proper urban features, often to an higher degree than the other larger settlement.

[138] Negev was convinced that several indicia suggested an abandonment of the site by the mid-6th century. The mid-6th century *terminus ante quem* was offered to him mainly by the churches. First of all, both the religious complex in Mampsis did not undertake several modifications from the moment they were built; then, there are no funerary inscriptions graves, a practice attested in the Negev starting with the great pestilence in December 541. According to Negev, by this time, the churches in Mampsis were already out of use. Furthermore, ashes layers - surely indicating a fire and preceding normally the collapse of some buildings and of their roofs - are connected with local struggles with the semi-pastoral groups living around the site. In particular, "Justinian's failure to support the *limitanei* in Mampsis, who probably constituted the bulk of the rather small population, weakened the town". Thus, it would have been the absence of the central support and the concomitant dissatisfaction of the Arab tribesmen to determine the abandonment of the site.

The major peculiarity of Mampsis in respect to most of the other rural sites considered in this analysis is the presence of a clear defensive structure delimiting the borders of the town (Figure 73b). There was a clear effort to include all the Late Nabatean structures within city-walls, and this is probably the main reason for its irregular plan, which that continued to be used also in the Byzantine period (Negev 1988b: 9).[139] The specificity of this wall is that it is not simply an enclosure, but a proper military structure, even if probably built with re-used building materials. In particular, the presence of the towers along the walls is a rarity in non-urban contexts and is normally seen only in purely military sites like military forts or main urban centres. As seen in Shivta, normally no defensive structure delimited the site, and that function is upheld by a particular structural solution of the more external buildings, creating a blind and continuous front to the outside.

One further aspect clearly distinguishes Mampsis' organisation from all the other case studies and is also a development contemporary to the city-wall, is the wall delimiting the south-eastern quarter (Figure 70f). It is rare in fact to find intentionally enclosed quarters, and the only possible parallels are the ones reusing earlier structures like military posts or acropoleis. Topographically, the south-eastern portion of Mampsis is not separated, since there isn't a proper acropolis in the site. Therefore, the reason has to be related to its socio-economical orientation, and it is not surprising that all the structures included within this second inner wall have a functional specialisation: on the one hand the Market represents a clear and structured commercial centre, and on the other hand the larger Church of the settlement representing the main religious heart (see infra). The quarter is not simply delimited, but its access is managed through clearly marked entrances, which do not recall the same spontaneous development of the rest of the settlement (Figure 73c). From this it can be deduced that this part of the site was managed differently. Therefore, I would suggest that the proximity of the Market with the Church is not merely physical, but their close connection could also have had administrative elements, and the inner wall could have physically marked a sort of administrative boundary.

The particular relevance of the church itself is also suggested by its isolated position, as opposed to the only other religious complex in the site, which is embedded with a residential unit. There are unfortunately no written sources supporting such hypotheses, but the spatial organisation of this part, clearly different from the rest of the site, cannot simply suggest a physical and casual separation of quarters. The huge Building XII is also included in this quarter, which according to Negev could possibly be understood in the Byzantine period as residential quarters for monks (Negev 1988b: 146f), along with a series of structures east of it and the Church (Figure 70f). Unfortunately, this south-eastern portion of the settlement was heavily damaged by the construction of the Police Station during the Mandate period, so one has an incomplete picture of the intended use of this area. The only thing that can confidently be said is that all the structures included in the inner city-walls share some sort of functional and/or close social ties, defining a completely separated quarter in Mampsis.

[139] There is only one possible exception: the so-called "Caravanserai", Building VIII. We have to be cautious to this respect, since the structure was badly preserved and the limited excavations conducted here did not offer much information. It is not clear either the function of the structure nor its chronology, but it seems that it was demolished during the Byzantine period – or Late Roman – in order to re-use its building materials (Negev 1988b: 191-194).

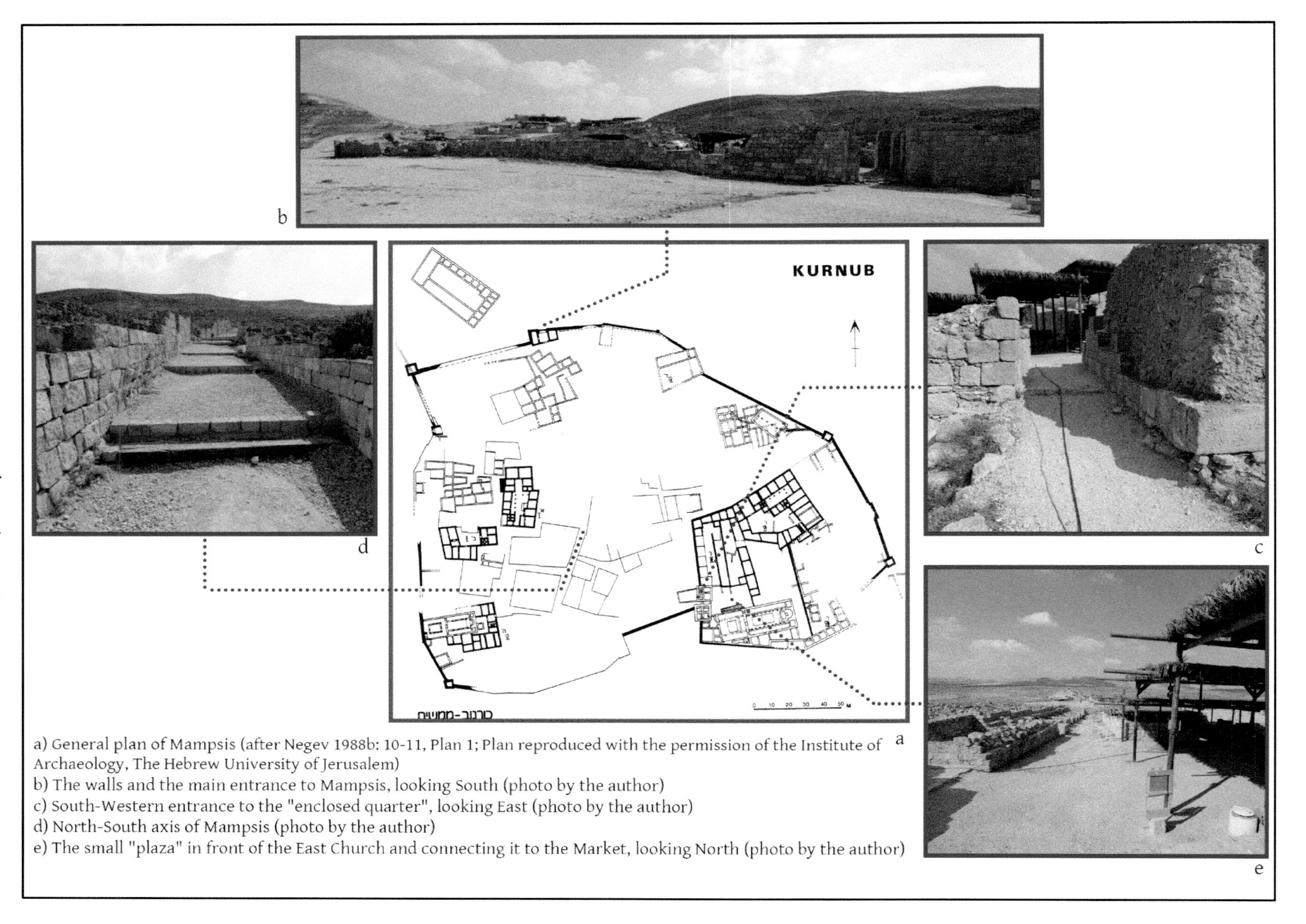

a) General plan of Mampsis (after Negev 1988b: 10-11, Plan 1; Plan reproduced with the permission of the Institute of Archaeology, The Hebrew University of Jerusalem)
b) The walls and the main entrance to Mampsis, looking South (photo by the author)
c) South-Western entrance to the "enclosed quarter", looking East (photo by the author)
d) North-South axis of Mampsis (photo by the author)
e) The small "plaza" in front of the East Church and connecting it to the Market, looking North (photo by the author)

Figure 73 - Mampsis: settlement's organisation

Despite the construction of these two properly urban features, the rest of the settlement maintains the features of a common rural settlement. The most evident is the absence of clear street network, with narrow paths spontaneously developing through the different buildings. Only one larger path running North-South (Figure 73d) seems to split into two in the southern quarter, but it is still not so neatly or clearly separated in order to define two completely separated quarters; this feature is also present in Shivta. In general, with the exception of the south-eastern quarter, no other socio-economic or functional zoning is detectable in the rest of the settlement and the two most important complexes of the town, Building I and Building II that were perhaps linked in their early phases by some administrative functions, do not differ substantially from the other buildings of the quarter. Their prominence is not marked by any element. Similarly, the importance of Building XI and the connected Church in the south-western corner is in no way reinforced by their location, but appear to be normally included in the rest of the quarter. The north-eastern part of Mampsis somewhat compensates the heavily built-up quarters of the settlement, being a large empty area. Its direct proximity to the Northern Gate and to the public Bathhouse and the connected reservoir seems to suggest its role as a communal space, a sort of unplanned plaza allowing the gathering of people. No other gathering areas could be clearly identified in the settlement; some junctions of different paths could form some sort of small plazas (like in front of the Palace), but these can hardly be seen as comparable to the north-eastern portion of the town. The direct parallel on a larger scale is with Umm el-Jimal, where a similar "blank" area is left in proximity of the main gate to the town, the "Commodus Gate", and other clearly communal buildings, like the so-called "Cathedral".

It is also interesting to note the coincidence with the location of the only public reservoir within the city-walls. It was supplied by an aqueduct built shortly after the realisation of the walls, showing that probably in an early phase, the reservoir was filled with water carried by hand from outside the site (Figure 70e). Surely, one of the functions of this structure was to supply the bathhouse nearby, interestingly also one of the few isolated buildings in the urban texture. It is nonetheless clear that this was not the only water-infrastructure present in the settlement, but was surely the only clearly public one. In fact, there is not a proper centralised water-supply system, and most of the water required by the inhabitants was guaranteed by the reservoir outside the settlement and the private cisterns in the single houses or churches. Furthermore, water management in Mampsis appears to be a private enterprise: each single dwelling had its own storage facility that was supplied by channels collecting the water falling from the roofs or that was directly poured in by the inhabitants. In the last case, the water was collected in a series of structures built outside the settlement, and most notably the dams and their connected reservoirs located across from the wadi Mampsis and controlled by the town.

The peculiarity of Mampsis in respect to the other case studies is the walled quarter, separated from the rest of the settlement and reflecting a clear functional and possibly administrative difference in the ownership of the area (Figure 70f). It does not resemble the situation where for instance military forts were converted, and which separation was somehow consequent to the characteristics of the previous structure. In these circumstances, instead of dismantling the entire buildings, it is simply reused and reorganised internally and often no clear functional or socio-economical distinctions from the rest of the settlement are detectable. Moreover, the new inner spatial patterns do share several features with the completely new quarters that sometimes surround the old *castra*. The walled quarter in Mampsis cannot be compared with an acropolis either, which is also always physically separated in the settlements: in the latter case, though, a clear topographical difference between the upper and the lower part of the settlement causes and clearly justifies a distinction functional distinction. This is definitely not the case for the quarter in Mampsis: there was a clear effort in marking a distinction between the walled area and the other quarter. The wall itself is not a previous structure reemployed or modified for its use, but is a brand-new construction, its purpose only being to separate the complexes in the south-eastern portion of the settlement. This separation would not otherwise be possible, because of the similar topographical context, the density of the structures, especially in this part of the town, and last but not least the new wall embracing the entire settlement.

Our general knowledge of the quarter is surely limited by the heavy destruction that followed the building up of the police station, erasing all traces of the entire easternmost part of it. Nonetheless, it is interesting to note that all the structures preserved present some peculiarities that are not directly comparable with other buildings in the settlement and are even rarely seen in other sites. If the presence of the large and important East church, conserving also reliquaries, is not in itself surprising, its direct connection with the so-called "Market" (Building IV) is even more impressive (Negev 1988a: 163-167). The two complexes are probably the only two in the quarter that were physically embedded: the main entrance of the church opens on the small plaza on which the two axes of the Market end. Their spatial relationship is clearly parenthetical, and their conceptual continuity is reinforced by the absence of any "buffer zone" or intermediate gate. A small "plaza" regulated the transition between the church and the Market (Figure 73e); it is also the same space on which one of the gates to the walled quarter opens. Its function is clearly a central redistributing place, working exclusively for this dichotomic complex, conceived as separate from the rest of the quarter.

The difference is even more evident seeing the situation on the opposite side of the Market, the northern side opening to the large empty area within the settlement. Here the entrance is composed of a series of small chambers, some probably functioning as guardrooms or at least areas to control the access to the Market and distributing the flow of visitors between the two axes constituting the complex. These two entrances serve only the Market-Church complex, and the other buildings included in the walled area apparently only opened to the outside of the quarter. Despite this clearly differentiated inner circulation, there are no clear indications for a subdivision into different blocks. All the structures enclosed by the wall share dividing walls and are adjacent one to the other; at least this is the case for the ones still standing. Some remains of other structures are also visible and directly abutting against the eastern walls of the three complexes still conserved, but their chronology and function is unknown. It is difficult to say what kind of relationship was established between the buildings of the quarters. On the one hand, belonging to the same walled quarter suggests some sort of functional or socio-economic proximity; on the other hand, the fact that apparently only the Market and the Church had a direct physical connection, reinforces the impression that each unit was somehow independent from one other.

Not much can be said for the complexes in the eastern part of this quarter, but one may get some information looking at Building XII. As mentioned earlier, no direct passage has been found between this complex and the Market, with which it nonetheless shares a wall. The only sure entrance opens directly to the empty area north of the quarter. Another passage to the south, through a "tower" room gave access to an area, which is also not clearly understood, extending to the east of the Market and the Church. From this "courtyard", no evidence for further openings or throughways to other structures is available. Despite this physical "isolation", Negev suggests a strong functional connection between the East Church and Building XII (Negev 1988a: 146f). In fact, he understands the new inner organisation of the house – indeed realized by the Byzantine period – as a possible conversion of an earlier private dwelling into a monastery or at least a dormitory of the monks serving in the Church. He deduced this interpretation by the presence of several small non-communicating rooms and the engraving of several crosses on the lintels of the house. Although no clear indication in support or against this hypothesis are available, I am not entirely convinced by this explanation. In fact, the phenomenon of creating single and non-communicating rooms is common within the settlement and the engraving of crosses also is in itself not a definitive indicator for a monastery or dormitory. Moreover, if it was a monastery, the absence of a direct passage between the complex and the East Church or any other religious complex would be unusual: a fact that could have been partially compensated, though, by the fact that they are enclosed by the wall into a distinguished unit. However, it is clear that the inhabitants of the dwelling benefited from some sort of status or had some clout that justified their inclusion within the walled quarter.

The different spatial organisation of the Eastern quarter is moreover not just evident because of its marked border. Its inner structure is unique from the rest of the town, where one can see a clear division in blocks (few and composed maximum by 2-3 units) and isolated complexes. As previously mentioned, this spatial organisation is already an achievement of the Nabatean period, and no substantial modifications on this intermediate level occurred in the Byzantine period. The inner circulation is guaranteed by the typical spontaneous path-network, developing without any plan through the buildings. The few proper blocks seem to be organised in clusters of courtyard houses, sharing dividing walls but having each unit an independent access from the paths. A particularly interesting aspect is that these clusters of houses are numerically less than the isolated buildings, that on the other hand, tendentially have large dimensions. However, both kinds of blocks do not seem to occupy any particular location within the settlement, thus one cannot gain an impression of any clear socio-economic or functional zoning.[140] The distance left between the isolated buildings and the different blocks was sufficient to guarantee the privacy of the single social group. That is probably the reason why there are no *cul-de-sacs* or intermediate courtyards – as opposite to Shivta – to ensure the privacy of the single dwelling. This absence could also suggest a major social homogeneity in the local community, with the only exception being the South-Eastern Quarter, without any marked division among different groups and the absence of people willing to distinguish themselves from the others, either for ethnic or socio-economic reasons.

One of the most impressive aspects of Mampsis is the extremely high quality of the architecture. The care in the realisation of the structures reaches standards hardly comparable with the other sites of the Negev, and even abroad. It is especially remarkable considering that it is not limited to a few buildings, but is a common feature of all the structures remaining and some of the structural elements. The Gamma-porticos or the tower-stairways, for example, are systematically used in their definitive form and reinforce the impression of highly specialized workers, likely trained in a properly urban context. Nevertheless, this concerns the Late Nabatean period, when most of the buildings were first realised: afterwards the structures had been heavily modified, and the internal organisation changed radically with the Byzantine period even if the earlier structures are normally maintained. The reasons for such drastic changes are difficult to understand and were likely determined by a set of causes; a shift in the economic context possibly played one major role, such as an increased population that required more living spaces. Almost all the Late Nabatean buildings were heavily modified, independent of the typology of the house and its original functional specialization. It was rare, however, that earlier structures were completely removed in order to rebuild new ones: the rooms may have undergone structural and functional modifications, but always limiting demolition as much as possible. A trend that can be observed in several complexes, especially in the larger ones, is the conversion of some

[140] Their relevance is marked by some architectural features (see infra). Also the compresence of large units – normally isolated – and smaller dwellings – normally clustered in blocks - confirms the impression (as noticed also by Hirschfeld 2003: 404).

functionally specialised rooms, most notably the stables, and an augmentation of the number of single ambiences in the dwellings.

Generally speaking, the typology of houses did not undergo radical changes: a clear predominance of "conjunctive" buildings persisted, most notably "Complex houses". In particular, Building II recalls closely some of the tower-houses seen in Shivta. As previously mentioned, the tower already appears to be in the original phase of the structure and connected to a courtyard and a series of dependences. Gradually, as in a typical "conjunctive" building, other rooms were then progressively added around the courtyard, giving the irregular plan of the unit preserved today. One further aspect of interest is the unit developing immediately southwards, which has not yet been excavated, but that seems to have been directly connected to Building II, ultimately forming a single larger dwelling. Here the only area identified is an installation interpreted as a kitchen. For the rest of the unit, the apparent coexistence of function in the same space can be confirmed, even if the specialisation of the complex itself is impossible to determine clearly.

In addition, Building XII (Figure 71), the largest complex in the site, follows the residential conjunctive model, even if the development is clearly more complex than all the other buildings in Mampsis. Furthermore, an area of the unit, interpreted by Negev as the residential one, closely resembles a "Courtyard House", while the rest of the unit is clearly closer to the "Complex Houses". It seems that Building XII is a direct parallel for Houses XVII and XVIII in Umm el-Jimal. In both the cases, there are two different units put into communication, even if the development seems to be different. In Umm el-Jimal two apparently separate dwellings were connected in a later moment, while Mampsis' Building XII seems to have always been a single unit, undergoing several and major expansions.[141] Differences are also clear in the functional use of some rooms in the dwellings, since some spaces in Building XII have a marked specialisation, understandable by the presence of installations. The most evident is the huge stable (Figure 71d), which even if it was not used anymore as such in later phases, kept a semi-public orientation. The large vestibule (Figure 71e), sided by a probable guardroom (Locus 417), also had mangers or a trough for animals, probably of the guests. Although, the rarest installation in the building to find is the so-called "Lavatory" (Loci 415 and 416), that is an *unicum* in the region[142]: it's peripherical and isolated position in the dwelling, with the rooms of this installation accessible through a staircase-tower, seems to be suited to meet hygienic requirements.

In light of these data and when considering this complex as a residential unit, one should be careful in suggesting its private character, at least in its final phases. More precisely, I think a possibility could be its use as caravanserai.[143] In fact, the several and small cells of the north-eastern section could have answered the residential needs of the guests (Figure 71f), combined with several units to house animals and included storage and hygienic facilities directly to the south. At the same time, the other cells on the outer courtyard could be easily understood as storage rooms (Figure 71c). The presence of a new caravanserai would not be illogical since the earlier one outside the city-walls was already demolished.

[141] As mentioned above, I am not entirely convinced of the chronological development proposed by Negev; nevertheless, both the hypothesis sees the Building XII as a single complex, and not two distinguished units connected in a later phase.
[142] It also finds few parallels in the Near East. It was in fact compensating the absence of a public sewer by a system of *anforae*.
[143] Like Building 3 in Shivta, which is normally considered a Khan, another consideration for the function of Building XII could have been for military use, such as a barracks or storage. The semi-public dimension of the building is nevertheless quite clear to me and makes a caravanserai a more plausible interpretation.

Moreover, the proximity of the market and its inclusion in the walled quarter suggests a particular status and/or function of this complex, which clearly did not merely have a residential function.

Fewer examples of "Courtyard Houses" can be found and are normally not integrated in blocks. Building I ("the Palace"; see Figure 70b, 70c and 74) is surely one example, and is also the largest in this typology. In its final configuration, the original administrative character was completely lost in favour of a purely residential function for an extended family. The typical functional ambiguity of the spaces can also be found. Apart from an installation opening to the courtyard (Figure 74) and interpreted as kitchen[144], no other structure suggests a clear zoning in the complex. It is clear that also in this example, the upper storeys (at least two, but likely three) were the more private and residential areas in the house, according to the common paradigm in the residential architecture in the region, similar to the other areas considered in this analysis. Despite the inner reorganisation of the spaces, the house perfectly fits the features of a "Courtyard house" and requires no further explanation here.

Another example is Building XI (Figure 72), the so-called "Nilus's House", in light of its particular development. Before the construction of the West Church, in fact, it was clearly an example of "Courtyard House", quite typical in its conformation, with a *vestibulum* opening on the one side to the street and to the other to the central courtyard (Figure 72c), a second storey accessible through a staircase-tower (like in Building I) and also a clearly specialised area, namely the stables. In the

Figure 74 - The courtyard of Building I in Mampsis (photo by the author)

[144] Two more rooms in the northern wings had been interpreted as a latrine and/or washroom (since it's served by a channel coming from the courtyard) and a "refrigerator" by Negev (1980: 15; 1988a: 63-64), but their ruined status did not allow further investigation. Moreover, it is not sure that they continued to work also in the later phase of occupation.

Byzantine period, the entire western wing was demolished by the construction of the church and a clear change in the spatial and functional organisation of the dwelling took place. The stables clearly went out of use and were divided into smaller rooms, and a direct access to the religious complex was assured (Figure 72d).[145] No clear zoning is present anymore, since no particular installations can be found. If the function of the house continued to be indeed residential, or at least to work as a private dwelling, is uncertain. Direct access to the church is not uncommon also for private houses, as seen in some complexes in Umm el-Jimal. Though one cannot exclude that it began to house some monks, the data are nonetheless insufficient to definitively call it a monastery or a dormitory. However, it seems logical the suggestion by Negev that the Nilus mentioned in the dedicational inscription as client of the Church, could have been the original owner of Building XI (Negev 1988a: 88).[146]

Other unexcavated structures that are also isolated could fit the "Courtyard House" model, mainly in light of their regular plan; though at present there is no certainty until they are studied in more detail. In any case, this type of residence occurs more infrequently. That is not surprising, since it is normally a properly urban typology, related to more Hellenised élites in the settlement. It is also not hard to believe that finding these houses in a semi-rural context like Mampsis, frequently associated with important complexes, could mark a socio-economical stratification in the local community. Even if one can argue on the proper function of Building I, identified by Negev as the Governor's house, it is without a doubt that at least in its earlier phases it was one of the major administrative centres of the settlement. Its function apparently changed considerably in the Byzantine period, also involving important changes in its structure. Accordingly, its physical modifications also reflect administrative and economic changes on a settlement level, which do find a direct confirmation by the widespread conversion of the stables.

Nonetheless, it is important to underline one further aspect that is ultimately shared by houses of both the typologies in Mampsis: the indirect access from the outside. In no case, any courtyard opened directly on the street, but the access was always mediated by a *vestibulum*. This suggests that despite an increase in the urban population, demonstrated by the addition and extension of buildings with new living quarters, there was no radical change in the social structure of the site. The patterns in privacy appear to be kept more or less similar to the earlier phases. Thus, no new discrete groups seem to have settled, but what can be documented by the built environment is likely a simple shift from more nuclear to a more extended family, requiring more living space within individual dwellings. The presence of larger groups - like clans or tribes - is evidently not reflected in the site by a new quarter or block organisation, as possibly seen for instance in Shivta. The only aspect that could be different is the presence fewer marks of socio-economic differentiation, possibly reflecting a more "homogeneous" society: the "normalisation" of the Governor's house may be considered a clue for such phenomenon. The prominent role in the settlement's administration might have been taken by the religious authorities or families connected to the religious hierarchies, which is possibly demonstrated by the

[145] In the text, Negev denies the direct communication between the church and Building XI (Negev 1988a: 88). Nevertheless, in the description of Locus 325 (idem, 107), he does not mention any blocked door avoiding the passage, and also in the plan offered in the publication do not show any separation between the two units.
[146] Moreover, a room (Locus 326) with a large niche, belonging to the house already in its early construction, is interpreted tentatively by Negev as a House-Shrine, suggesting that Nilus was descendent of a Nabataean priest. The concession to demolish part of the house to build a church would have been therefore a sort of continuation of the religious commitment of the family and sanctity of the dwelling itself (Negev 1988a: 107-108). Although a surely suggestive hypothesis, I think it is not supported by the evidence available now.

direct connection between the Nilus's House and the West Church. This shift of power finds perhaps reflection also in the new organisation of the eastern quarter.[147]

[147] Without any documentary information, this is a hardly demonstrable hypothesis, and the effective role of Nilus – for instance – in the local community is unknown. Comparison with other sites in the Negev suggests some family traditions in the religious administration, but again, their weight in determining the *res publica* of the local community is yet to be determined.

Chapter 5: Regionalism and Transregional patterns

After having considered each case study individually, it is necessary to confront all of them together in order to identify regional differences emerging based on the common morphological patterns described in the first section (see Table 1). Of course, the reasons behind such patterns can be different and this chapter will present possible explanations, which will also be revisited during the concluding part of the thesis.

Site	Settlement	Quarter/Block	House	Functions present in the settlement (apart from residential)
Umm el-jimal	Close / Scattered	Large and Compact Clusters and Liminal clusters; rare *insulae*; one converted fort	Almost exclusively large " Complex Houses" ; few	Administrative (?), military, commercial, agricultural, religious
Umm es-Surab	Open / Scattered	Compact Clusters and Liminal clusters	Almost exclusively large " Complex Houses" ; few	Agricultural, religious
Sharah	Closed /Scattered	Loose concentrations of farmhouses; few small clusters	Exclusively " Complex Houses"	Military, agricultural, religious, commercial
Mseikeh	Open /Scattered	Clusters and Liminal clusters	Exclusively " Complex Houses"	Agricultural, religious
Shivta	Close / Compact	Extremely compact urban structures; *Few Insulae; few clusters with " privacy devices"*	Mainly " Courtyard Houses" ; fewer " Complex	Aministrative (?), military (?), agricultural, religious, productive,
Mampsis	Close / Compact	Mainly isolated buildings; few clusters; one enclosed quarter	" Complex" and " Courtyard" houses	Administrative (?), Military (?), religious, commercial

Table 1 - Synthesis of the Typologies of the built environment and of the settlement's functions of the case studies in the Byzantine period

First of all, the sites described until now share one main aspect: they all testify a considerable development during the Byzantine period, especially during the latest phase between the 5th and 6th centuries. This development, credibly triggered by an increase in the population, appears to determine a drastic change in the organisation of the sites from the earlier phases, on all the levels though to different degrees and varying forms. With few exceptions, a total dismantling of the Hellenistic and Late Roman spatial frame can be documented for different reasons and to different degrees depending on the case study. Generally speaking, the sites in the southern Hauran appear to be the ones most heavily modified during the Byzantine era among the case studies considered here. The comparison with the other example is quite clear on this aspect. In sites like Umm el-Jimal, the Hellenistic and Roman town was almost completely erased, save for few buildings, which nonetheless underwent radical changes that completely distorted their original form. Even though the circumstances that triggered this change are different, the Byzantine settlement does not seem to resemble in any manner the previous site. A similar history is shared by the village of Umm es-Surab, where nothing but few stones under the central church and some reused building material survived from a probable Nabatean site. On the contrary, in the central Hauran, only limited ground-destruction has been documented, and the Byzantine period bears witness to a densification and reoccupation of the earlier living spaces: a process that continues also in the later phases and is a common feature to almost all the case studies, even if with chronological differences. The same trend can be observed for the Negev settlements, where Hellenistic phases are still identifiable under the later over structures. This is more applicable to Mampsis than Shivta, where the original nucleus has not yet been clearly found, but where the Byzantine village clearly extend into earlier unoccupied areas. Still, the Byzantine occupation also in these two sites does not imply any complete destruction of the earlier settlement. Like in the Syrian context, one can simply record a notable expansion of the villages and towns and a densification of the occupation of the space that is most evident on the house level.

The change in the urban patterns that was often elevated as cultural proof for the "Islamic conquest" is not visible in the case studies. In fact, it appears that the Early Islamic settlements follow the same spatial frames of earlier phases, at least continuing processes that had already begun in the Byzantine period, such as the densification of the living spaces. In this regard, the example of Mampsis is particularly important: it is indeed the only site where the conservation of the Hellenistic structures, though heavily modified, allows to track clearly the major changes made in the following phases, occurring between the Hellenistic/Roman and Byzantine period. This process is indeed recognisable in the other sites also occupied in later periods, showing that the presumed shift in the settlements' organisation already took place at the end of Hellenistic and Roman periods, and not in the Islamic one. The morphological differences between the sites analysed appear to be determined by regional peculiarities, which are clearly the consequence of local techniques and of a specific local development. Nevertheless, the phenomenon described is the same, taking different forms on different levels. A similar situation is also described for later phases of occupation, as in the Syrian examples: here in fact, it seems that the sites do not have any clearly identified Early Islamic phase, but a consistent Middle Islamic development. It is interesting to note that in spite of a densification of the buildings, detectable also in the other case studies in earlier periods, no substantial difference in the way of structuring the urban space can be seen. This later development can be seen ultimately as the final result of the same phenomenon seen in Mampsis in the Byzantine period or in Umm el-Jimal in the Late Byzantine-Early Islamic phases.

It is needless to say that the chronology of the settlement is the most delicate and hypothetical aspect of the present analysis and until other excavations will shed some additional light, caution is required. The architectural remains alone, even where cleared by the collapsed stones and exposed in their entirety, do not always offer reliable or certain chronological coordinates. The frequent reuse of building materials is surely one of the major obstacles for a precise recognition of the specific techniques in the different phases. It follows that based on what evidence has been described, especially for the changes between the Byzantine and the Early Islamic period, is a matter under major review. Nonetheless, aspects of continuity between the two periods have been demonstrated in other contexts and in different regions, and the diversity of contexts offering more or less the same indications supports this view (see Part 3).

The second interesting aspect to consider here is the regional differences between the case studies. In order to summarise the evidence, it is first necessary to consider some factors influencing the site appearance to different degrees and on several levels. The most evident element is the topography of the site, especially on the settlement and quarter level. It is quite obvious that a settlement developed on a hill or a mountain region must adapt to the context, adopting organisational pattern accordingly to the morphology of the land and the practicable space. In contrast with this more "constricted" situation, the sites occupying plains or more even surfaces are freer to develop and to follow whatever spatial solution they prefer. In these contexts, the social imprint is perhaps more evident than in hilly or mountainous topographies, without being a universal rule, in determining the spatial features. The other element, more influential on architectural techniques and therefore more evident on the house level, is the available building material that is later used in the sites. In the examples considered, the use of stone is almost exclusive, and only in the Negev both local and imported wood beams were employed in roofing. On the contrary, no wood can be found in the Hauran area. Bricks is totally absence in all the case studies considered here and are in general extremely rare in these areas. The different kind of stones is therefore critical and ultimately the factor determining the different architectural solutions. There are actually two possibilities with basalt on the one hand (Clauss-Balty 2008: 52) and limestone on the other. Clearly, the harder and more resistant basalt allows different architectural solutions than softer limestone: it is especially evident in the roofing techniques and in the dimensions of the rooms covered with stone, with those in the Hauran being normally larger than in the Negev, indeed because it was possible to have longer basaltic slabs without the risk of breakage.

One of the most problematic issues of the present analysis is indeed to distinguish which features can be explained exclusively by these more "technical" or "geographical" elements, rather than social elements. For instance, do the recurring transregional features – on all of the levels of analysis considered – reflect the use of the same building materials and possibly the sharing of the same techniques? Is there also a potential consequence of a common social structure or the way of thinking about space in terms of social ties and proximity? Probably, both components play an important role on different degrees depending on the level of analysis considered. On the most general settlement level, variability is extreme, at least morphologically. It is quite easy to attribute the different geographical context as a major cause: not surprisingly the examples belonging to the same region share more features on this level, than the sites from different regions. In the southern Hauran, for example, Umm el-Jimal and Umm es-Surab, despite their evident different role in the local context, have the same "opened" structure, with isolated (and in a way scattered) quarters and few isolated communal single buildings. This kind of urban organisation is probably explainable to a large degree by the availability of land easily suitable for building, i.e. the flat basaltic plateau.

In contrast, the situation is slightly different in the Leja, the central region of the Hauran in Syria, where the two sites considered, Sharah and Mseikeh, do to a certain extent closely resemble the spatial patterns seen in the above-mentioned settlements. Nevertheless, the hillier geographical context represents an important feature, especially in the demarcation of the borders of the settlements and in the structure and distribution of the single quarters. In Sharah, for example, the peaks present within the walled area are an obstacle to the creation of a continuous urban frame or larger compact quarters, and there is a higher frequency of isolated or small clusters of houses. On the other hand, in Mseikeh, despite the deep slopes surrounding the village, the grouping of larger quarters is made possible by the plain plateau occupied by it, that nonetheless limited its spread and/or expansion.

The Negev highlands are again something different, but despite the presence of a hilly landscape, it does not seem that this topographical aspect was a major element influencing settlement morphology. In fact, both Shivta and Mampsis could have occupied a larger surface area and clearly the more compact inner organisation is not determined by the lack of space available for urban expansion. The development of the sites themselves seems to be the cause of a higher density and compactness, which is indeed an "urban" characteristic. Nonetheless, some clear differences are visible between the two Negev case studies, and may be explained in terms of a different economic orientation and administrative role. In the case of Mampsis, its northern side is opened to a plain, and could have allowed a more widespread settlement, which is likely for earlier phases. The Late Roman city-walls mark a limit that would not have been present otherwise and represents a clear social and administrative choice, which nonetheless does not have any direct parallel in Shivta. This aspect could ultimately confirm the fact that since the very first stages of its history, Mampsis played a more important role in the local administration. This impression is strengthened by two further elements: the extremely high quality of the architecture and more importantly the presence of buildings clearly related to a civic authority in the Middle Nabatean period until at least the Late Roman period, and later during the Late Roman – Byzantine phase the enclosed inner quarter. Moreover, the wide "empty" areas left between the different quarters is also reminiscent of the urban organisation seen in the Syrian and Jordan sites and may be a sign both of communal activities and exchange areas destined to non-locals, for instance for regional markets or guest caravans. These elements could possibly explain why a settlement with limited to no agricultural hinterland was able to survive so long and even to finance major public enterprises as the walls, the "Suq" and the churches.

Shivta clearly shows an opposite conception. Despite the absence of an "official" demarcation of the settlement by the means of walls, its compactness and marked closeness from the outside suggest a higher degree of socio-economic "autonomy", also allowed by the extensively exploited agricultural hinterland. The morphological features of the settlement seem to reflect a strong sense of community, separated from the outside, and the only facilities and infrastructures that may have also been opened to other people, such the supposed caravanserai and likely the large northern plaza, who were relegated to marginal areas. The village might have had a certain local civic administration and a certain level of socio-economic stratification, likely attested by some larger buildings across the settlement, explaining the more "planned"-looking extension of the Byzantine period. In any case it is clear that the economic prosperity of the local community was mainly due to the agricultural production and eventually the exchange with pastoral nomadic groups living on the fringe of the desert or in regional markets, taking place somewhere nearby the village. Therefore, the properly urban character of Shivta is not much on a functional level, but mainly on a morphological one, especially if compared with some later *medinas*. This aspect especially involves the presence of determined "privacy" systems that would have little reason to be in more scattered settlements. This aspect is particularly evident on the quarter and block levels of analysis, the "highest" level where the combined influence of morphological and socially related features have left a clear imprint.

Ethnographic research confirms the tendency of having members from the same familial group – more or less extended – living as much close as possible to the other family members (see Chapter 2.2): in general, the higher the spatial proximity is, the tighter the social bond is. Of course, there cannot be absolute certainty that this was also the case for the ancient periods and there are clear exceptions that can be found. Nonetheless, this phenomenon can be also detected in many historical sites, including some of the current case studies. For instance, in a quarter made of a cluster of farmhouses, which courtyards are simply divided by shared walls, their inhabitants could have belonged to the same social group, that being an extended family, a clan or even a tribe. Nonetheless, besides proximity, one must be aware of these aspects of privacy. In a case like the one just described, the absence of particular devices reinforces the impression of strong social ties shared by the inhabitants of the quarter or of the block. In fact, a family member was no stranger to whom the sight of the private quarters had to be kept eluded. A further element not to be underestimated is that the absence of physical obstacles offered the possibility of a more effective control over the other family members.

In contrast, the presence or not of *cul-de-sac* and intermediate courtyards or any other sort of solution creating a sort of "buffer zones" between private and public areas is ultimately one of the major differences in the way of organising quarters and blocks and it is closely related to a perceived need of an incremented protection of the privacy. It is interesting, even if not entirely surprising, to see that the presence of these spatial solution is most recurrent in only one single example among the six sites we considered: Shivta, the most compact settlement. Why is that the case? Why are these devices rare if not totally absent in the other five more scattered and opened settlements? One probable reason may lie in the combination of a varied social composition of the local community or the presence of discrete social groups, like different extended families, clans or tribes, a feature in common ultimately with the other case studies, and the compactness of the site, which is clearly a peculiarity of Shivta. If one compares, for instance, this organisation with Umm es-Surab – two settlements that apparently have a similar "rank" – the difference is quite clear. In front of one almost single compact quarter of the Negev's village, the Jordanian site shows a clearly marked division of the living spaces, with clusters of farmhouses independently isolated within the urban space. This last pattern both reflects a less bounded expansion of the village due to topographical characteristics of the region, the flat and wide basaltic plateau, and mainly a double social element. There is both the expression by some inhabitants that they belong to the same restricted social group – testified by the extreme compactness of the single quarter – and at the same time the lack of a developed communal membership on the village level – materialised in the separation of the quarters. The absence of "spatial devices" such as the ones mentioned above – intermediate courtyards or *cul-de-sac* – can therefore be explained by a lower "need" of privacy among the same family group or kinship, while at the same time implementing a more effective social control on their members. The structure of the houses then sufficiently guarantees the privacy of the smallest social group: the nuclear or an expanded family depending on the type of dwelling. There is no need for further "defences" from the prying eyes of outsiders, since the villagers that do not belong to the same discrete social group are living in a separate quarter in the settlement.

This impression is further supported by the position of the courtyard. It is not rare to find two houses that share a dividing wall between the two private open spaces. Since the houses often had more than one storey and anyway the roofs were also used, it would not be difficult to have a direct line of sight to the private quarters of the neighbours. The absence of structures that would avoid such inconveniences suggests that no worry was felt to this regard. Therefore, it is not absurd to think that the two families could have shared a strong social bound.

Once more the situation is different in settlements like Shivta, where the recurrent use of spatial devices like *cul-de-sac* and intermediate courtyards that a more urban character of the settlement can be found. This could be explained in terms of an augmented need for privacy, a reinforced urge to divide more markedly the private areas from the neighbours and the properly public sphere. If in the previous context, there is substantial duality between the private and public dominions, then in sites like Shivta a triadic spatial organisation had likely been developed: the intermediate courtyard and the *cul-de-sac* create a semi-private area shared by families linked to one another by social bonds. The main reason for the adoption of such solutions can be seen in the compactness of the settlement itself. In sites like Umm es-Surab, as already mentioned, the opened structure makes them useless, since close social groups can easily build separate quarters far and isolated by the ones belonging to others. In more urban contexts or in compact settlements, this is not possible, and the "privacy-devices" are the more logical solution to guarantee the same level of privacy. The way the houses are structured within the blocks created by intermediate courtyards and *cul-de-sac* then closely resembles the same spatial organisation seen for the other kinds of settlements: smaller courtyards or fenced areas facing directly the shared spaces, sometimes sharing dividing walls with the neighbours' house. Once again, the social proximity of the inhabitants could be quite clearly confirmed.

A further indirect confirmation of close social proximity is given by the places where such devices are not employed: in particular, the blocks similar to the Hellenistic *insulae*, geometrically planned and offering direct access to the public streets even in a more densely-constructed context. In these less frequent cases, the spatial conception of the house is different, and it alone is sufficient to meet the privacy standards that are elsewhere guaranteed by the block and quarter organisation. Two factors were therefore developed. The first element is the vertical "isolation" of the building: on all four sides of the house, in fact, the roofed areas avoid a direct line of sight into the opened private areas from the neighbouring complexes, even from their roofs. The second element is a direct consequence of this: the position of the courtyard, occupying a central position in the dwelling. Generally speaking, the so-called "injunctive" type is normally the type of house that can be found here.

Each type of house reflects a certain degree of the inhabitants' independence: the different spatial organisation possibly indicates a different familiar structure, most importantly in terms of how many members the complex had to house. Therefore, a "single house" might reflect a nuclear family, while the larger "Complex houses" might reflect an extended family. Surely this last type allowed the greatest degree of flexibility, since a new building could have been added to the complex when the need arose. This need could follow either the addition of a new nucleus in the extended family or an economic improvement in the family situation – the two possibilities not necessarily unrelated to one another. At the same time, the choice of determined house typologies – and this is particularly true for the Courtyard house in the Negev – could simply reflect the maintenance of a local architectural tradition, technical know-how and building techniques, as well as an identarian socio-economic status. It is nonetheless true that the same process of densification of the living spaces is noticeable also in these types of structures, as demonstrated in Mampsis: this clearly shows a shift towards residential functions for more extended families, though particular socio-economic functions or statuses were difficult to determine.

This social interpretation refers of course to the later phases of the settlements. Moreover, it does not take into account the relationship between the development of structures and the socio-economic changes that might have occurred. These issues and a deeper social evaluation of the morphological features will be further investigated in the last chapter with a direct reference to the regional historical context. Nonetheless, it is worth considering few elements here. The topography of the site clearly plays an important role in determining some characteristics of the settlements. But if it is true that more dense settlements are more likely to be found in hillier or mountain contexts and their expansion is somehow linked to the morphological configuration of the terrain, why is the compact pattern still so frequent in more flat regions, where no obstacles are to be found? Which circumstance(s) triggered this development during the 5th-8th century phenomenon that radically modified the previous settlements' pattern, especially in the Hauran region? Even in cases of complete destruction determined by sudden and violent episodes, nothing would have prevented the inhabitants from re-building their villages or towns according to more "conventional" Hellenised models or at least with quarters closer one to the other, keeping them nonetheless clearly separated. This matter is difficult to address; however, what I think might be a plausible hypothesis - but far from being a general rule for all the open settlements - is the involvement of sedentarisation processes in the development of such settlements.

It is possible that under determined circumstances - political or economic or both - non-settled groups or segments of tribes were encouraged - or forced - to sedentarise. Normally, processes of this sort tend to take place in proximity of already existing settlements or - in particular cases - around communal sanctuaries (see Chapter 2.1).[148] A tendency that often takes place is that where there is a presence of already settled members of the same tribe or clan, the new comers set their own dwellings in the immediate surroundings. Sedentarisation processes are therefore the reflection of the social proximity in the spatial organisation that determines clusters of houses one next to the other (Whitcomb 2009: 243). This situation is indeed observable in many Syrian and Jordanian sites, like the ones of present analysis. Therefore, the relative isolation of quarters or blocks - a frequently recurring feature and probably their most characterising features - is a mark of their belonging to a different clan or even tribe.[149]

Although this is a plausible explanation, it is far from being a general rule applicable in every context, even within the same settlement. In fact, at least two further elements must be considered. First of all, it is often difficult to clearly distinguish morphologically between the original nucleus and the later products of sedentarisation, especially if no precise chronology based on excavations is available. This is particularly true in the rural context, where the planned settlements were almost absent.[150] This is mainly due to the fact that pastoral groups living in the desert - in Syria, Arabia and the Negev - always had a socio-economical relationship with settled communities. This similarity is indeed the second and most important aspect to keep in mind: as seen in the introduction and will be further investigated in the next part, the social structures of these two components are almost the same and direct ties between them are extremely frequent (see Chapters 2.1 and 2.2). It follows that this social acquaintance may also have determined almost indistinguishable spatial solutions. This is strikingly evident on the lower level of analysis, the house. The types of dwellings are more or less the same, either within the same settlement but also on a trans-regional level, recurring with few variants. This homogeneity is not entirely surprising if one considers the element that constitutes the basis of almost all the practical - and also ideological - aspects of the life of both the pastoral and settled communities: the family. The centrality of the family in the social structure of the inhabitants of these regions is clearly shown, independently from how the further levels of the segmentary paradigm develop in higher levels. Anyhow, the different types of houses may reflect the different types of families, or better, the different

[148] To quick reference to archaeological studies, see for instance: Dentzer 1999; Whitcomb 2009.

[149] It is extremely difficult, if not impossible, to clearly establish on which social level this spatial distance is operating: as we also have already seen, the terminology describing the segmentary societies is far from being standardized and clear. In some lucky examples, like in Umm el-Jimal, we have some epigraphical indications, but still we are not entirely sure which kind of social entity is described by the different terms used (see Part 3).

[150] The only exception may be the presence of a Roman fort, later modified and merged in the later settlements. In this case, it can have been the original nucleus, around and within which the later development took place. Also in this respect, though, no generalisation is possible and each example is to be considered singularly.

degree of extension a family may reach. It is also clear that the economic situation of such group plays an important role, possibly influencing the dimension of the dwelling.

The interpretation of social groups and divisions of such groups, however, has to be considered in light of the diachronic modification of the urban space, that was demonstrated had affected all the three levels considered. Their causes and the degree in which they happened may vary from case to case, but it begs to question if each spatial change necessarily implies a social change as well. And if so in which terms: different family structures, as suggested for the densification of the house-space, new identity pattern, or ethnic shifts? One aspect, however, is quite clear: the strong discontinuity in the spatial organisation of the settlements - not only the major properly urban centres - precedes the Byzantine period in most cases, and it is not to be connected - as was previously (and to an extent unjustifiably) suggested for a long period - with the arrival of Islam. For example, one of the major changes taking place from the beginning of the Byzantine period is the extreme densification of the buildings on a quarter and block level, and also on a house level. The earlier scattered and isolated houses start to be grouped around a common courtyard, starting to form the recurring clusters of dwellings. The process is gradual and seems to follow a similarly gradual increase of the need of living quarters and facilities. This phenomenon is also detectable in different form in the few cases of "Hellenistic" houses, like in Mampsis with the Governor's House and in Umm el-Jimal in the so-called *Praetorium*. In the first case, the earlier official and representative areas of the complex are divided into smaller independent rooms accessible normally only from the central courtyard. In a second phase, the upper storey - or storeys - were added, probably responding to a further demographic increase. In Umm el-Jimal a different version of the same process can be described: here, the earlier structure is only partially maintained - most notably in its most representative rooms - to be incorporated into the typical courtyard house, with the gradual addiction of other structures and facilities.

The different form just described may be explained in two ways, one not mutually exclusive the other. First, a conjunctive way of constructing the buildings, resulting in a more or less free aggregation of structures one next to the others. This is also possible since no clear demarcation of the private land property is to be found. This aspect is partially linked with the second aspect: the availability of suitable land. In more dense and compacted settlements is not possible to have larger courtyard houses, and the most practical answer to the need of more living space, is to extend vertically the building. The reasons that determined this change is not only a purely demographic one, or at least not in all the settlements. Clear economic changes have determined the functional change of several areas in earlier structures. A marked shift towards a more subsistence economy can be recognised for instance in the modification of the large stables and in the abandonment of the caravanserai in Mampsis.

In conclusion, I would like to point out a more methodologically oriented issue. If the purpose of the analysis is to offer some insights on the society of the local communities, it is absolutely necessary to consider the settlement in a holistic perspective. The three-level analysis applied here is one attempt to embrace the complex social patterns that characterised these sites. Though, not necessarily relating to issues of ethnicity, it is clear that a multi-faceted identity is somehow reflected in the spatial organisation: as previously discussed, several "identities" can be at play according to the context the individual find himself. Thus, in the material substance of a settlement one can possibly glimpse some physical markers separating different "contexts" in which each one of these multiple identities is operating. The house therefore can reflect the ground-level identity of an individual, the one that is activated in a situation where the most detailed definition is required: the family. It implies a clear separation from other dwellings belonging to other families. Similarly, the pattern of the blocks and the quarters may involve larger identity groups, like clans or even tribes, the definition of which can be extremely flexible. On the settlement level, one can at last see how much the sense of community is developed or at least the degree on which it is working as clearly distinguishing identity criteria.

The differences detected between the Negev settlements, for example, which show a more "urban" orientation, could possibly reflect a more developed sense of the local community. An indication of this can also be obtained looking at the incredible *corpus* of papyri found in Nessana, another village in the Negev, that offers a unique insight of the social and identity pattern of its inhabitants (see Chapter 7.4). The idea of community is firmly stated throughout all these documents, which is not only indicative of socio-economic stratification but also a certain degree of pride belonging to the "people of Nessana". Apart from the reason that may have determined such a cohesion – and in particular the ecological explanation suggested by one author of study on the papyri (Stroumsa 2008: 80) – what is interesting to note here is the striking contrast with the situation described in other regions. Here a major emphasis is placed on the segmentary social groups on different levels. The recurrence of inscriptions carrying information on the tribal membership of the individuals, as the many found in Umm el-Jimal, is not casual (see Chapter 7.4). The physical separation of quarters, at the same time extremely compacted in their inside, probably could have been determined by a similarly fragmented population. The inhabitants may have felt a communal belonging, as demonstrated by some public buildings serving quite clearly to a public life; nonetheless, familial ties were stronger, determining the clearer separation among the different social groups within the settlements than between the settlement itself and the outside.

Clearly, Nessana could represent an exception or have its own peculiarities even in respect to other settlement from the same region. It is nonetheless interesting that the settlements showing a clear unity on a settlement level, a higher density of the spaces and ultimately more urban features also on lower levels of analysis are located in the Negev: a region where it is clearly attested, at least in one settlement, that tribal identity wasn't defining the local identity patterns. The segmentary structure of the society, and namely the membership to a determined family, simply does not seem to play a key role in the inner dynamics of the settlement (Stroumsa 2008: 82). The population prefers to identify itself in terms of the belonging to a particular local community, even being proud of its own long-term familiar link with the specific site. It is not to say that the social and familial ties did not have any importance in the daily life of Nessana – and possibly in the other settlement in the region: it is clearly proved, for instance, that these were the guiding elements in the administration of the properties (Stroumsa 2008: 83). The major difference with the other areas is really the degree of development of the communal feeling of belonging to the settlement. This could explain why the major differences are to be found on the most general level of the analysis, namely the settlement as a whole and in the quarters' patterns, but not on the house level. The key social element throughout the region is always the family, being nuclear or extended; what is different is the importance played by the larger segments of the society, or better yet, which identity took the leading role: the local community or the "blood-based" tribal one. Investigating the causes of this different pattern is an extremely difficult task that requires the involvement of other sources alongside the archaeological dataset.

Chapter 6: The "test" case-studies - The Central Jordanian Plateau

Before considering the archaeological and anthropological information within the general historical and socio-economic contexts, two more case studies from the central region of present-day Jordan will be examined: Tall Hisban and Umm er-Rasas (Figure 75). The purpose of this last section is two-fold. The first is to verify if the three-level analysis can be also useful for sites where the state of conservation is not as good as in the examples considered until now. The second is the set of information that could offer, enriching and improving the record of typologies mentioned in Chapter 3. More specifically these two sites address many different topics that are critical in the final discussion.

Figure 75 - Localisation of Tall Hisban and Umm er-Rasas in Central Jordan (Google Earth)

First of all, Tall Hisban is one of the documented cases in which a large village becomes a proper urban centre already in the Late Roman period. This implies that the morphological features do not follow the standard Hellenistic urban form seen in the proper *poleis.* Nevertheless, the settlement surely functions as one of them, possibly on a smaller regional context and maintains this urban status also in the Byzantine and possibly Early Islamic periods. What is certain, however, is that in the Middle Islamic period it is mentioned as a *madina* in the historical sources. Tall Hisban ultimately provides a case study to investigate to which extent some morphological and social aspects known for so-called "Islamic" cities can be also applied to Late Antique cities, especially in the examples where a former rural or semi-urban settlement was elevated to *poleis*, as happened to the Late Roman Esbus.

Umm er-Rasas, on the other hand, opens up the important topic of the continuity of religious structures in the Early Islamic period, which can have various social and cultural implications. Establishing a precise chronology of abandonment or functional (and not necessarily religious) conversion for these sites and their structures complicates matters and has been subject of controversy. In fact, the mosaics are among the mostly used elements for dating, and the different reparations the churches in particular underwent over the course of time. The presence of inscriptions mentioning some interventions or dedication of entire parts of the buildings has offered in some cases important information to reconstruct the general development of the building in which they are found in. Nonetheless, Umm er-Rasas clearly represents a key case study to investigate how the presence of some important religious centres might have contributed to the shape and maintenance of the communal identity, as well as to the built environment. Moreover, from a more morphological point of view, it is also an excellent example of how a Roman fort was converted into residential quarter, functioning as the "nucleus" for the development of the Late Byzantine settlement: Umm er-Rasas ultimately completes the record of types of quarters mentioned earlier in this part (see Chapter 3.2.2).

Differing from the other case studies considered before, the state of conservation of Hisban and Umm er-Rasas does not allow the analysis for each level of the settlement. Therefore, a general analysis will be conducted, offering all the data available until now – as far as the material evidence is concerned – and more importantly, parallels with the other examples will be established, in order to tentatively reconstruct the missing dowels of the urban fabric. Therefore, as previously mentioned, they are to be understood as a testing ground for the methodology, and later the results from these case studies will prove their worth in the concluding part (see Part 3). This aspect ultimately compensates the poor conservation of the urban fabric, which nonetheless could be partially investigated though the comparison of with the other case studies. Needless to say, all the conclusions are highly hypothetical, and only further archaeological investigations will offer data to verify or falsify them, and consequently the validity of this methodological approach in such contexts.

6.1: Tall Hisban

The settlement of Hisban (Figure 76) – identified as the Roman Esbus[151] – bore witness to an almost continuous occupation at least from the Iron Age period, when, according to the early investigations

[151] The site is known also through some historical sources, see Ferch *et al.* 1989: 3-23. The name Esbus, together with some variation like Esbouta or Esebon, giving origin to various adjectives and *ethnica*, derives from the Hebrew Heshbon, appearing 38 times in the Old Testament. Particularly interesting is the account by Pliny, mentioning "Esbonitarum" Arabs (*Nat. Hist.* V 12). Also Josephus reports of Ἐσσέβων (*Ant.* XII 15.4) and its region/district – Ἐσεβώνιτιν (*JW* II 18.1), Ἐσεβωνίτιδι (*JW* III 3.3), Ἐσσεβωνίτιδος (*Ant.* XII 4.11). The name of the city also appears on some coinage produced in the site – *Aurelia Esbus* on a coin minted under Elagabalus (218-

Figure 76 - Aerial picture of the archaeological site of Tall Hisban (drone photos by I. LaBianca; courtesy of Prof. Dr B.J. Walker)

conducted here, it could had been the Biblical capital of the kingdom of Sion. As mentioned above, the extremely fragmented nature of the archaeological remains at Tall Hisban does not allow for a comprehensive analysis of the urban organisation of the site. Nevertheless, it may offer important hints, especially for the development of the idea of city in the later phases of the Graeco-Roman period and the Islamic occupation. In particular, the methodological setting of the projects studying Tall Hisban could be a useful theoretical reference for addressing the more social and anthropological questions, some of which had been already partially stated in the previous part and that will represent the main focus of the final conclusions of this thesis.

The study of the site of Tall Hisban began in 1969 and immediately after the first season of excavation, it was oriented towards a strong interdisciplinary approach. In particular, scholars from a wide range of disciplines from archaeology to geology to anthropology were involved in the project in order to yield a complete set of data not only for the site of Tall Hisban itself, but also for the region. From this, a combination of regional surveys and local excavations were developed. The general research was based on a strong theoretical background, specifically the Food System Configuration Theory proposed

222) – and two Roman milestones found near the site – Esb(unte) in the earlier one, dated to the AD 288, and Ἐσβοῦτος in the later, dated to 364-375. The site is mentioned also in some later sources, like Eusebius and Jerome, as well as in the lists of the participants to the Council of Ephesus (431) and Chalcedon (451), where the Bishop Zosus of Esbus is recorded. A variant Essmos appears in the Notitia Antiochena (570), even if in the 6th and 7th centuries the name Esbus appears to be most widespread. The first Arabic sources uses the name *Ḥesbân*, and the variants *Ḥesbân*, *Ḥusbān* and *Ḥisbān* spread only from the Middle Age onwards.

by LaBianca for interpreting and reconstructing the development of the region surrounding Tall Hisban and of the site itself through its long occupation. It has been a useful theoretical framework in which consider the material evidence from the site (LaBianca, Hubbard, and Running 1990: 131-133). This theory defines through careful analysis of the regional survey and archaeological data different circles of intensification and abatement of the settlement's occupation. Seeing the interdisciplinary backbone given to the study of the site since the very beginning of the archaeological investigations, five different types of conditions are taken into account: environmental, settlement, land use, operational, and diet. Each one of these is explored through different set of data – from the archaeological excavations through geological investigations to the archaeobotanical and faunal analysis – and ultimately their combined analysis could allow different food system configurations for a specific period to be determined, and therefore different occupational patterns. More specifically, LaBianca describes a three-degree range from low intensity to high intensity: the lower represents a "lighter" occupation, more oriented towards seasonal presence, temporary infrastructures and subsistence economy. On the contrary, the highest intensity reflects more stratified and "settled" contexts, with a clear delocalisation of food supply thanks to long-distance trade, a lower diversity of animals and plants and a minimal variation in the location and intensity settlement (LaBianca, Hubbard, and Running 1990: 131). Through this analysis, three broader cycles of intensification and abatement were identified for the region: from the 15th to the 5th centuries BC; from the 4th BC to the 7th AD; and finally, from the 8th to the 19th. Each one of these cycles ultimately testifies different food system configurations, which involves the entire region of Hisban; furthermore, it is also to a certain extent reflected by the transformations the settlement itself underwent through time. For the purposes of the present research, only the diachronic development of the site will be considered in detailed, with a limited reference to the regional trends, which are nonetheless an important component in yielding significant historical conclusions for Hisban itself.[152]

After an important development during the Iron Age, with one peak of the settled activity in the 7th - 6th centuries, the first phases of the Hellenistic period continue to represent a low intensity occupation of the region, and consequently the site itself do not have consistent remains testifying occupation during the periods before the 1st century BC. The site reacquires a primary importance during the Late Hellenistic period, between the end of the 2nd and the 1st centuries BC, especially after the establishment on the summit of the acropolis of a fortress, a construction probably urged by Herod the Great in order to prevent revolts and control an area contended with the rising Nabatean Kingdom.[153] The original military nature of the settlement is confirmed by the excavations, with several objects of military nature been found.[154]

The exact moment in which the slopes beneath the Hellenistic fort were occupied is not yet clear. According to the archaeological remains, it appears that for the entire Hellenistic and even Early Roman periods the occupation of the surrounding area, and by extension the entire region, was mainly oriented

---

[152] For the general regional trends, see LaBianca, Hubbard, and Running 1990: Chapters 4 to 7, and in particular chapters 6 and 7 for the chronological horizons more relevant for the present research (from the Hellenistic to the Modern periods).

[153] Josephus refers to the effort by Herod the Great to rebuild several fortress and fortified cities in order to be prepared for a possible revolt. In *Ant.* XV 8.5 in particular he refers to establishment of a group from his elite cavalry in Esebonite in Paraea. The site was already mentioned as Essebon in *Ant.* XIII 15.4, for being captured by Alexander Jannaeus from the Nabataean. According to Josephus, the control over the site was short lived, since already under the reign of Herod Antipas it may had been fallen back under Nabataean control. It is interesting to note though, that no Nabataean influx – for instance inscriptions – had been found so far, contrarily to the site for instance of Umm er-Rasas.

[154] The Stratum 15 is the one dated to the Late Hellenistic period (Mitchel 1992: 17-39).

towards a more seasonal frequentation and pastoral economy. This was also due to the unstable political context or a low-to-medium intensity configuration, as LaBianca defines it (LaBianca, Hubbard, and Running 1990: 168), where only a small population was settled around fort, dedicated mainly to subsistence economy and to a certain extent a domestic textile production.[155] Architectural remains from this phase are hardly identifiable, with few reported walls of field stone or semi-dressed stones and the use of some of the caves. As suggested by LaBianca, the settlement underwent a gradual but constant shift towards more sedentary forms of occupation, reaching a medium intensity configuration, with a more intensive use of the caves and possibly even with the construction of some public building, that nevertheless does not show any abrupt change in the archaeological stratification before the Late Roman period.[156]

The Roman control over the province of *Arabia* in AD 106 and the end of this land being a contested region offered the site a good opportunity to prosper. Its strategic position close to two important commercial routes, namely the north-south axis represented by the *Via Nova Traiana* and another east-west street connecting the Transjordan area with the Jordan Valley and Palestine, through Livias, Jericho and Jerusalem, played an important role in the development of the settlement. After the destruction documented in Early Roman Stratum 14, it appears that the reconstruction of the settlement was immediate and its commercial vocation clear (Stratum 13; Mitchel 1992: 75-91). In particular, the building activity in Area B and D appears to be consistent, with the creation of the so-called "Plaza" (Area B) and the filling of the Iron Age cistern with debris coming from the summit of the acropolis, likely indicating the end of its military use as fort, and the following establishment of the "Inn". Ultimately, Mitchel considers it more plausible to see these remains as part of a fortified Roman road-station and village: the "Plaza" at this time simply would have been an enclosure, probably connected to the summit of the acropolis, still hosting a small Roman garrison,[157] and the series of rooms forming the "Inn" were possibly designated to house and provide food to travellers and caravaners and their animals (Mitchel 1992: 86f).

The acceleration of urbanisation appears to be markedly increasing in the later phases of the Late Roman period, especially in the 3rd and 4th centuries (Strata 12 and 11), which ultimately represent a fundamental stage in the development of the site. In general, the entire site greatly expanded, with the construction of several domestic buildings and the reuse of some of the underground spaces. Economically the relevance of transhumance and pastoral economies seems to have reduced drastically in favour of a more intensive – and presumably more economically valuable – agricultural exploitation

[155] As suggested by the presence of spinning and weaving implements on the southern slope of the acropolis (Mitchel 1992: 39).

[156] In this phase, it appears that a consistent transhumant population was still present and living in the caves, together with a more settled group, possibly the veterans moved by Herod living in the fort. Still a trend to sedentarisation is supposed by the scholar in the light of the increase in the percentage of cattle and donkeys, and also the emergence of raising pigs (LaBianca, Hubbard, and Running 1990: 198). According to Mitchel, the transition from Stratum 15 to Stratum 14 was smooth and not "traumatic" (Mitchel 1992: 39), in contrast with to the transition to the Late Roman Stratum 13 is in fact a bit more disrupted than the previous one, with several and widespread layers of collapse, maybe due to an earthquake (Mitchel 1992: 75). Moreover, the underground structures fell out of use in Stratum 13.

[157] The nature of the occupation of the acropolis at this point is extremely problematic, and it is not clear from the archaeological reports if the fort really ceased to end in this phase or in a later moment immediately before the construction of the temple in Stratum 11, see below.

of the region.[158] Some elements though suggest also a real change in the status of the settlement, with the official recognition of an urban role.

The monumentalising of the settlement is the first hint pointing to a renewed importance in the regional context. The temple, of which only few architectural remains are left under the church, but represented on the *recto* of the locally minted coinage and in two mosaics,[159] is dated to this period along with some other buildings occupying the acropolis and also the monumental stairway conducting to it from the so-called "Plaza". This area also underwent some changes, contextually with the monumentalising of the access to the acropolis, being expanded to the detriment of the "Inn", which was therefore dismantled during Strata 11 (end 3rd -first half 4th centuries). Even more important is the presence of a mint, active at least under the reign of Elagabalus between 218 and 222, who made Esbus a *municipium* called *Aurelia Esbus.*[160] This prosperous condition seems to continue into the Early Byzantine period in the first half of the 4th century, making the transition between the two phases quite difficult to distinguish clearly archaeologically.[161]

The transition between Strata 11 and 10 is marked by some destruction layers across the site, including the temple on the acropolis, probably determined by a seismic event (ca. AD 363 or 365). Nevertheless, no large-scale construction activity is described by Storfjell for the middle 4th century, making him suspect that the damage for the entire settlement was not so severe to represent a proper breakthrough for its occupation.[162] Building activity continues to be limited to few areas of the site, namely the further expansion of the "Plaza" - which third and last resurfacing has a *terminus ante quem* of the 375 (Storfjell 1983: 47) - and a hypothetically reoccupation of a possible Roman military installation in area C, where the presence of six groups of loci may be associated - without any new construction - with either domestic and gardening activities (Storfjell 1983: 42). The heavy use of the Plaza is considered by LaBianca and Storfjell as clear evidence for the complete shift towards a trade economy, with a local production of crops that produced a surplus and forced the settlement to open to the market (Storfjell 1983: 53).

[158] LaBianca suggest a clear process of polarisation in the population, with some groups migrating into the larger regional settlement like Esbus, while others preferring to a return to pastoralism (LaBianca, Hubbard, and Running 1990: 200)

[159] The first one is in Umm er-Rasas, in the famous St. Stephen Church; and the second one is in Ma'in. The features in all these representation of the temple - including the coins - are similar: the building is set on a central pillared platform with four columns, with a façade which incorporated an arch between the innermost columns ("Syrian" Arch, according to the definition of Boethius and Ward-Perkins or "arcuate lintel" after Prince and Trell).

[160] In this period, it is documented a general trend in the increase of local mints in this region (Mitchel 1992: 104f). During the 2nd century only Philadelphia and Gerasa had right to mint coinage, replaced afterwards by Adraa an Bosra - even if their activity overlaps for a certain period. In coincidence with Stratum 12 in Esbus - end of the 2nd and 3rd centuries - just within 100 km radius from the site six mints are active (Esbus, Dium, Philippopolis, Madaba, Rabbathmoba and Charachmoba).

[161] One potentially disorienting aspect of the local chronology based on Strata is its missing coincidence with the general chronological division for the region. Therefore, the proper Early Byzantine period should begin - according to the widely accepted - in the middle of Stratum 11. As Storfjell (1983: 26) affirms: "Late Roman Stratum 11 extended into the historical and archaeological Byzantine period, but without a change which would warrant a new stratum. There was no new Esbus which appeared in AD 324".

[162] The passage to Stratum 10 is nevertheless justified. There was in fact "enough of a break due to earthquake activity to justify a new stratum, although it is a new stratum where the culture and life of the previous period seems to continue almost uninterrupted" (Storfjell 1983: 42).

Despite the evident prosperity of the site, no attempt to reconstruct the temple on the acropolis was made, and in the Stratum 9 (408-527) some of its remains were used to build the Christian basilica.[163] In general, this represents the most intensely occupied phases of the site's history, with a further occupation of the area of the Plaza, and the construction of new structures and possibly storage facilities. The expansion of the population also required the creation of a large reservoir in the direct proximity of the today's enclosed site – even if its dating is uncertain between Strata 10 and 9.

The same pattern of urban expansion characterises Stratum 8 (427-614), even if some areas of the site might have undergone some destruction, possibly still related to an earthquake in the first quarter of the 6th century. Once again, the summit of the acropolis is one of these areas, and an intensive rebuilding activity is documented for the church. Nonetheless, a change in the economic orientation of the site may be indicated by the ceased activity in the "Plaza", or at least a shift in the functional zoning within the settlement, with new areas dedicated to the commercial life of the site (Storfjell 1983: 171f). The relative prosperity of the site is still sustainable, as demonstrated by the construction of a second Byzantine church and possibly the construction of a third.[164] On the contrary, one can argue that a shift in the economic assets in the site and the region occurs, with a "revival" of a more subsistence economy and a return to pastoralism, or at least a less trade-oriented production.[165]

The status of the Late Byzantine site is quite difficult to determine, even if a prominence in the regional context can be quite confidently stated, especially for the first part of this period – corresponding more or less to the Stratum 9. In fact, the presence of a bishop in Esbus is documented at least for the 5th century,[166] and also Eusebius of Caesarea mentions the site as capital (*'epismos polis*) of a district (Eusebius, On. 84, 1-6; see also Ferch *et al.* 1989: 14). The later phases signify a change, though it cannot be determined if the economic shift also resulted in the reduced importance of Esbus in the regional administrative framework. Surely the hypothesis suggested by Storfjell that the Persian invasion determined a gap in the occupation of the site is not supported by the material evidence, since no destruction levels can be identified[167], but simply a slowing down or a virtual absence of building activity – documented only in the south-eastern quadrant of the Acropolis. Stratum 7 (AD 614-636) therefore does not represent a decisive change from the previous phase, and in addition the transition to the Early Islamic period does not appear to be destructive or marking a cultural shift, simply bearing

[163] "The lack of significant building activity at Tell Hesban during Stratum 10 period can be interpreted as reflecting the existent tension between the Christian Church and the Graeco-Roman religions" (Storfjell 1983: 53). Even if plausible, there are in my opinion too few evidences to totally support such hypothesis. Surely enough, though, the site benefitted of an economically florid situation, as mentioned before.

[164] In this case, a mosaic floor poorly repaired was found in a probe in area G. The interpretation of the structures is therefore not entirely certain (Storfjell 1983: 176). Storfjell also suggests that the poor reparations of these mosaics are a further prove of a weakened economy in the city, but I would argue it is not a decisive prove in this direction.

[165] The creation of the large reservoir is in itself not a clear mark of a predominant pastoral strategy: such structures may be found also in heavily exploited agricultural lands, especially if inserted into a larger irrigation system. Nevertheless, LaBianca and Storfjell are confident in suggesting that this moment represents a shift "from a cash-crop economy necessitating trade to a more self-reliant semi-pastoral economy of sheep and goat raising" (Storfjell 1983: 176).

[166] Bishop Gennadius of Esbus appears in the list of the participants at the Council of Nicaea in 325 and Zosus of Esbus in the ones of Ephesus in 431 and Chalcedon in 451 (Ferch *et al.* 1989: 5).

[167] Storfjell suggests actually that the takeover of the city by the Persian did not imply a widespread destruction of the site, but their destructive activity focused exclusively to the ecclesiastical structures (Storfjell 1983: 183). I have not found any evidence in the archaeological reports of a widespread destruction at least on the Acropolis that may justify such interpretation.

witness to a possible continuation in the reduction of the size of the settlement (Storfjell 1983: 182-184).[168]

In general, the Early Islamic phase is difficult to understand from the available publications for the site, and therefore the role of the settlement in the larger administrative context is not surely identifiable: the continuity of occupation, though bearing witness to a contraction, seems to be confirmed by the archaeological investigations through ceramic finds. If the continuity was also in the administrative status as "urban centre" is not verifiable, though it seems unlikely. The occupation of the summit of the acropolis is confirmed, even if undergoing a clear functional and physical change. In particular, it seems that it was converted to an outpost once again, where later in the Abbasid period, even a small *hammam* was erected.[169]

One of the reasons why both the Early Islamic structures and those from earlier phases are not so easily identified is the overwhelming amount of buildings constructed during the Mamluk period, especially in the 14th century, which represents the second phase of high intensity configuration as described by LaBianca. Even if not considering the development of the site in this phase in detail, it would be worth mentioning two elements that have emerged from the most recent archaeological investigation. First is the development of the settlement following an official recognition as urban centre. In fact, Hisban appears on the historical and documentary sources as *madina*, equipped with a mosque and *madrasa* and being for a short period even a district capital with a small garrison, a governor, and a market. After the transfer of the governor, garrison and the market to Amman, the citadel is described as "ruined", however, the settlement still seems to have maintained its urban status ruling over 300 villages (Walker and LaBianca 2003: 446f; Walker 2013: 7). This new administrative role implied the reoccupation of the acropolis (Figure 77), which appears to have been abandoned in the Abbasid period, the phase to which the *hammam* is dated (Walker and LaBianca 2003: 454; Walker 2013: 6f). Once again, this area of the site was elected as an area to erect buildings of the power, in this case more political than spiritual, since the garrison and the governor were based here. It is interesting to note the close relationship that this distinct group of people, clearly belonging to a non-local elite that was established with the rest of the population. Archaeozoological investigation, for instance, clearly demonstrated that the food supplies were gathered from the village, where the worst cuts of meat were discarded. It is even possible though that two separated flocks – at least in the case of the predominantly consumed animal, the sheep/goat – were kept in the surrounding of the site: this could be suggested by the major percentage of pathologies due to intensive breeding in the samples from the citadel than in the village (Walker *et al.* 2014; Walker *et al.* 2017).

Therefore, it is quite clear that a sharp social distinction between the two groups in the population was established and remarked by a different physical location in the settlement and by a different quality in diet and ceramics that were used. It is not to say that all the wealth and luxury products were limited to the consumption of the elite settled in the acropolis, as the finding of other imported ceramics and fine glasses found in the 2016 season in one of the farmhouses seems to point to (Walker *et al.* 2017; Walker *et al.* LaBianca 2018). It would be interesting to verify if such a social configuration was active in the Roman and Byzantine period, when the city of Esbus was clearly an administrative regional centre

---

[168] In general, in all the four Strata no clear cultural shift is identifiable (Storfjell 1983: 185-193).

[169] "Medieval Islamic "Ḥisbān" was a fortified administrative center that was transformed physically and functionally from the Umayyad period on. What now appears to be an Abbasid-era bathhouse may have constituted part of a much larger complex that included a building located along the northern sally gate to the summit. These may be the remains of the fortified Umayyad outpost that was the scene of a revolt against the Abbasid family, according to medieval Arab sources" (Walker 2013: 6f).

Figure 77 - The acropolis of Tall Hisban, with the rest of the Church on the left and the Mamluk citadel on the right (photo by the author)

of a certain importance, with a distinguished Hellenised elite group and an indigenous population still connected with the more pastoral component of the society or at least able to reconvert its economic strategy and switch back to less settled solutions. The constant flux between different grades of sedentarism and pastoralism in the course of Hisban's history is a clear indication that the ties with the transhumant strategies were never totally forgotten by a part of the local population, supposedly the one less engaged in the administration or less assimilated by the urban elite.

However, just as for the Roman and Byzantine period, the entire settlement benefited by the renewed centrality in the regional context in the Mamluk period, undergoing a clear intensification in the occupation of the site. This determined an almost systematic reuse of older structures, in some instances simply restoring collapsed or out-of-use buildings, while in some other cases building them anew. Recent excavations in the village at the foot of the hill of the acropolis showed that the series of farmhouses dated to the Mamluk period insist on previous buildings, keeping almost an identical orientation. It is therefore not unreasonable to think that the spatial conformation documented for the Mamluk period, with some clear clusters of houses, recalls to a certain extent, and on a more reduced scale, the organisation of the Byzantine city.

Given this historical background, what can be ultimately being said about the organisation of Esbus in the Byzantine and Early Islamic period? Despite the limited extension of the architectural remains of the site, it will be still possible to draw some conclusions, even if highly hypothetical, combining the few indications gathered by the spatial analysis of Hisban with the clues offered by the other case studies and the Food System theory. There are no clues on the exact extension of the settlement in any of its occupational phases (Figure 78a). At the same time, it is certain that at least in the Byzantine period it was much more extensive than the fenced site today, as confirmed by the presence of a church among the houses of the modern village to the north. Moreover, several remains of structures are still visible on the eastern slope of the wadi. Likewise, Qasr Nabulsi is most probably reoccupying older structures (Figures 79 and 80). The presence of Late Roman and Byzantine tombs on the slope opposite to the citadel suggest that the main settlement was developing only on one side of the wadi.

It is then needless to say then that our understanding of the general urban frame and spatial organisation is highly partial, and ultimately, it is limited to the areas re-occupied during the Mamluk and Ottoman periods, namely the citadel and the immediate surroundings.

Despite this drawback, the morphology of the site necessarily influenced both the shape and the organisation of the settlement. As far as the surviving architectural remains currently indicate, it was at least based on the dichotomic presence of one acropolis and a larger settlement at the bottom of its slopes. As mentioned above, these areas had been heavily reoccupied during the Mamluk period, when the site experienced its second occupational apex, though not reaching the level of the Byzantine phase. What it's extremely interesting not only for Hisban in particular, but for the sites that testifies a later medieval occupation, is the tendency of reoccupy the exact same structures built in earlier periods, with limited changes in the rooms' arrangements and in the orientation of some of the buildings. Thus, based on this assumption, the town of Esbus was also likely formed by a series of cluster of farmhouses, similar to the ones seen in other case studies. One peculiarity of this settlement, though, it that the hilly morphology of the site forced to create a series of terraces, particularly visible looking at the section of the Tell and in particular in Area C (Figure 78b and 78c). Nonetheless, this area showed that probably a series of "empty spaces" were spared. This use of apparently blank areas in the urban area of Hisban could possibly recall the presence of enclosure and "blank spaces" seen for instance in the southern Hauran cases: spaces that today may look empty, but that nonetheless show from aerial photos forms of enclosures and subdivision, suggesting some sort of functional specialisation. It would not be surprising in Hisban either to document the presence of some small spaces for cultivation or even for the gathering of small flocks, especially if considered the presence of pastoral groups in the area.

One further aspect that surely had an important influx on the settlement organisation, is a complex system of caves, which had been alternatively dedicated to different uses through time, such as animal keeping, water collection, storage, and even residential purposes. Seeing the continuity in their use and the complexity of this underground system – especially if considered also the investment in keeping it working and useable – it is clear that these caves played an important role in the life of Esbus. In some cases, they are ultimately combined with standing structures on the ground, with which they were clearly integrated within a unique settlement frame.

The Byzantine portion of the settlement still visible today also seems to suggest a sort of functional differentiation of the spaces, not limited to the obvious specialisation of the acropolis, occupied either in the Roman and in the Byzantine periods by the religious complexes. The top of the acropolis and its southern slope clearly show signs of monumentalising, representing the public heart of the city. The large stairways leading to the summit starts from the so-called "Plaza", a large space which function is not clearly understood, but that probably functioned as commercial hub. Some structures that could have functioned as storage facilities face this area, which in the earlier phases of Hisban was occupied

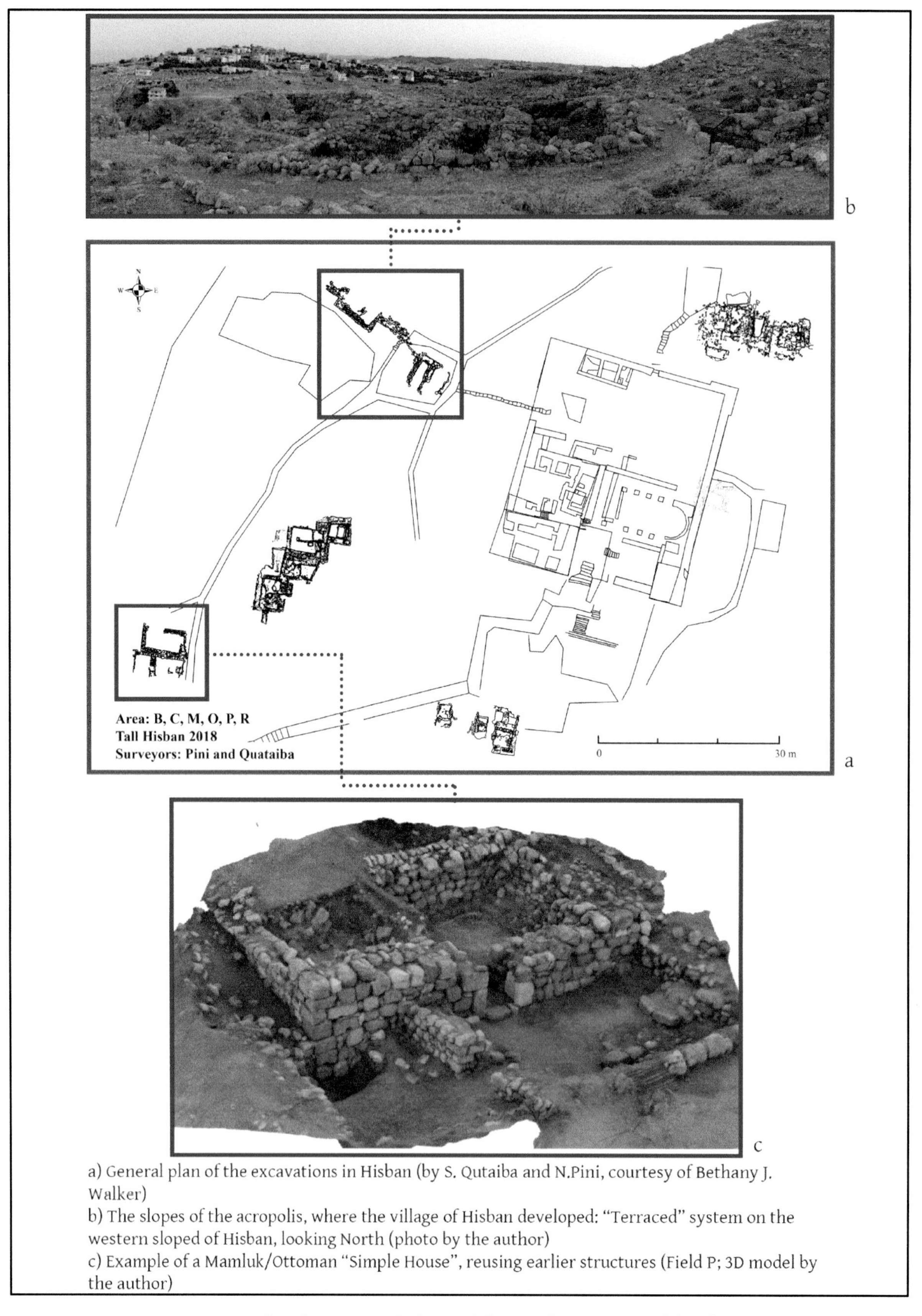

a) General plan of the excavations in Hisban (by S. Qutaiba and N.Pini, courtesy of Bethany J. Walker)
b) The slopes of the acropolis, where the village of Hisban developed: "Terraced" system on the western sloped of Hisban, looking North (photo by the author)
c) Example of a Mamluk/Ottoman "Simple House", reusing earlier structures (Field P; 3D model by the author)

Figure 78 - Tall Hisban: general plan and forms of occupation of the slopes

Figure 79 - Nabulsi Qasr, from the summit of Tall Hisban, looking South-West (photo by the author)

Figure 80 - Reused building materials in Nabulsi Qasr (photo by the author)

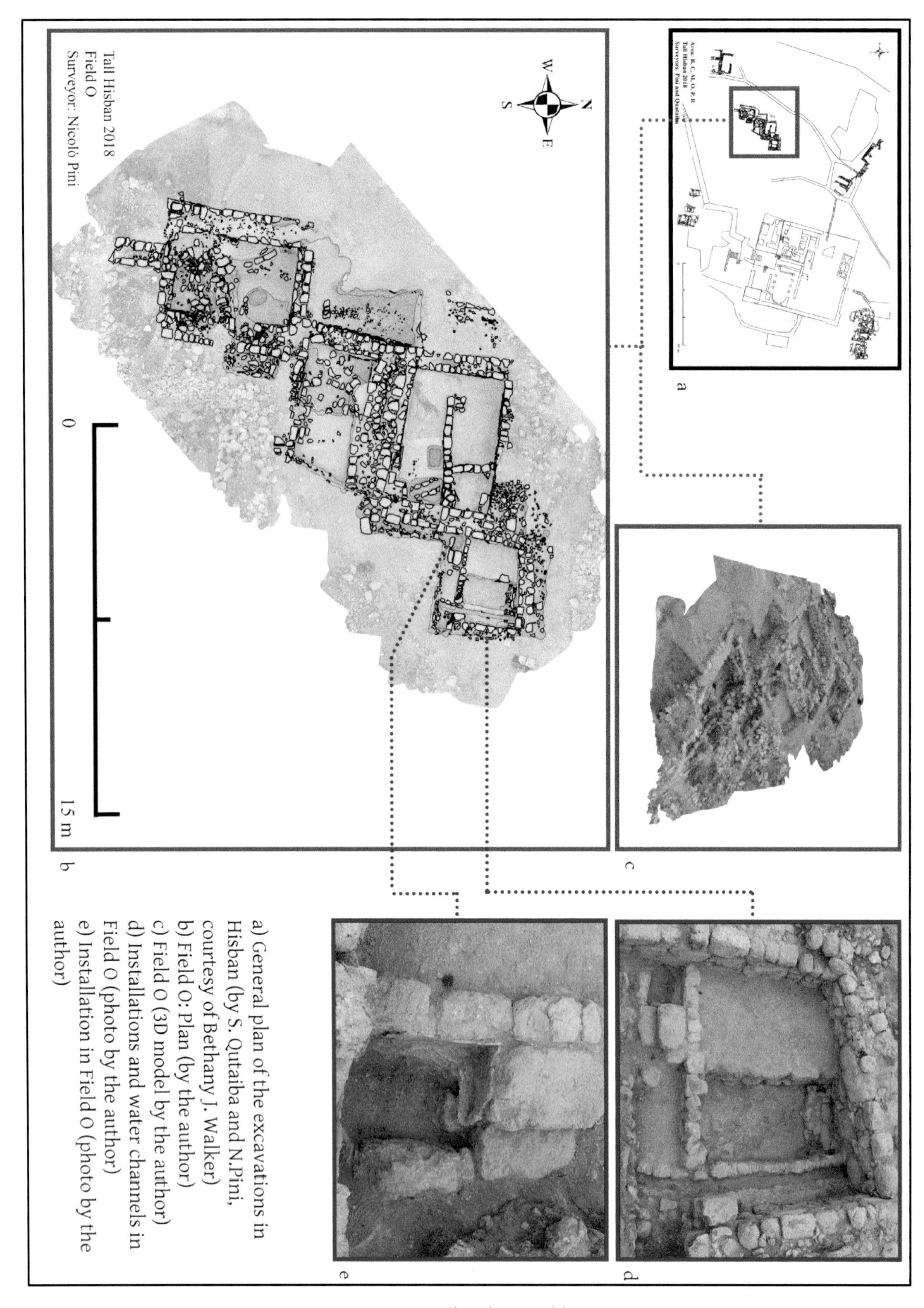

Figure 81 - Tall Hisban: Field O

by a *statio*, confirming the tendency of having a public use of this part of the settlement. Not much more can be said, apart from the fact that the recurring mix of functions seems to characterise the rest of the settlement. It is likely that only a series of structures directly connected with the acropolis served also as storage. For the rest, it seems that the house fulfilled most of the needs of their inhabitants, and that it is probably also true for other periods when the economical orientation of the settlement was more probably set towards surplus-production and trade.

The houses themselves - at least from the scanty information available and especially supposing that the Mamluk structures resembled a bit the earlier structures - appear to have been heavily influenced by the topography of the site, while never occupying extremely large surfaces, but at the same time benefitting of the huge underground system of spaces. The most relevant implication of the anthropological framework suggested by LaBianca on this reconstruction of Hisban organisation is considering it as a possible mark for "tribal structures" in the society, with the settlement following a spatial frame defined by blood-related or kinship groups. In particular, the clusters of houses that appear to pattern the Mamluk village (for instance Field O, Figure 81), but almost certainly also at least part of the Byzantine city, point in that direction. The centrality in the daily life of the family groups - more or less extended - as shown also for the other case studies, is clearly not determined exclusively by the sedentarisation processes involving the villagers,[170] but indeed is one further clue for the fact that the pastoral and settled components are indeed "enmeshed in a single system" (Cole, quoted in LaBianca, Hubbard, and Running 1990: 40).[171] The sharing of this social structure was ultimately the element guaranteeing the almost continuous shift from more settled to more pastoral solutions and the opposite, as material evidence, above all from the faunal and botanical evidence, seem to suggest for entire Hisban's entire history. It follows - and that also applies to the other case studies - that it would be highly limiting to consider the two groups as separated or eventually interacting only in light of purely economic and political interests. At the same time, it is also highly difficult to identify a single propelling force determining the development of a site or even its abandonment. The reference to the larger socio-economic and political context in which the settlement is included is therefore a key-element for any sort of historical conclusion.

In conclusion, despite the fact that the more visible remains today on the site are from a later period than the main focus of the present research, especially as far as the residential quarters of the settlement are concerned, some of the observations gathered could offer interesting insights for the organisation of the settlement and its socio-economic and cultural implications. First of all, the anthropological approach proposed by LaBianca underlines the constant oscillation between different

[170] "In other words, the presence of this form of organization among settled villagers may not necessarily be taken as a sign of their nomadic descent, as has been suggested by Patai (1970: 191). Instead, as Gulick (1971) has emphasized, farming and herding are symbiotically related *ecologies* between whom exist certain common forms of organization and beliefs, such as adherence to the practice of patrilineal descent" (LaBianca, Hubbard, and Running 1990: 39).

[171] According to the American scholar, this interaction between the different social components act of at least three levels: the village level, where "villagers interacts mainly with nomads at the subtribal grouping of the linage"; the tribal level, on which "nomadic tribe, as a grouping in itself, is involved in dealings with regional urban centres"; and finally, a national level, where "nomadic tribes provide the major military support of the nation-state". Even if this contribute clearly show its age, especially in terms of terminology, and I would say in particular for the conception of National level, it surely spots out one key-feature of the issue, namely the importance of the multi-level analysis. Different dynamics act accordingly to the context, involving different portions of the society and consequently determining a diversified range outcome in several aspects of the material culture.

economic strategies, which implied and determined changes in the settlement organisation itself. In particular, archaeozoological and botanical finds offers important hints to help reconstruct these dynamics, since the different proportion of sheep/goat – always the most frequent among the faunal rests – may indicate important changes in the pastoral component of the society and/or a more or less accentuated stress on subsistence in the agricultural and breeding production. In addition, some structures, not only in Hisban but also in the hinterland, may be associated with seasonal occupation in some periods, which has often been suggested and is likely the case for some of the caves. On the contrary, a more economically valuable agricultural production, in particular for the Roman period, is not only suggested by the new status and level of monumentalising the settlement reached at this moment, but also a relative reduction in the absolute importance of sheep/goat, in favour of other species possibly employed also for work in the arable fields.

All of these analyses, especially if with a strong statistical component, should be handled carefully, seeing the highly fragmentary picture got from the excavation and the limited portion of the site that was investigated until this point. Nevertheless, important information can be yielded by more recent periods, especially from the Mamluk period, as seen for instance in Hisban with the combined interpretation of architectural and excavation's data and the new archaeozoological and faunal analysis. The latter offered a confirmation of the agricultural vocation of the settlement, even if the mamluk strata yielded materials that are normally not so automatically associated with rural settlements, like imported wares and glasses that refer surely to more urban contexts. Moreover, during 2018 season a series of installations, probably related to a complex system of water channels, was brought to light in one of the clusters of farmhouses (Field O; Figures 81d and 81e): even though the chronology of the first construction of such features is still unclear, it seems that during the mamluk period an earlier basin was substitute by a second one, suggesting still a producive specialisation for the area. The relative prosperity of the site was reinforced by the presence of the garrison, with which the villagers surely were closely engaged either economically or socially: a situation that could had direct precedents in the history of Hisban. Moreover, the presence not only of urban elite but also of social formations or "tribes" with both settled and pastoral groups have been confirmed for this period, erasing the big question of if or how they can be possibly distinguished in the material culture and in the architectonical/spatial organisation of the spaces. If the two social groups occupied separated areas in the settlement in light of this different economic specialisation, or if the social ties prevailed – as I would suggest is more plausible, also considering the other case studies – is difficult to prove.

Moreover, in the case of Hisban there is a further demonstration of how frequently building materials were stolen from ancient structures in order to build the new houses. Even if this habit makes it more difficult to date some structures with certainty and simultaneously creates a lacuna in the earlier material remains, it also shows the tendency to reoccupy the same structures, possibly maintaining most of their original conformation. The recent excavations of some of the Mamluk farmhouses confirm that these structures are built on top of earlier structures, showing few differences on the orientation and tracks of the walls. As seen in the other sites, then, it appears that Late Byzantine structures were reoccupied or restored in this later phase; therefore, this does not represent a radical change in the organisation of the site. While caution should be taken, and the phenomenon must not be considered universally valid, it is simply a possibility that can be verified on a case-by-case basis. This especially applies to Tall Hisban, where a larger portion of the still "undisturbed" site is not excavated.

6.2: Umm er-Rasas

The case of Umm er-Rasas[172] is consistently different from the previous one for two main reasons: the typology of site and its development and its state of conservation. It is a village that never played an administrative role in the regional context despite a significant increase in size during the Byzantine period. The interest in the site resides in its particularly unique development, that ultimately determines one of the most interesting spatial characteristics of the Late Antique settlement. The original nucleus is in fact a Roman fort, *castron Mafaa*, where its defensive walls (150 x 120 m) are still clearly standing and dividing the inner quarter from the outer clusters of houses and churches, which radically changed the original configuration of the fort (Figure 82). Only a few indicators of chronology are available for reconstructing the history of the site. The record of the Nabatean funerary inscription of the *Strategos* Abd-Maliku, dated to the AD 41, points to an early occupation of the area, at least with a military presence that coincided with the later Roman fort (Kennedy 2004: 138). It would not be surprising, since the site is located in a contested area during the Hellenistic-Early Roman period like at Tall Hisban. Nonetheless, the excavations suggest that the original establishment of the fort visible today is dated to the 3rd-4th century, and the presence of the fort is confirmed also by contemporary

Figure 82 - Aerial picture of the fort and the later village of Umm er-Rasas (APAAME_20170920_RHB-0090. Photographer: Robert Bewley, courtesy of APAAME)

[172] The site was included in 2004 in the World Heritage List by UNESCO (Official web site: http://whc.unesco.org/en/list/1093), with the official designation: "Um er-Rasas (Kastrom Mefa'a)". In particular, "the Outstanding Universal Value of the site resides in the extensive settlement of the Byzantine/Umayyad period" and in particular for the artistic value of the mosaics found in the site and for the role of monasticism in development of the settlement.

historical sources, like Eusebius (*On.* 128.21-3 and 129.20), the Notitia Dignitatum (Or. 37.8 and 19: "*Equites promoti indigenae, Mefa*", "Advanced native Cavalry, at Mefa), and Palladios.

For later phases, important information is obtained from the inscriptions on the mosaics from the churches. This kind of chronological evidence, though, often represents a more general period of time, since they rarely identify the exact moment of construction, restoration or abandonment of the building they are found in.[173] Therefore, many of the religious complexes are still far from having a well-established (or even relative) chronology. On top of that, the use of relative dates from buildings constructed after the "Islamic conquest" (or at least after the last dates available through the mosaics) has been an extremely controversial issue. In fact, according to the dated mosaics, most of the churches in Umm er-Rasas were erected starting from the 6th century and continued construction well into the Early Islamic period, with restorations and additions at least until the 8th century (Gatier 2011; Michel 2011). While the last inscriptions in the mosaics provide at least a *terminus post quem*, one would be mistaken in treating them as the exact date of the renovations in the structures. The inscriptions date exclusively to the restoring and modifications of the mosaic floors, and are separate from the rest of the building, which therefore could have been modified also much earlier or later that the last interventions on the decorative apparatus. A further complication is the common phenomenon where the earlier churches were converted into residential spaces during the last occupational phase in the 9th-10th centuries before having the definitive abandonment of the structures and their subsequent collapse.[174] When considering that the conversion of churches, not only into mosques (like in Umm es-Surab), but also into residential or productive sites is fairly common (like in Tall Hisban) precisely dating these changes nevertheless becomes quite complicated.

The case of the complex of St. Stephen – the largest religious complex at Umm er-Rasas, counting at least four interconnected churches – is particularly interesting not only in demonstrating the issue of chronology despite the presence of absolute dates offered by mosaics, but in general for the development of the site (Figure 83). A precise relative chronology was established by analysing the architectural remains of the complex through the presence of at least three well-dated inscriptions; however, these inscriptions do not offer any consistent help in reconstructing the different stages of construction any more precisely (Michel 2001: 100-102 and 381-397). However, one earlier unidentified structure was clearly identified distinguished below the churches *de l'évêque Serge* and of St. Stephen, which later reused some of its walls. This is the first documented architectural activity in the area, which apparently was designated as the necropolis of the Byzantine quarter – developed outside the former Roman fort (Michel 2001: 101). In fact, several tombs were found underneath the earlier mosaic floors, opening to the possibility, as stated by Michel, of the presence of some "saints' tombs" justifying the development of not only the first structure, but also of the later complex. The connection with the cult of the saints is clearly attested by the presence of reliquaries.

The first church to be built in this space is the *Église de la Niche* (Michel 2001: 383-384), which is also ca. 2 m elevated in respect to the rest of the complex and is tentatively dated to the 6th century and remained in use until at least the 7th-8th centuries. No mosaic or inscription has been found in relation to this church. The second church to be added to the religious complex was the *Église de l'évêque Serge*

[173] As stated by Gatier (2011: 16-18), in regard specifically of the mosaics of the Islamic period, but extendible also to earlier phases: "Nous avons vu que les inscriptions des mosaïques d'époque islamique ne concernaient pas l'architecture, à quelques exceptions près [...]. Rien ne nous dit leur ampleur, la nature des travaux qui les concernent – construction, réparation ou adjonction – ni leur généalogie, la phase à laquelle ils se rattachent".

[174] This is true either for the churches inside the older fort and for the one in the so-called Byzantine quarter outside. For the analyses of the single churches, see Michel 2001: 379-418.

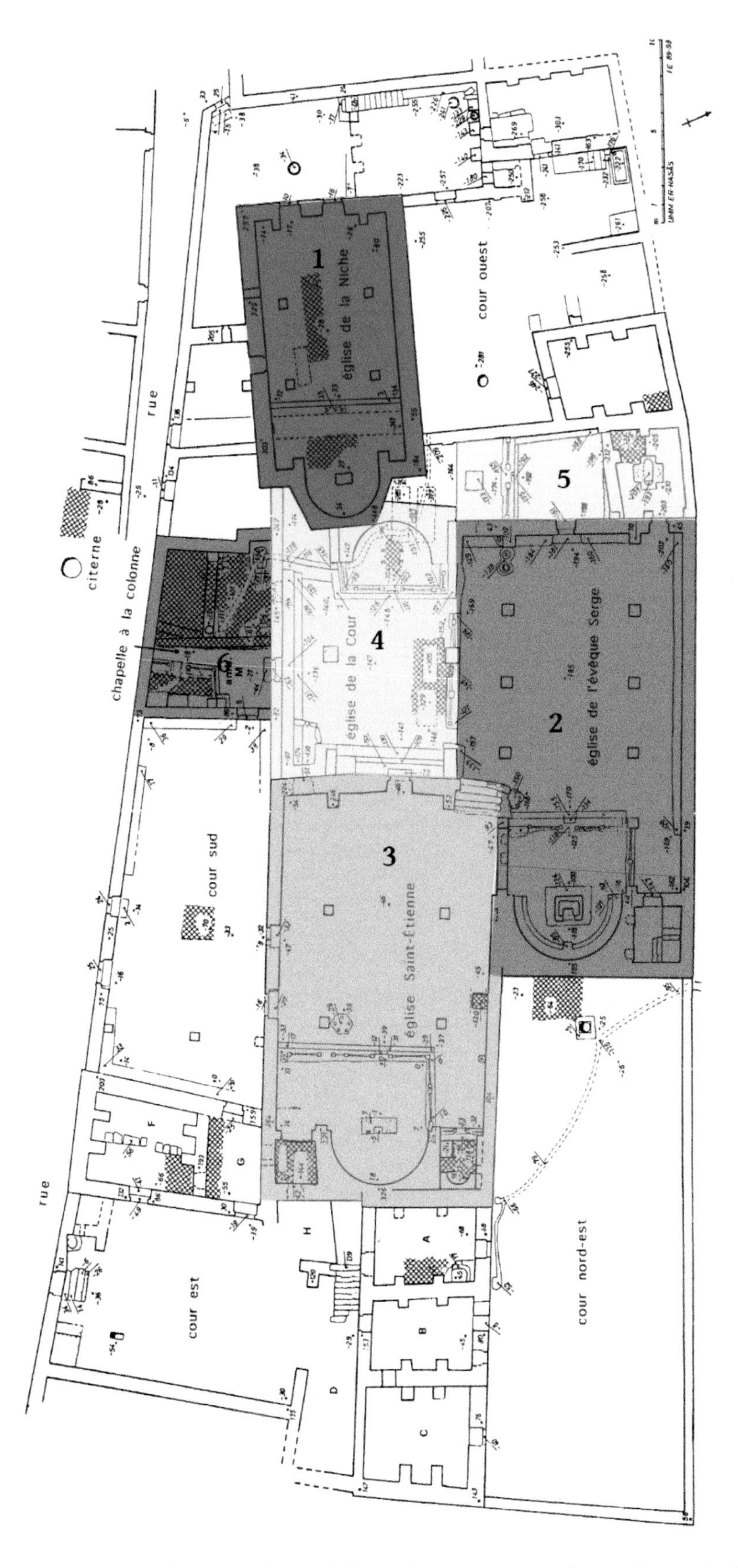

Figure 83 - Plan of the St. Stephen complex and chronologic sequence of the religious buildings (elaboration by the author on Michel 2001: 382, Figure 358)

(Michel 2001: 384-388), partially integrating the first structures. One of the inscriptions in the decoration offers the earliest absolute date of the entire complex, with donation of the first mosaic floor by the Bishop Sergius in AD 587, providing also a *terminus ante quem* for the construction and first development of the church itself. All of the constructions that follow are relatively dated by the ceramics and the relationship between the different units in the complex.

The first important change in the structure was the destruction of the southern corner of the apsis area, following the construction of St. Stephen church, with which the complex was connected through stairways. Even if no clear date for this activity is available, it postdates AD 587 and precedes AD 718, the date of the earliest mosaic in St. Stephen. Further major changes were made between the 7th and the 8th, though the precise order is not yet possible to determine. On the one side, the dependences at the western extremity of the building – the baptistery and the chapel – were added; on the other side, the construction of the *Église de la Cour* (see infra) resulted in the complete rearrangement of the inner circulation, with the opening of a direct passage with the new unit on the southern side of the building. The entire building appears to be occupied for the last time during the Abbasid period, with the evidence of reparations of the mosaics after iconoclastic damages (dated in the region to the 8th century) and of a small wall associated to 8th-9th centuries ceramics in the Baptistery.

The third church that was constructed, being the largest and possibly the most well-known one, is the St. Stephen's church (*Saint-Étienne* in the French literature; see Michel 2001: 394-396), where the famous mosaic representing 16 cities of the Byzantine Near East occupies the intercolumniation.[175]This building is set upon the first structure built in the area and largely reuses its structures in the northern and eastern part. The precise moment of construction of the new church is uncertain, but as previously mentioned, has to be dated between the 587 and the 718,[176] when the earliest dated mosaic floor was put in place. The creation of the mosaic floor does not seem to follow immediately or be contemporary with the first phases of the church, plausibly dating to the end 6th-7th centuries. A further consistent modification of the church is dated to the 756, with the elevation of the choir and the realisation of another mosaic floor, indeed dated to that year. Evidence of reconstruction of part of the decorations following the iconoclasts of the 8th century, and also changes to the inner circulation of the building (with the blockage of southernmost entrance in the façade), suggesting an abandonment of the church only in the 9th century, as is confirmed also by ceramic finds.

The last church to be built, the *Église de la Cour* (Michel 2001: 388-394), has a peculiar conformation, mainly due to its position within the complex, communicating directly to the north with the *Église de l'évêque Serge* and to the east with St. Stephen's, while simultaneously flanked to the west by the *Église de la Niche*. Its later date, namely the end of the Umayyad period, is suggested not only by the chronological relationship with the surrounding buildings, but also by the ceramic evidence. Moreover, the recovery of destroyed rests of decorations from the other churches, possibly result from the iconoclastic phase, also found in the "Chapel M" (called "*à la colonne*"), as suggested by Michel to propose a construction date later than the first quarter of the 8th century. The abandonment of the structure wasn't much later, taking place possibly at the end of the same century or at the beginning of the following one.

---

[175] Among the settlement represented, there are Umm er-Rasas itself and Esbus.

[176] A debate on the reading of the inscription recording such date had been going on, since the mosaic itself surely underwent modifications at an earlier stage. The hypothesis of the dedication of the inscription in 718 – therefore implying a wrong restoration of the original mosaic into AD 785, the other date proposed for the mosaic – seems to be the most likely: see Michel 2001: 392.

Other than the four churches, the complex was also joined by other structures than the four churches, such as by a series of courtyards, dependences and smaller chapels, such as the aforementioned "Chapel M", which was possibly one of the last occupied areas in the entire complex, even if not for a religious purpose but as a stable.[177] A similar pattern emerges from other religious buildings spread throughout the entire settlement, attesting a construction in the 6th century, with an almost constant renovation in the 7th and 8th centuries, and a final conversion into residential units before the abandonment in the 9th and 10th centuries.

How much this development of this complex is representative of the most general history of Umm er-Rasas is still yet to be determined precisely. One can nonetheless confidently state that the settlement underwent an important development during the 6th and 7th centuries, as testified by the widespread construction activities and the dedication of several mosaics. The economic implication of pilgrimage and the martyr cult should not to be underestimated in this context. Two aspects though are particularly relevant, even if only partially related to the purely urban development of the site. One is the unusually late start of the phenomenon: dedication of mosaics and most importantly the construction of a large number of churches, even in small settlements, is a common process already at the end of the 5th century, almost one century before Umm er-Rasas (Michel 2001: 100-104). The second aspect is its perdurance during the first phases of the Islamic period, with the continuation of the Christian use of the religious complex well into the 8th century. This is surely an important factor to be kept into consideration for the following discussion, especially as far as the identities of the local communities are concerned, and in particular their role in shaping, maintaining and transforming the built environment.

As far as spatial analysis is concerned, this complex is favoured by a better state of conservation of the site, at least outside the perimeter of the original castrum, for which a survey showing also the precise configuration of its internal space is nonetheless available (Figure 84). The presence of these two clearly distinguishable areas – the former castrum and the village around it – is the most visible feature of the settlement. Beyond this feature, the rest of the settlement does not present many similar characteristics to other case studies considered here, on all the three levels of analysis. Nonetheless, a clear distinction can be made between the two areas that do present clearly a different organisation, mainly because of the previous structures.

In the castrum, the original configuration is completely revolutionised, with only a section of the *via principalis* still visible and possibly the space occupied by the *praetorium* recognisable near the twin churches. For the rest of the structure, the spatial organisation is completely new, and features of the former castrum are no longer visible, apart from the outer enclosing wall (Figure 85). Surely, the presence of an earlier structure and most importantly a clearly defined and enclosed space are the two elements determining the peculiar new organisation; these elements do not have any direct parallel in the other case studies. Two case studies have some similar elements. At Mampsis a sector of the settlement was enclosed and separated physically from the rest of the town, while at Umm el-Jimal possibly another example of former fort converted into residential or productive area. In Umm er-Rasas, however, at least two key features are different: the dimensions of the quarter and the inner compactness.

[177] A coin date to the 9th-10th century was found in the abandonment level in the room (Michel 2001: 394-397).

Figure 84 - Plan of Umm er-Rasas fort (elaboration by the author on Wirth 2002: Figure 17)

Figure 85 - The northern wall of the Roman fort, facing the Byzantine village; looking west (photo by the author)

Only one clear street, assuming almost the form of an inner ring, is left free, with few *cul-de-sacs* developing from it and leading to the private accesses of some buildings. Furthermore, no clear distinction in functional compartments is recognisable: a series of units can be possibly interpreted as shops, only because of their single-room plan. No excavation has been conducted inside the castrum apart from some of the Churches, immediately in front of a visitor entrance in front of the quarter from the western gate. Next, a large religious complex, with the twin churches in the centre of it, develops in the northern side. For the rest, all the residential and commercial/productive functions seem to be mixed as seen in the other case studies.

It is also clear that the compactness and the peculiar conformation of the quarter determined a different spatial setting of the single buildings, not entirely matching any of the typologies described above. As far as it is possible to understand from the only plan available, it could be said that there is a tendency to have "courtyard houses" – following the conjunctive models – all with relatively small dimensions. The courtyard of the castrum occupied the centre of the complex and was accessible through a *vestibulum* or a series of rooms opening to the street. Still there are some examples of houses – or otherwise functionally destined buildings – that have their court opening directly to the public pathway. In general, the castrum of Umm er-Rasas shares some features and spatial models seen also in the Negev settlement, which is not surprising since compactness was one of the key features of those case studies, which ultimately makes them the closest examples to what had been long defined as the archetypical "Islamic" city.

On the other hand, the quarter outside the castrum is directly reminiscent of the other form of settlements found in the Hauran, especially the southern part. It is spontaneous and less compacted, with clusters of buildings grouping up without forming proper quarters. One element is striking, and it's probably to be seen in connection with the economic context of the settlement itself: the large concentration of churches and monastic complexes probably occupies a larger portion of the outer village than the purely residential units (Figure 86). It is clear that the village had a valuable economic income thanks to pilgrims visiting the site. At the same time, the presence close to the village of the "Stylite" tower used by eremites, which is located at the centre of the courtyard in front of the "Church of the Tower", reinforces the impression of a deep influence of the religious component in the daily life of the entire settlement.

As far as the spatial organisation is concerned, the features in the outer village are likened to those seen in the villages in the Hauran on all levels, especially when the house structures are concerned. The farmhouses, in some cases are clustered in blocks of two or three, entirely match the examples described for the other case studies. The differences between the two quarters in Umm er-Rasas depend to a large degree on the condition of their development. It is not unusual for forts or even monasteries to stimulate the development of settlements around them, likely as a matter of higher security ensured by the local garrison, but most likely for economic opportunities. The change in the function of the castrum, the precise date of which is unknown, did not involve a change in this pattern, and ultimately the presence of the large religious complexes in the outer village, contributed in keeping both the quarters alive and socio-economically interconnected.

Figure 86 - Major religious complexes in the Byzantine village of Umm er-Rasas (elaboration by the author on APAAME_20170920_RHB-0090. Photographer: Robert Bewley, courtesy of APAAME)

Part 3: Archaeological data in the bigger picture

Chapter 7: The regional historical context

The comparison of the case studies brings focus to the organisation of the sites from a purely architectonical perspective and highlights their unique urban development, and in particular the influence that different social structures might have had on the various levels of those settlements. Through this analysis, marked regional particularism could be revealed; for instance, a clear distinction on the most general level of analysis (the entire settlement) could be made between the Negev and the Hauran settlements, the former being a compacted and more urban form and the latter being open and scattered. The Central Jordanian case studies, by contrast, represent a more intermediate position on the spectrum between them. These differences can be seen as reflections of different ideas and choices made by respective local communities. Whereas communal and civic identity are more evidently attested in the Negev sites, other evidence supports the separation and nucleation of single quarters in the Hauran that could be indicative of a predominant role of family groups in daily life. Although, at the same time, similar family structures are reflected in the house plans of all case studies discussed in this study. In order to explain the different spatial solutions and to shed light on the processes that determined their developments, it is crucial to contextualise the sites and look at the bigger picture.

In order to fully understand the context in which the general Byzantine and Early Islamic settlement pattern and likewise the specific spatial organisation of the case studies developed, it is essential to consider the *longue durée* of some of the social phenomena. Many of the processes, such as the interaction between mobile groups and settled communities, might be more evident in the findings of this period, but are also still recurrent in other chronological phases.[178] It has become clearer that, as is also evident from the case studies presented here, in the first half of the 6th century the Byzantine Near East experienced an astonishing expansion of settlements, both in number and their dimensions. Nonetheless, the regions where these centres developed were rarely newly settled: both the Negev and the Hauran, not to mention the Madaba area, show an extremely long occupation history, starting at least from the Hellenistic period – and in several cases even from proto-history onward.

Likely, the region of Bosra benefitted from the establishment of the new capital of the Nabatean Kingdom in AD 93, probably triggering the development of rural settlements. Similarly, the presence of the Nabateans is attested elsewhere in Jordan and in the Negev in relation to trade routes. Especially in the Negev, some settlements are normally considered as Nabatean foundations linked to the so-called "spice route" from the Arabian Peninsula towards Egypt and the Mediterranean, developing in some cases from forts and/or caravanserai, like in Oboda or possibly Mampsis. On the contrary, the Madaba region was a border area between the Hesmonian and the Nabatean Kingdoms that contended for the direct control over the area. Hisban and its citadel (but also the clear reference to the Nabatean military corps in Umm er-Rasas) are in some ways a reflection of this political instability. The arrival of the Romans encouraged the further develop rural areas, being facilitated by a smooth transition from Nabatean rule, which were left to Rome in the will of Rabbel II after his death in AD 106. The effects of Roman control are particularly visible with the implementation of a widespread road network.[179] The

[178] Indeed, the elements necessary to grasp entirely the *long durée* of some phenomena will be considered later in detail in this chapter. For a comprehensive and purely historical overview for the region in the Hellenistic and Roman periods, some useful references include: Bowersock 1994; Ball 2000; Sartre 2005.

[179] The better-known case is the *Via Nova Traiana* that nonetheless follows the track of the North-South axis of the Nabatean Kingdom – the "Kings' highway". The major development though can be seen on the margins of the steppe, in particular from the 3rd century AD: the foundation of new forts like in Khirbet as-Samra and also Umm

development of rural areas, parallel to the widespread monumentalising of major cities in the 2nd century, was probably accelerated by the newly assured security guaranteed over these regions: the cases of Hisban and the Negev towns seem to testify that a clear conversion towards a monetary economy, with the new focus on the production of cash crops (in particular olive and grape, most likely also grain in Hisban).

The Byzantine expansion in the hinterland is not quite that simple to explain: it is likely that a coincidence of positive economic, political and religious factors had a central role in this phenomenon, together with favourable environmental conditions in the 5th century with a sequence of shorter droughts and longer wetter and warmer periods (McCormick *et al.* 2012: 188). Nonetheless, recent paleoclimatic studies (Büntgen *et al.* 2011; McCormick *et al.* 2012; Büntgen *et al.* 2016; Fuks *et al.* 2017) reconstruct a particularly complex environmental picture, also potentially affecting settlements patterns, especially in the 6th century. Severe droughts are documented (also by written sources) between 523 and 538, consistently outnumbering episodes of precipitation; this could have had an effect on the movement of pastoral groups from outside the Empire (McCormick *et al.* 2012: 197). More dramatically, a series of volcanic eruptions (in particular, an "explosive near-equatorial" eruption in AD 536; Büntgen *et al.* 2011: 580[180]) appear to have had serious consequences on the local climate in the Near East, causing a drastic fall in temperature due to the decline in solar irradiance (especially between AD 536 and 545) and a significant rise in precipitation (Büntgen *et al.* 2016: 233-234; Fuks *et al.* 2017: 5-6). Even if the consequence of this "Late Antique Little Ice Age", spanning from ca. AD 536 until 660 (Büntgen *et al.* 2016: 233), could appear positive for desert or semi-desertic regions, the paleoclimatic analyses suggest that on the contrary these climatic conditions, because of their stormy nature, negatively affected agricultural infrastructures and the cultivation of crops (testified through the reduction of the pollen of cash crops: Fuks *et al.* 2017: 8-9) and simultaneously triggered severe cases of soil erosion (Fuks *et al.* 2017: 11-13). Clearly, this environmental stress was combined with other factors, such as epidemics, such as the famous Justinian Plague in 541-542, and famines, a total of eight occurring between the 536 eruption and the Justinian pandemic (McCormick *et al.* 2012: 198), in a tragic cycle that was well known to ancient and pre-industrial societies (Büntgen *et al.* 2016: 234; Fuks *et al.* 2017: 6). The local population also suffered the consequences of political instability, notably following the Persian Invasion and the Islamic Conquest, which in turn may have been aided by the fact that opponents were debilitated by the sequence of famines and epidemics (Büntgen *et al.* 2016: 234-235).

Yet, architectural remains are not always precisely dated in their entirety. When considered together with the archaeological record, the evidence tends to offer a less stark and catastrophic picture: it is surely possible that some settlements suffered a contraction in their population, although they continued to be occupied albeit on a reduced scale. In general, archaeological reports for the study region still suggest a continuity in settlement even after these "natural" catastrophes, showing that the local communities were resilient enough to cope and recover from these environmental stresses. For instance, recent investigations in the Negev demonstrated that most of the terrace systems and agricultural interventions are dated to the Byzantine period and underwent constant repair and

er-Rasas are possibly related to these investments, aiming also to secure better the liminal regions of the Empire. For a comprehensive summary of the evidence of the Roman army in Jordan, see Kennedy 2004.

[180] Ashes associated with this eruption have been documented archeologically on several sites as well deposed ash layers in the stratigraphy (Fuks et al. 2017: 5-6); written sources also report solar veiling and consequent crop failure in 536 and 537 (McCormick et al. 2012: 195).

maintenance over a long period, being kept in use well into the Islamic period, and some even until the 11th century (G. Avni, Porat, and Y. Avni 2013: 340-344).[181]

Even if strong regional patterns can be recognised (Fuks *et al.* 2017: 13), the Near East as a whole appears to be one of the most (culturally) resilient areas in the Mediterranean (Harper 2017: 270): when considering parallels in other geographical contexts, it seems that when faced with "unpreceded environmental stresses, cultures can shift to lower subsistence levels by reducing social complexity, abandoning urban centres, and reorganising systems of supply and production" (deMenocal 2001: 672). This could to a certain extent explain the parallel increase in rural settlements and the changes taking place in the cities, as well as an increased relevance of pastoralism and more mobile strategies, involving at least sections of local communities (see infra).

A similar argument can be made for the transition to the Early Islamic period. Until thirty years ago, it was common belief that the establishment of Islamic rule marked a clear break in the history of the Middle East, following the idea (based mainly on written sources) that the Late Antique and the beginning of the Middle Ages represented the moment in which the "Old Roman" world finally collapsed and was radically replaced by new cultures, societies, and political systems. Specifically, for the Near East the so-called "Arab Conquest" of the Eastern part of the empire determined not only a new political administration, but also a change in ethnicity. Consequently, the original population was either forced to migrate elsewhere or slaughtered, one way or another leading to the complete abandonment of several settlements. This historical reconstruction had been heavily challenged by recent studies, pointing out several aspects of continuity, thanks to the combined increase in regional surveys and better chrono-typologies of pottery and other classes of findings (Walmsley 2007: 48-70).[182] A general and widespread tendency towards a continuity in the occupation of many settlements had been confirmed: sites continued to be occupied with few changes and without undergoing destruction identifiable as a consequence of the so-called "violent" conquest by the "Arabs" (with few exceptions, as is the case in Jerusalem). Similarly, material culture neither attests these transitional phases nor suggests a newly arrived population that was culturally distinguishable from the "native" one.[183]

---

[181] In some cases, a reuse of the same installations is documented for the Mamluk period and more recent times by nomadic pastoral groups. In general, it appears that the abandonment of fields and installations was caused by a series of political and economic factors rather than exclusively environmental agents. Perhaps a prolonged exposure to environmental stress combined with other anthropogenic causes might had brought the local communities to the end of their resilience capability.

[182] The issue related to the availability of refined chrono-typologies not only concerns the 7th century, but also the 8th and 9th centuries: "The compression of important diagnostic ceramics groups into earlier periods, and especially the failure to recognise post-Umayyad sequences, has had a deleterious and long-lasting impact on accurately understanding the social history of early Islamic Syria-Palestine from an archaeological perspective. Because of these chronological errors, the Abbasid and Fatimid periods have been almost entirely written out of the archaeology of southern Syria-Palestine, thereby creating a false 'dark age' that has been difficult to dispel and replace with a more considered evaluation of socio-economic developments after the mid-eight century" (Walmsley 2007: 55).

[183] Walmsley describes a "modest development from pre-Islamic styles during the seventh century, a period of accelerated change from the end of the 7th century (coinage especially) extending into the first half of the 8th (ceramics), continuing development until the mid-8th century and then a period of rapid and systemic change, involving elements of cultural discontinuity, in the later eight and throughout the ninth century" (Walmsley 2007: 69). Palestine, and particularly the Negev, is probably the region for which the most systematic evidence is available, while for the Transjordanian regions it is more scattered. For the purpose of this brief overview, I will mainly refer to two recent publications, one mainly focused on the Syria and Jordanian area, and one on Palestine (either the works though embrace a broader geographical area through punctual comparisons between site):

In general, three phenomena had been recognised for this period (Whitcomb 1994; Walmsley 2007: 104-107). The first is the reoccupation of abandoned and derelict areas, especially in large urban centres like Damascus and Hims. Even though the occupation of several sites is confirmed, it is not always easy to determine whether after the Byzantine period such occupation involved the entire settlement or only a portion of it.[184] The second phenomenon is the foundation of new settlements, normally described as *amsar* (sing. *misr*), often developing next to already existing sites, like in the case of Tiberias or Aila.[185] The third phenomenon is the establishment of "dynastic foundations", which are often associated with land estates.[186] Moreover, the foundation of entirely new settlements, such as Ramla, are also attested.

While the previous examples represent more or less urban sites, proper rural settlements also show similar evidence of either reoccupation or full continuity in habitation, dislocation, and new foundations. In particular, Walmsley interprets the evidence of the creation of new industrial settlements as the "deliberate development of unproductive land" (Walmsley 2007: 111). The reasons and the ways by (or through) which these developments were accomplished varied considerably from case to case. However, in general a heavy investment of resources was necessary, through the implementation of new infrastructures. Avni (2014: 288) further underlines the great regional variability in the settlement patterns, since while some sites "continued uninterrupted and even intensified during the seventh and eight centuries, other areas showed a temporary decline followed by renewal and expansion of settlements in the eighth and ninth centuries". In the case of modern Northern Jordan and the southern region of Syria, strong patterns of continuity emerge, with a rapid recovery from traumatic events such as the devastating earthquake in AD 749 and an renewed intense occupation until at least the 9th-10th centuries.[187] On the other hand, the Negev Highlands presents a more diversified pattern (Avni 2014: 289), with a clear difference between the eastern part of the region, where the settlements (e.g. Mampsis) seem to survive only until the 7th and 8th centuries, and the western part of the region, where sites testify a strong continuity until the 9th-10th centuries is documented (e.g. Shivta).

Generally speaking, both Avni and Walmsley ultimately describe the same development either for settlement patterns and for the material culture: a strong continuity is well-attested between the Byzantine and Early Islamic period, but the second half of the 8th century (and even more clearly the 9th century) saw some important changes. In the settlements, these changes are largely due to already ongoing transformations in the society (as suggested by Walmsley in 2007: 112), likely stimulating the creation of new identities or the transformation of already existing ones.

Walmsley 2007; Avni 2014. Other earlier important regional studies are available and quite often quoted in the aforementioned works, most notably, for Palestine, Magness 2003.

[184] Walmsley (2007: 108-109) admits that there is the possibility that a reduction in the effectual occupation of some sites took place. He nonetheless considers one of the principal reasons why it is in general so difficult to examine the Early Islamic settlements patterns the fact that "obviously, the archaeological evidence created by three centuries of Byzantine occupation will be considerably greater than that coming from much shorter 90 Umayyad years". It is clear, that this is an aspect to be kept into account in any analysis.

[185] This typology of settlement had been considered as originated by the pre-Islamic phenomenon of the *hadir* ('settlement'), normally characterised by a seasonal or not entirely continuous occupation (Walmsley 2007: 105; for specific examples, see also pp. 94-98).

[186] fall For instance, the so-called and much debated "desert castles" (*qasr*, pl. *qusur*) falls into this category, subdivided also into other typologies (Walmsley 2007: 99-104).

[187] In this context of continuity, Avni notices certain cases of dislocation of the centres of the settlements, matching one of the phenomena mentioned already by Walmsley. In the rural context, he also describes a progressive change from "large and spacious residential buildings to smaller, denser ones" (Avni 2014: 288).

Still, it is difficult to identify the precise causes determining the shifts in regional settlement patterns. All of the "major crises" during the 6th and 7th centuries had been demonstrated as not so influential[188] on altering such patterns, especially if compared with the 11th century, where marked signs of widespread discontinuity and abatement are documented. This eleventh century breakdown may be seen as the consequence of a combination of several naturally and anthropogenically related factors that finally stretched the limits of the resilience and adaptability of the local communities, finally resulting in the abandonment of several sites (Avni 2014: 341-343). On the contrary, the 6th-9th centuries events might have had an impact on the local population, but not in the long term. The damage done to settlements (if any in most cases) were probably repaired quick enough not to leave any consistent archaeological evidence.

Finally, one of the most important aspects of this study is making the link between the evolution and continuity of settlement patterns and the creation and reinforcement of identities. Or more specifically, could the creation, maintenance and change of identity patterns have had a role in determining the continuity of the settlement in these historical phases, which had been for so long considered as one of the most dramatic periods in the Near East history? In my opinion, it is worth focusing on three aspects to help answer this question: firstly, the economic investment in the liminal and semi-desert regions, mainly in relation to agricultural exploitation and the trade routes; secondly, the role of the army and the involvement of local social groups in the defence of the borders, especially as far as the recruitment of "Arabs" in the military organisation; and thirdly, the conversion to Christianity, and in particular the spread of the martyr cult and of monasticism, which impact both had deep implications on the socio-economic development of settlement on the political sphere.

7.1: The Desert and the Sown

Economic factors are possibly among the major elements which both influenced the development of settlements and helped shape identities, in particular in areas on the border of the Empire. The local climate closely related to this and clearly had a major impact on settlements, which would not have had a chance to exist without some major investments in supportive infrastructure, for example as far as water management is concerned. Indeed, all of the case studies considered may be connected to an increased extension and exploitation of the agricultural land, requiring a considerable effort in creating systems to collect, store and distribute water to both the inhabitants of the site and for sustaining the arable fields. The similarities in these features are particularly strong between the southern Hauran and the Negev, in the sense that an augmentation in the settlement in these areas can be seen as coinciding with the expansion of the Nabatean Kingdom. In particular, the foundation of the two cities of Elusa and Borsa are normally seen in connection to a series of major investments in the development of their agricultural hinterland as well. Yet, if for the Hauran archaeological studies seem to confirm such hypotheses (even if the dating is normally based on "circumstantial evidence": G. Avni, Porat, and Y. Avni 2013: 343), recent studies in the Negev offer a different picture, dating the development of fields

[188] Avni summarises all of the supposed decisive moments that would have determined the end of the settlement in the Near East (Avni 2014: 300-331): military actions, starting with the Persian invasion between 614 and 628 and after the even more dramatic "Arab conquest" in 634-640, both mostly invisible in the archaeological record; crisis like the bubonic plague in 542 or 744 or famines; similarly catastrophic events like earthquakes (particularly severe seismic events are reported to be in 551, 659 and 749, with less dramatic episodes in 633, 756 and 859); and finally, a climate change.

and agricultural installations not before the 3rd-4th centuries (Ashkenazi, Y. Avni, and G. Avni 2012; G. Avni, Porat, and Y. Avni 2013).[189]

Likely, the agricultural exploitation of new lands was not the only factor playing a role, but the commercial potentiality of establishing settlements on (or in proximity of) the major trade routes: the north-south axis, namely the Kings' Highway in Transjordan; the east-west from the city of Aila to the Mediterranean coast, the so-called "Spice route" passing through the Negev; and also to a certain extent the route passing through Wadi Sirḥan, coming directly from the Arabic Peninsula and ending in proximity of Umm el-Jimal. A side effect of the development of these trade routes is the establishment of military posts to secure them that in turn also stimulated the foundation of civilian settlements. This phenomenon had been documented for several towns in the Negev, like Oboda and possibly Mampsis itself, but also in Esbus and other sites in Transjordan.

The production of wheat, wine and oil that would be sold on the Mediterranean market seemed to have been one of the principal drivers for the development of rural settlements already during the Roman period, but it is with the Byzantine period that the investment in the hinterland appears to have been much greater, and as a consequence also the increase in number of settlements (Avni 2014: 204-207). At the same time, some of the settlements (especially in the Hauran) show clear evidence for animal breeding, not necessarily for the market (as suggested for the horse commerce of the Nabatean and Early Roman Mampsis in the Negev), but for the production of secondary products and the subsistence of the local population. The frequency of stables in the single houses, some of which also of considerable dimensions (for instance the ones in Sharah) is likely a hint for such economic specialisation. A similar impression is given by the widespread presence of pens found in the urban space, without neglecting the possibility of keeping some animals in the houses' courtyards.

Despite this increase of investments in agriculture and the consequent accentuation of some settlements' agricultural specialisation attested for the Byzantine period, scholars generally agree in seeing a trend towards regionalisation and the decline of the long-distance trade, from a "global" Mediterranean circulation of goods to a more local and regional trade, especially from the 7th century onward (Avni 2014: 290-294). Likely, the focus on local production might have led to an increased involvement of rural agricultural communities, which were apparently responsible for the development of field systems and agricultural installations in the first place (G. Avni, Porat, and Y. Avni 2013: 343-344). As mentioned earlier, investments in these areas continued during the early stages of the Islamic period, and specifically for the 8th and 9th centuries, the presence of specific technological solutions like the *qanat* had been considered features that provide evidence for the presence of specific "ethnic", non-local groups.[190] Investments like the construction of a *qanat* and establishment of entirely new settlements are rarely driven from the local communities, but more likely result from a direct

[189] Though, Avni does not exclude the possibility that some earlier structures were completely replaced by the Byzantine developments, similarly to what seen in some of the case studies considered in the present study (G. Avni, Porat, and Y. Avni 2013: 342).

[190] This hypothesis follows the long-debated theory by Wilson of the so-called Islamic "Green Revolution", according to which the Islamic conquest of the Near East brought together new crops and new technologies until then unknown (Watson 1974). Several aspects of Wilson's theory had been questioned and I will not discuss it here further discuss them here. Still, one should be cautious on the "ethnical" relevance of such elements in studying settlements. In general, the diffusion of these technologies and crops follows the gradual shift of cultural and economic interest from the West towards East, visible in the increasing influence of Iraqi models in many aspects of the material culture in the 8th and especially from the 9th century. As seen, ceramic and glass from this period show clearly such phenomenon, not only in terms of import of extra-regional vessels, but also the adoption of new forms and decorations for local productions (Walmsley 2007: 57-58).

involvement of the elite class, that likely came from other regions than the previous Byzantine provinces (Walmsley 2007: 110-111). Nonetheless, these new settlements did not necessarily exclude the local population; on the contrary these populations could have moved from other sites following new opportunities represented by these sites. These movements might have been boosted as well by phenomena as erosion and degradation of soil quality (due to deforestation and excessive land use), attested in different locations starting from the 6th century (Walmsley 2007: 112).

Similar movements of people are not but one possible manifestation of mobility patterns in these areas, and possibly took place for earlier periods. Ultimately, such horizontal shifts of settled people from one site to another, considered better fitting their needs or offering more opportunities, do not clearly differ in the material remains from the sedentarisation of nomads. Probably, the increasing involvement of pastoral groups (with different degrees of mobility) had a strong impact on the development of the settled rural sites. One result of the interaction with settled communities is the sedentarisation of these pastoral groups in proximity or even inside the already existing settlement. Still, it is important to keep in mind that these two groups (the settled and "sedentarised") cannot necessarily be considered as separated societies: in fact, their integration within the same economic system most likely determined the establishment of social ties. Similarly, it is not unusual for empires or state-formations to encourage or even force the sedentarisation of sections of nomadic populations in order to better control them and at the same time have a mobile force for patrolling the desert borderlands (See intra, chap. 7.2).[191]

Sedentarisation of groups had been well documented in the Negev, with temporary settlement progressively developing into more stable sites, forming within close proximity of the Negev towns. Avni suggests that the diffusion of agricultural farms in the most "remote corners of the Negev Highlands" can ultimately be seen as a product of the spontaneous sedentarisation of nomads.[192] This process is the opposite of what is usually described for the Early Islamic Wadi 'Arabah, where the direct involvement of the Islamic political power in the development of the agricultural hinterland might have determined the "forced" introduction of people, maybe coming from other regions of the Empire (Avni 2014: 282-283).

Social ties between the "settled" and "nomads" possibly played a major role in determining certain spatial features within some settlements. In the case of sedentarisation processes, the tendency is to have members of the same family or extended blood-related group cluster together, forming well defined physical areas. This is ultimately one impression that can be gained by looking at urban morphologies like the one seen in the Hauran case studies, such as at Umm el-Jimal and Sharah. Sedentarisation processes could result in the development of clearly distinguished quarters or blocks, as is possibly seen in Shivta, where the belonging to different social groups is well reflected by their spatial organisation within the settlement. It expresses the desire to underscore the "alterity" of the newly arrived groups from the rest of the local community, but at the same time underlines those groups' mutual social ties within the same settlement. Nevertheless, whether this alterity is expressed by distinctive sedentarised "nomads" or simply by groups moving to and from other settlements is difficult to determine.

---

[191] The tight connections are also visible in some papyri studied by Sijpesteijn, where the simultaneous presence in a document of "village" (*qarya*) and "tribe" (*qabīla*) might appear to contradict each other. Nevertheless, "nomadic tribes could very well be associated with a village even if residing there only a part of the year" (Sijpesteijn 2013: 262-263).

[192] It was noticed that a shift from rounded structures to squared dwellings can testify a completed process of sedentarisation of nomads (Avni 2014: 281-283).

A useful comparison can be made with Mamluk Egypt, and more specifically with the Fayyum. As described by Rapoport (2004)[193], in al-Nābulusī's *Tārīkh al-Fayyūm wa-Bilādihi* (dated to the 1240s) the dichotomy between *badw* (Beduins) and *ḥaḍar* villages was exclusively based on two features. The first is purely economic, distinguishing between two different agricultural specialisations (the *badw* specializing in cereal cultivation and the *ḥaḍar* specializing in profitable cash-crops[194]) and irrigation methods. In particular, one element that appears to be particularly relevant for the present discussion is the fact that the *badw's* "specialisation" had a direct consequence on the social structure of these groups. According to Rapoport (2004: 10), "tribal social organization was linked to the organization of cereal cultivation, and, especially, to the local irrigation system. In cereal-growing villages, the amount of arable land was subject to drastic annual fluctuations, and the peasants may have looked for ways of sharing the risk of a low Nile level". Being aware of the evident differences between Egypt and the regions considered here, one could wonder to what degree such a remark is also applicable to the Late Byzantine and Early Islamic periods. If so, one might suggest that differences in the spatial structures of the villages between the Hauran (especially the southern part, where the main cultivation appeared to be cereals) and the Negev (where in turn olive and wine were more intensively cultivated) can be a reflection of a different economic importance of the tribal structures. Consequently, this different built environment could be a consequence of the importance of these social structures in the local identity patterns.

The other discriminating factor, that ultimately results in a dichotomy, which is discussed in al-Nābulusī's *Tārīkh al-Fayyūm wa-Bilādihi,* is the element of religious identity, namely the distinction between Muslims (the *badw*) and non-Muslims.[195] The social and identity consequences of this situation are interesting, especially in light of the continuity in occupation of some villages from the Late Antiquity into the Mamluk period, as is well attested in the sources. Some sites mentioned by al-Nābulusī as *badw* villages were formerly Christian villages. Their process of "Bedouinisation" however is not connected to the real sedentarisation of Bedouins, but is determined by the Islamisation of the local Christian farmers, which during this process also assimilated the tribal social organisation and the *badw* identity (Rapoport 2004: 13). This assimilation of a "made-up" Bedouin, the "nomadic past" origin and identity was rectified by modifying their genealogies respectively. Together with the political and fiscal benefits of being part of the Muslim community, the adoption of the "Nomadic origin" had an important ideological consequence: the notorious "independence of the Nomads" functioned as a justification on which the opposition against the State and the taxes were later justified (Rapoport 2004: 20).

[193] As far as the terminology is concerned, Rapaport offers a crystal-clear statement on the approach he used for his study, that I tried to demonstrate perfectly applies also for the chronological context considered for this research: "I would contend here that a clear distinction between pastoral nomadism as an economic option, tribalism as a form of social organization, and bedouin-ness as a cultural identity allows for a richer interpretation of the resistance of the *'urbān* [a non-classical plural of *'arab*] to Mamluk rule" (Rapoport 2004: 4).

[194] To be noted though is that during the Mamluk period, cereals were also cash-crops. Therefore, the economic distinction of the two groups would need further investigations (thanks to personal conversation with Prof. Walker of the University of Bonn).

[195] "The Arab tribesmen of the Fayyum lived, for the most part, in settled communities. al-Nābulusī's fundamental dichotomy of *badw* and *ḥaḍar* is not between nomads and settled cultivators, not between the desert and the sown, but rather between Muslim tribesmen and non-Muslims" (Rapoport 2004: 11-12).

7.2: Brothers in arms

Settlements might be also associated with the urgency to protect the border or nearby commercial routes, developing around an existing fort or tower, like in the cases of Esbus and Umm er-Rasas. Therefore, the presence of the army could have influenced both settlement patterns and local identities. In the regions considered here, these developments are closely related to the defensive strategy that was first put into action by the Romans and later more intensively by the Byzantines (in particular, from the 6th century onwards). Namely, the integration of "Arab" groups as *foederati*, or even proper allies, in the regular army, was followed by the (self-)identification of strong personalities or influential local groups.

Determining the direct consequences of this strategy on the regional context in Syria and Arabia before the 6th century, especially as far as the social background of the *foederati* is concerned, is difficult. Contacts with "Bedouins" are clearly attested in sources already in the earliest phases of the Roman presence in the East, even though at the beginning the task of dealing with them was mainly in the hands of the local client states (Schmitt 2005: 275). Safaitic inscriptions attest contacts with nomads already before the 3rd century, engaged in the Nabatean army in particular as mounted units (al-Housan 2017: 35 and 42-43).[196] Most importantly, Rome continued and encouraged a "tradition" of engaging "nomads" in the army after the annexation of the Nabatean kingdom and the incorporation of its army into the Roman one (Nehmé 2017: 142-155). Apparently, this led to a deep cultural exchange (Schmitt 2005: 276-278), possibly manifested through the adoption of the Greek language as is attested by some authors of Safaitic inscriptions (see infra). Beginning in the 4th century, the engagement of local and nomadic groups appears to intensify, even if the exact nature of the *foederati* and their respective social structure is not always clear. It appears that the policy of integrating nomadic groups continued as it had in the earlier phases; this time, however, they were more enmeshed within the political dynamics of the Byzantine-Sasanian confrontation (Schmitt 2005: 278-283; Nehmé 2017: 142-155). The development is particularly evident from the 5th century onward, with the integration of the "tribal" leaders into the local Roman hierarchy and at a higher level in the provincial Roman society.[197] In general, the implication of progressively more intense contact between populations in the peripheries and central authorities through the army are extremely relevant, and possibly led to the creation of new socio-political identities (bringing together groups from different backgrounds) and therefore impacting settlement patterns. "Tribal" leaders could have become interested in at first seeking Roman and later on Byzantine support:[198] this resulted in movement and settling towards into the territories directly controlled by the Empire, although not yet full sedendarisation at this stage, and therefore a more stable presence on the border. Similarly, they might have determined the creation of

[196] The Nabateans appear to not only integrate nomads in their army, but also are well-aware of the tribal patterns and were engaged in the surrounding regions of their kingdom, especially the Hijaz, as the support to the Nabateans expressed by some "nomads" in Safaitic inscriptions suggest (Norris and Al-Manaser 2018: 16-17).

[197] "Another clue to increased levels of contact and cooperation is provided by the prominence of the term *phylarch* in the literacy and epigraphic sources. First appearing as a neutral description of local tribal chiefs, its meaning evolved as frontier relationship grew in intricacy, eventually acquiring a level of administrative meaning with connotations of authority within the local Roman military hierarchy, over a specific provincial region or areas" (Fisher 2011: 79). A significant change was the equalisation in the administrative hierarchies of the *duces* and "Saracen" phylarches, requiring a more direct cooperation between leaders of the *foederati* and Roman authorities (Schmitt 2005: 280).

[198] "The Roman attitude towards tribes also involved attempts to identify and support leaders against whom they could apply leverage, backed up with punitive expeditions and gifts of money where it suited. [...] The Arabs whom Rome encountered in the east in the second, third and fourth centuries were subject to similar initiatives, which varied in their nature and the levels of coercion that were needed" (Fisher 2011: 74).

larger tribal formations, as suggested by the "*Thamudeni*", a group mentioned in a Greek-Nabatean inscription in Riwwafa in the Hijaz and later listed in the *Notitia Dignitatum* (Macdonald 2009c; Fisher 2011: 75). It is possible that the creation of a new identity was facilitated by the presence of groups sharing a similar cultural and social background, more or less integrated into the Roman army, but still clearly distinct from the rest of the official Roman army[199]; this intermediate position represented a political opportunity for these leaders and their followers to reinforce their position in respect to other tribes that were not under the Roman influence. This political opportunism might have ultimately resulted in the development of new social identities, though not yet distinguishable as an ethnicity, that were based on socio-cultural factors and potentially geographic provenance (Haldon and Kennedy 2012: 340-341). The reference in the Namara inscription to "Arabs" as opposed to other tribal groups or possibly confederations may reflect a distinction between groups under the Roman influence (the "Arabs" directly led by Imru 'l Qays) from others that were not (see infra). These sort of statements that declare a specific identity (with a clear political intention) are rare before the 4th century. In general, the absence of evidence documenting a recurrent broader consciousness resulting in an identity is manifested by the preferred reference to single "tribal" affiliations rather than the reference to a proper confederation of tribes.

The Christianisation of the Empire also had a major impact on the above-mentioned socio-political dynamics in the Near East during the Byzantine period (see Chapter 7.3). Firstly, belonging to the Byzantine army meant being immersed in Christian cultic ritual and surrounded by Christian symbolism: oaths from the period testify several Christian formulas, reflecting dramatic social changes.[200] In fact, oaths can be seen often as a way used by a central authority to indirectly impose obligations and in a way to bind individuals despite the presence of other possibly conflicting identities or loyalties (Esders 2012: 359). The integration of the late Roman Arab *foederati*, or at least their leaders, is indeed one of these cases, since it was "centred around a Christianized idea of military loyalty" (Esders 2012: 365). In practise, however, taking an oath was not always a guarantee of the "neutralisation of other social relationships": this context has a multiplicity of concomitant factors having all high resonance on the single individuals. In fact, belonging to the same religion was an element that both integrated the Arab elites into the Byzantine administrative framework, and at the same time allowed them to keep a well-established influence on the region and keep a noticeable degree of independence from the central hierarchies (see Chapter 7.3). Similarly, the increasing involvement of religion in the military sphere is testified by the role of specific bishops in dealing with the army, who more and more frequently filled the void left by the imperial authorities. The cases of Evagrius and Gregory of Antioch are renowned in the historical sources for their role as intermediaries between the lacking imperial authorities and the armies on the verge of mutiny (Leppin 2012: 241-258). Clearly, the involvement of bishops in the army is a clear sign of the increasing importance of religious authorities in filling the void left by the imperial hierarchies. At the same time, it also demonstrates the importance of the Christianisation of the Empire in providing a foundation for a common identity for a population with extremely diversified social and geographical origins. In this context, their consistent use of the term "*Romanoi*" to indicate that common identity is particularly, as it can be understood as the unshaking loyalty to the Empire and its ruler.[201] Furthermore, historical sources also offer examples of clergymen

[199] This finds a direct parallel in the interpretation of an inscription from Palmyra, where the dedicand defines himself as a Nabatean: Al-Otaibi (2015: 297-299), which suggests that the statement of such identity can explained in light of the contact with "other" groups within the Roman army.

[200] The Ghassanids are reported to have been converted to Christianity in AD 502, under emperor Anastasius I (Shahîd 2009: 364-365).

[201] "Romanness, as it is understood here, was a matter of performance. The label 'Roman' was very flexible at this time, because it could include people of different ethnic origins and most of the soldiers at this time were anything

intermediating for the settlements during the period of the Persian invasion and Arab conquest, making agreements with the "invaders" in order to spare the population from a violent subjugation (Leppin 2012; Avni 2014: 302-319; an interesting parallel, for Egypt during the Arab conquest: Sijpesteijn 2013: 49-58).

As mentioned earlier, a phenomenon parallel to the Christianisation of the population was the acceleration of the integration of "Arab tribes" into the defensive system of the Empire, resulting in a consolidated political power on the local and provincial levels, which represented the final development of the defensive strategy mentioned above (Schmitt 2005: 280-283; Fisher 2011: 80). The Jafnids (and their counterpart under the Sasanian sphere of influence, the Nafnids) are the most noticeable example for the 6th century, especially after the official recognition of Al-Harith (Arethas in Greek) as an ally by Emperor Justinian in 528/9.[202] It is likely that a combination of factors determined the success of the Jafnids. The frequent conflicts between different Christian factions, allowed the rapid development of both a new social identity (with the official integration of "Arab" leaders into the imperial elite) and of an established political entity, who were not officially recognised by the imperial authorities. The Jafnids and their leaders were able to maintain their independence while becoming more and more integrated into the central administration. Their role was twofold: as military commanders of the "Arab" tribes (either settled and nomadic) in the Eastern provinces and as mediators within religious conflicts, acting in certain situations (in particular in the first half of the 6th century) as the imperial diplomats for the miaphysite Church. Acting in these roles simultaneously increased their credibility at the court of Constantinople and their local support (Fisher and Wood *et al.* 2015: 281).

Similarly, the increasing tensions with the Sasanians to the East, played a major role in the rise of the Jafnids', as they were able to access important positions within the imperial hierarchy.[203] A similar development is documented for the Nasrids (actually developing into a political entity itself, based in al-Hirah, modern Iraq) and is a precise indication of the new importance of such "tribal" formations in Late Antique political dynamics (Fisher 2011: 92).

---

but Roman by origin" (Leppin 2012: 252). This broad concept of the Roman identity was also possible in light of the strong Christian characterisation of the Empire itself: "Roman identity could be detached from the political organisation of the empire without becoming merely a religious identity. The guarantor of the Roman Empire [...] was not the emperor, but the bishop and the army when commanded and exhorted by the right people. The Roman Empire was based on Christianity – the Christian organisation for its part did not depend on the Roman Empire" (Leppin 2012: 258).

[202] In this context the idea of a proper confederation of tribes, namely the Ghassanids, under the leadership of the Jafnids, has been challenged by several scholars, since no direct evidence from that period for such a type of formation is available. See for instance Fisher 2011: 83: "Unfortunately, the lack of information at our disposal about the people under Jafnids or Nasrid control makes it virtually impossible to ascertain if any type of 'confederation' – for example, the popular idea of the 'Ghassanid confederation', for which there is no contemporary evidence, and which may in fact be something of a historical chimera – was evolving".

[203] The change in the terminology used in the Greek diplomatic language to refer to the Jafnids is particularly interesting: from the description as *foederati* (ὑπόσπονδοι), they are later defined as "allies" (σύμμαχοι). It is still an ambiguous recognition, since in other sources these Arab "allies" are described in extremely degrading terms. This also has a parallel on the Sasanian side, with their relationship to the Nasrids. Generally speaking, it appears to witness proper attempts by the Byzantine Emperor to undermine the Jafnids, once he became aware of their increased territorial political role. These attempts, however, were unsuccessful as is demonstrated by the reception of Al-Harith's son Al-Mundhir (Alamundarus) in Constantinople in 580 (Schmitt 2005: 282; Fisher 2011: 122).

It is not clear if there was a direct cause and effect relationship between the augmented power of the Jafnids and the settlement boom in the Syrian and Arabian regions between the 5th and 6th centuries.[204] Nevertheless, the political presence and possibly direct territorial control of the Jafnids[205] is documented by some of the epigraphic evidence, mainly attesting the foundation of religious complexes rather than military structures or "palaces" (Genequand 2015). It could be suggested that the security provided by the Jafnids to the region encouraged the development of settlements, investments in agriculture and trade, alongside the growing number of pilgrimages made to the region following the "Christianisation" of the landscape. It is likely that at this time sedentarisation of other nomadic groups also took place, being that this process also occurred within the pastoral-nomadic and settled sections of the Jafnid tribe, as is more clearly indicated by written and archaeological evidence (Fisher 2011: 108-116). It however remains difficult to determine to which degree these various sedentarisation processes impacted the settlement patterns and how these could be distinguished from movements of other people. As Fisher points out, the importance of the Late Antique development of the Byzantine-Arab relationship in the periphery of the Empire, and in the specific case of the Jafnids, is not only the actual integration of new groups - at least their leaders - into the imperial elite, but also and at the same time the formation of an independent political identity, with deep roots into a determined social and cultural context (Fisher 2011: 116). To a certain extent, the social phenomena described for Trans-Jordanian and Syria may also be identified in the Negev, with one major difference: there are no hints for the presence of a politically and territorially active "tribal" presence like that of the Jafnids. Although at the same time, the presence of *foederati* in the Negev settlements is widely documented for the Roman and Byzantine periods as well as the involvement of nomadic tribes around the settlements, as the Nessana papyri strongly attests (see infra).

The Early Islamic period does not seem to represent a particular break in these social patterns. The new rulers maintained a similar pattern of interaction that the Byzantines had used with the Jafnids and that the Jafnids had with the other local tribes, however with one substantial difference: the social composition of the Umayyad elite was closer to the one of the Jafnids, sharing the same social structures also the people they were ruling over, determining an increased political importance of family and tribal connections. Moreover, the fact that the Umayyads were now ruling a newly established Empire put them in a completely different position than the Jafnids, despite their relatively strong territorial control.[206]

---

[204] If Shahîd was firmly convinced that the Ghassan (of which the Jafnids are presumed to be the leading tribe) were already an entirely settled population, occupying specifically the Hauran region, no direct proofs are available. Parallelly, attempts to identify a proper Jafnid architecture had to face an avoidable lack of strong evidence to sustain the hypothesis of the existence of a distinctive tradition (Fisher 2011: 108-116; Genequand 2015: 173-207).

[205] The precise geographical area over which it was exerted is also still debated. Generally speaking, it is quite confidently demonstrated their direct control over Mesopotamia (as seen in Rasafa), the Hauran, and part of the province of Arabia, at least until the region of Madaba; but might as well extended northwards (Hoyland 2009: 117-139; Fisher 2011: 34-71).

[206] "It is obvious that the Umayyads had to rely heavily on the tribes of Greater Syria. It is also highly possible that the way they dealt with these tribes was not fundamentally different from the way the Jafnids interacted with the peoples under their control. Nevertheless, the scale involved was certainly different. The formers were at the head of an empire and had all the power in their hands (political, military, economic ...), while the latter had a mainly military role and to a lesser degree some political functions; they never really replaced provincial governors or *duces*. Another point is that after *c.* 582 the Jafnids lost much of their leadership" (Genequand 2015: 187).

One possible direct consequence on the settlement patterns is that the new "Arab" elite began "reusing specific places and acting like the Jafnids", occurring in the first half of the 8th century (Genequand 2015: 187). Furthermore, several "Arab" Christian tribes also assimilated into the Islamic armies (Avni 2014: 312-313) may explain why there are no indications of a radical change in the social components in the local communities. This may offer one further explanation into the survival of the majority of the settlements. In general, the newly established Islamic administration did not impose a mass-conversion or the abduction of large communities, contributing to the survival of several Christian communities, as demonstrated for instance by Umm er-Rasas.

One main aspect that characterised the social structure during the Early Islamic period was the progressive separation of the military section of the population (ultimately made up of the Umayyad elites) and the civilian one, marking a clear difference with the preceding periods (Haldon and Kennedy 2012: 341). The distinction between the military and non-military sections of the Muslim *umma* is clearly a factor that had major consequences for the (new) social structure, and may have also influenced new organisational patterns within the settlements and their built environments. That being said, it is rarely possible to identify distinguished areas in the settlements (not even in properly urban centres) that were reserved for the military elite, with only a few significant exceptions, such as Aila, where a separated camp for the newly arrived "Arabs" was established close to the pre-existing city (Walmsley 2007: 94-95).

One further interesting development within this social structure, specifically within the ranks of the army, was the presence of two clearly distinguished and competing identities groups based on geographical grounds, namely the people of Syria (*ahl al-shām*) and the people of Iraq (*ahl al-'irāq*) (Haldon and Kennedy 2012: 342-353). These two new identities formed based on their regional origin that superseded their local tribal affiliations in importance, and the difference socially between these groups was ultimately determined by the distribution of land and spoils following the conquest of the Byzantine and Sasanian territories.[207] During the Umayyad Caliphate in particular, this contraposition and rivalry escalated, which later led to the Iraqis' rebellion, who often protested against the privileged conditions the Syrians benefitted from.

From an ideological point of view, the "regional identity became as important as kinship (real or imagined) as a marker" (Haldon and Kennedy 2012: 351) at the end of the 7th century, while at the same time daily life in general (especially in the rural setting), and the choices made in the built environment more specifically, seems to have continued following early trends. The fact that many of the tribes already integrated into the Byzantine social system, were subsequently included into the Early Islamic armies and later progressively converted to Islam, possible resulted in the continuity of these regional settlement patterns. For this aspect, it would be once again useful to refer to the oaths of allegiance, which showed a more neutral religious identification during the early phases of the Muslim Empire than in the Byzantine period. The integration of the Christian or at least non-Muslim groups into the Islamic army was simplified by the fact that these groups were already monotheistic; and this link made up part of the terms on with their allegiance was based: "the *bay'a* [understood as pledge for 'loyalty in

[207] "The most important reason for the emergence of this new sense of community must lie in the exploitation of resources in the aftermath of the Muslim conquest and settlement. As has been known for a long time, the conquerors and their progeny were entitled to pensions from tax revenues of the conquered lands, Syrians to the revenues of Syria, Iraqis to the revenues of Iraq. This in turn meant that the *ahl*, while they may not have lived in the villages and farmed the land, nonetheless had a clear identification with the area which produced the incomes on which they lived and a clear sense of community and common interest with those with whom they shared these revenues. Above all they sought to prevent the *ahl* of Iraq from appropriating the revenues of Syria and vice versa, even if they shared bonds of kinship with them" (Haldon and Kennedy 2012: 343).

war'] not only meant accepting the legitimacy of the caliph and fighting for him, but in a way also adapting monotheism, whatever it may have looked like at this time. Defined in such broad terms, it could be accepted by Christian Arabs, too, who became part of the movement" (Esders 2012: 370).

7.3: Defended by the saints

Religion increasingly gained importance in defining local identities in the Byzantine period, and parallelly in the integration of groups defined as "Arabs" within the Empire; therefore, this had an impact on the regional settlements' organisation and in the spatial organisation of villages and towns. The astonishing number of churches and monasteries across the landscape cannot be simply described in terms of numerical account, nor be considered a simple testimony of devote local communities, but as a key feature in the formation of community identities and a provision of the socio-political structure. A particularly relevant role was played by monks and rural monasteries.

According to Fowden (2015: 175-196), two factors worked together in the conversion of the local population: the establishment of "shared holy sites" (*haram*) and the social features characterising the "Arabs", especially concerning mobility and kinship relationships. Some of the sites considered in the present analysis show continuity in the presence of religious buildings before and after the conversion to Christianity, sometimes located in the same place and reusing the same building materials of earlier complexes (for instance in Umm es-Surab and Umm el-Jimal, but possibly also in Sharah and Mseikeh). The importance of establishing and maintaining the status of a "holy site" must be seen in terms of "opportunities for observation of others which might result in the cross-fertilization of linguistic, economic and social practices or, in the case of religious behaviour, even conversion" (Fowden 2015: 180).

On the other hand, the heterogeneous group of people described as "Arab" is crucial in the conversion of local communities in light of two features; two features that ultimately explain why they were chosen by the Byzantine Imperial administration and by the ecclesiastical hierarchies as an intermediary with other local communities: their mobility and flexibility. After playing an important role within the Roman/Byzantine army , their potential as a vector for the Christianisation of the countryside was also understood by the ecclesiastical authorities. Fowden provides an example of a miaphysite priest named Ahudemmeh in Syria, who is particularly relevant to understand this strategy (Fowden 2015: 183-184). After being able to convert some leading members of local tribes in the northern mountain area and having consecrated priests and built churches, he was able to exploit these groups' mobility in order to guarantee a stronger control and defence over the monasteries in central Mesopotamia. The idea that this kind of conversion of tribes led to the "forced" sedentarization of these tribes is not entirely accurate as it reinforces the misconception of a clear separation between nomads and settled communities, where the "Arabs" are exclusively nomads. On the contrary, the flexibility of these "Arab" groups, easily shifting from more mobile to more settled strategies, guaranteed a greater effect of the conversion on a wider geographical scale and on different social, religious, economic and political levels. The social and religious consequences for these groups are surely the most evident results of the mass conversion, leading to the conversion and the establishment of new Christian communities. Although the consequences are also extended to the economic and politic spheres, since the movement of these groups and the contextual foundation of new shrines and monasteries guaranteed an income for their residents from increasing pilgrimage to these new sites and generated and supported local trade. It guaranteed more stability from a cultural perspective as well: even after the frequent conflicts from different groups within Christianity subsided, with the diffusion of different sects and the creation of what Fowden defines "the variegated spectrum of religious cultures" in the Near East, the presence of these places of confrontation generated a local population "accustomed to operating in a wide variety of linguistic and cultural registers" (Fowden 2015: 188-189). This element is especially relevant for explaining social continuity in the population after the arrival of the Islam, where the groups later defined ethnically as "Arabs" because of their flexible social structure quickly adapted to the arrival of the new "Arab" Muslims, a group which they also identified with. Furthermore, Fowden suggests that the smooth Islamisation of these tribes can be explained by the fact that Islam itself was put in terms of "restoring an unperverted monotheism and starting afresh", after the religious strife taking place in the later phases of the Byzantine periods (Fowden 2015: 196).

The political consequences of the conversion of "Arab" groups to Christianity (and particularly the Jafnids of the Ghassan) resulted in a particularly ambiguous relationship between these new allies and the Byzantine Empire (at least before the definitive conflict between al-Mundhir and Emperor Mauritius in 584). The conversion to Christianity and belonging to the Christian community was the way the Jafnids progressively became more integrated into the imperial elite and politic, guaranteeing the inclusion of their leaders of the Ghassan in the provincial administration. The risk was that the large majority of the "Arab" population following the Jafnids could have eventually become alienated, as consequence of its exclusion from the Byzantine elite and the possible cultural difference following the progressive "acculturation" of their leaders. The backup of the Miaphysite church was the solution to this potentially volatile situation, allowing them to maintain a certain degree of independence from the central administration (that was supporting the Chalcedonian church) and to establish a sort of independent territorial control in the regions where the members of their confederation were present.[208] This choice led to the creation of a parallel identity, partially in contrast with the Byzantine elite and thus ensuring them a certain degree of autonomy (cultural, social and political), but more importantly strengthening the support of their followers (Fisher 2011: 63). At the same time, the benefits of this strong connection with the Miaphysites guaranteed this church the protection of the Ghassans and the Jafnids (Hoyland 2009: 129; ter Haar Romeny 2012: 199).

The social consequences of this "in-between" condition is reflected in both the settlement pattern and cultural sphere: for instance, many have suggested that the foundation of monasteries and churches (especially connected with the cult of St. Sergius[209]) are potentially evidence for the sedentarisation of a large section of the Ghassan confederation.[210]

---

[208] Both Hoyland (2009) and Fisher (2011: 95-116) suggest a sort of direct control over some areas in Syria and Transjordan, particularly over the Hauran, but to different degrees. Hoyland, for instance, does not exclude the presence of a proper direct territorial control of the Ghassan on a presumed "Monophysite church State of Arabia", as already suggested by Nöldeke (1875: 419-443). On the over hand, Fisher does not exactly suggest the development of a territorial control, but underscores that the ambiguity in the Jafnids' religious position in the 6th century sectarian conflict, ultimately guaranteed them the access to the political imperial elite and at the same time the maintenance of a direct and autonomous control over the groups of the confederation, in particular the more pastoral one.

[209] The importance of the cult of St. Sergius as identity-element had been clearly established, and the saint himself represents the perfect synthesis of the "pastoral" and "Christian" identities: "The cult of St. Sergius, popular amongst the Arabs of both steppe and settled areas such as the Hauran, contained an intrinsic connection to the steppe itself. Recast as a rider, Sergius could be conceptualised as a 'saint on horseback, a speedy guardian of cameleers', who, as Elizabeth Fowden speculates, perhaps held an 'immediate appeal' to those who lived in or travelled through the steppe, acting as their 'guide and divine protector'" (Fisher 2011: 47).

[210] Shahîd actually refutes the idea that the Ghassanids are nomads, suggesting that they were already entirely settled by the time they became *foederati* of the Byzantine Empire (Shahîd 2009: 1-20).

Even if at the present stage of research it is difficult to determine the effective extension of the sedentarisation process as consequence of the Christianisation of the general population, it is well evidenced by settlement patterns that at least some groups took the opportunity to settle, potentially also in the proximity of the "holy places". These places offered particularly valuable opportunities, not necessarily for religious reasons but especially for the economic benefits that they could provide (Fisher 2011: 41-42). Moreover, the recurring link between *harams* and water is particularly significant in regions on the border of the desert.[211]

The cultural consequences of both sedentarisation and Christianisation on nomadic pastoral groups is similarly hard to determine precisely. Considering the often long-existing contact between these groups and other communities that might have encouraged the sedentarisation and the conversion processes, it appears too extreme to suggest that they could have suffered an identity crisis, due to a complete change in uses and re-arrangement of daily life (Fisher 2011: 42). If settling down is potentially a dramatic change in traditional lifeways, this process is not always forced and prompted, though rather a gradual adaptation of several daily customs helped by the already mentioned flexibility of the pastoral communities. The organisation of settlements could be indeed one example of such an adaptation. Ethnographic studies have shown that in many circumstances, settlements developing through sedentarisation often have a direct parallel the pitch camp from which they could have originated from (Saidel 2009; Whitcomb 2009: 242-245).

Moreover, a newly acquired identity do not necessarily erase the previous ones but often absorbs them, being active in different contexts and particular situations. This is a phenomenon that can be directly compared to (evolution of) the different languages spoken in the region, as documented also by a variety of written sources (especially papyri and epigraphic finds): the co-occurrence of several different idioms reflects a differentiated use according to the social context (Hoyland 2009: 129-132). Still, each idiom could have been spoken by the same individuals and possibly understood by the entire population (see infra). For instance, the emergence of Syriac has been considered by some as correlated to the diffusion of the Monophysites and more importantly to the progressive inclusion of "Arabs" into the miaphysite hierarchy (ter Haar Romeny 2012). One could suggest that with the use and uptake of lingual constructions like idioms into other languages (Greek and Aramaic) might be applicable for other identity features as well.

Similarly, it is not entirely true that the stratification between elite and non-elite social levels among the pastoral groups was an entirely unknown phenomenon or a development following the sedentarisation. The presence of more prominent segments in the tribes and of more influential tribes within confederations (as in the case of the Jafnids) is well attested and did not necessarily alter or destabilise the social structure. The manipulation of the genealogies is a proof of the adaptability of the segmented societies, such as the one later defined as Arab.

The second half of the 6th century is more evidently characterised by the diffusion of monachism, another religious phenomenon that might have directly impacted the shape of the settlement themselves. The centrality of monks and hermits has been shown by Fowden in particular relation to

[211] A continuity in the association between dedication and water is well attested even between Nabatean and Islamic inscriptions, either thematically (al-Salameen and Hazza 2018: 95-97) and physically (al-Manaser and Ellis 2018: 70). In particular, Al-Manaser and Ellis (2018: 70) point out the fact that while Safaitic inscriptions are widespread across the entire *harra*, Islamic inscriptions and simple mosques of the region are always close to permanent or temporary water sources.

their "magical" powers as healers and exorcists (Fowden 2015: 182-183; see also Fisher 2011: 36-37). Even if the accounts might had been the result of a rhetorical style used by the sources to underline the "dramatic changes brought about by a new faith in God, and the rejection of a pagan past" (Fisher 2011: 37), it is nonetheless evident that a new importance was now played by religion in shaping identities. It is not only testified by the changes in the oaths of the soldiers, as seen earlier but also by the astonishing number of churches and chapels even in small villages, as seen in Umm er-Rasas. More likely, the magical and apotropaic power of relics belonged to saints and martyrs was highly considered by the local communities,[212] especially in a period of extreme political insecurity. In this context, the great vector for the diffusion of these beliefs, even in the most peripherical sites, were indeed monks and hermits (Haldon and Kennedy 2012: 325). Therefore, it is not entirely surprising to find many references to "divine interventions in in the defence of towns" (Leppin 2012: 248) in written sources, especially related to actions of bishops. The gradual collapse of Byzantine control over the Eastern territories, starting with the Persian invasion between 614 and 628 and culminating with the arrival of the Arab armies in 634, determined a clear shift towards more religious features in these communities' identity, and at the same time a more evident political role of the Christian authorities within the local administration. The example of Gregory of Antioch mentioned before is only one of several cases in which clergymen take on roles in political authority; also, other regions (like Egypt) attest other examples of treaties agreed for villages and town by the ecclesiastical authorities.

Furthermore, the combination of the broader Christian identity and the regional "sectarian" one had a role in shaping the regional and urban landscape. The huge increase of the number of churches and martyria are one direct consequence, also in remote settlements, as shown by some of the case studies, like Umm er-Rasas and Umm el-Jimal. But even more importantly, it could have been crucial also after the "Arab conquest" in maintaining these identity and settlement patterns for several reasons. The example of Evagrius proves that the Church itself was able to guarantee a continuous presence even if it was strongly connected with the Byzantine Empire: "The Roman Empire was based on Christianity – but the Christian organisation for its part did not depend on the Roman Empire" (Leppin 2012: 258). It was not only the persistence of the Christian hierarchies that were guaranteed in the first phases of the Early Islamic period, but also the universal Christian identity across the broader region. This can help explain why local communities continued to invest in churches and mosaics at least until the Abbasid period, as demonstrated by the case of Umm er-Rasas. Parallelly, this continuity in identities and settlement patterns was also facilitated by that fact that the first Caliphs and the Umayyad rulers never imposed a mass conversion upon their new subjects, not even on the nomadic groups who joined their armies. It is important to remember that some of these groups were part of the Byzantine forces in the preceding. In this context, the role of the "tribal" ties could have played a major role, surpassing political and religious differences.

Nonetheless, if the belonging to one of the different sects of Christianism did not determine a complete exclusion from the society, this status likely represented the basis on which the new administration superimposed a new taxation and legal system. This ended the definition clearly social groups, benefitting of different conditions, also resulting in new identity boundaries with clearer religious connotations. At the same time, cases of conversion to the new religion were not rare and increased over time, most likely stimulated by the legal and fiscal benefits that belonging to the *umma* guaranteed.

[212] In general, building churches in the rural context appears to have been a local enterprise that was desired by the villagers and not stimulated by the central authority. Possibly, this is confirmed by the diversified distribution and location of churches and monasteries in historic Palestine that is indicative of a villager-driven process of Christianisation as described by Bar (2003; 2005).

This fluidity of the religious context should also be also understood in light of the adaptability of the local populations.

7.4: A *"Kaleidoscope of identities"*

The socio-economic backdrop, the military and political engagement and religious conversion of the "Arab" population contributed to shape, change and maintain the different levels of the local identities. Though, attesting these phenomena through archaeological data is highly problematic. Even when historical sources are available, they are rarely reliable for the daily life and social structures of local communities, reflecting a more "city-centred" or elite perspective in most instances. All of the semi-urban or entirely urban contexts are normally cut off from the available accounts, and especially Classical and Late Antique sources quite often reverted to stereotypes when describing nomads and using terms as Arabs and Saracens (see Chapter 2.3). In general, these accounts can hardly capture the multi-faced and sometimes contradictory nature of identity of these groups.

With this in mind, epigraphic findings are precious and particularly useful, as they can yield information on the social structures that would not be available otherwise through historical sources or archaeological studies. Even if they do not directly mention and describe the actors who play a role in determining spatial patterns, this type of documentation is still informative in the reconstruction of which groups were the most important in impacting identity in region. One of the most problematic issues is the identification of the groups mentioned in these epigraphic sources. This is particularly true relevant for the recurrence of the term "Arab". Often the term had been referred to the presence of nomads in a specific area, since the dichotomy in the historical sources and in particular their generalised identification with the "Arabs" was particularly stressed.

In general, for the Hellenistic and Roman period at least, the epigraphic evidence suggests the presence of nomadic tribes, for which the diffusion of the so-called "Safaitic" inscriptions across the entire Ḥarra (al-Manaser and Ellis 2018: 70), with rare examples also found inside settlements, has long been considered a proof (Macdonald *et al.* 2015: 22-28).[213] One issue needs to be addressed first before considering these inscriptions: the mention of *mlk'* in Nabatean inscriptions or *basileus* in Greek is translated either as "king" or "chief". If the involvement of local groups in the protection of the Eastern border of the Roman Empire is also well documented by historical sources (see Chapter 7.2), the presence of such a denomination suggests a segmented social structure in which some lineages or individuals were able to access a privileged position within the broader societal hierarchy. For instance, a bilingual Nabatean-Greek inscription, tentatively dated to the mid-3rd century AD, mentions a "tutor of Gadhīmat king of Tanūkh" (Littmann, Magie, and Stuart 1930: 138; Littmann 1930: 38, Figure 87).[214] Since no "tribal" formation appears to have gained territorial control before the Late Byzantine period, the term "king" would likely refer to a social predominance among a specific social group.

---

[213] Macdonald also suggested that the presence of the Safaitic inscriptions inside settlements can be seen as a proof of contact between some sections of nomadic groups and settled communities already in the Hellenistic and Roman period (Macdonald 1993: 303-403).

[214] Both Greek and Nabatean inscriptions have been translated by Littmann as follows: "This (is) the stele of Fihr, (son) of Shullai, tutor of Gadhzmat, king of (the) Tanukh". If this stele is an interesting evidence for the presence of individuals recognised as "tribal" chief at Umm el-Jimal already in an early date, it is questionable who exactly the "Tanūkhs" are. Likely, they were formal foederati allies of Rome in the 4th and 5th century, succeeding to the Salīḥ (Fisher 2011: 79).

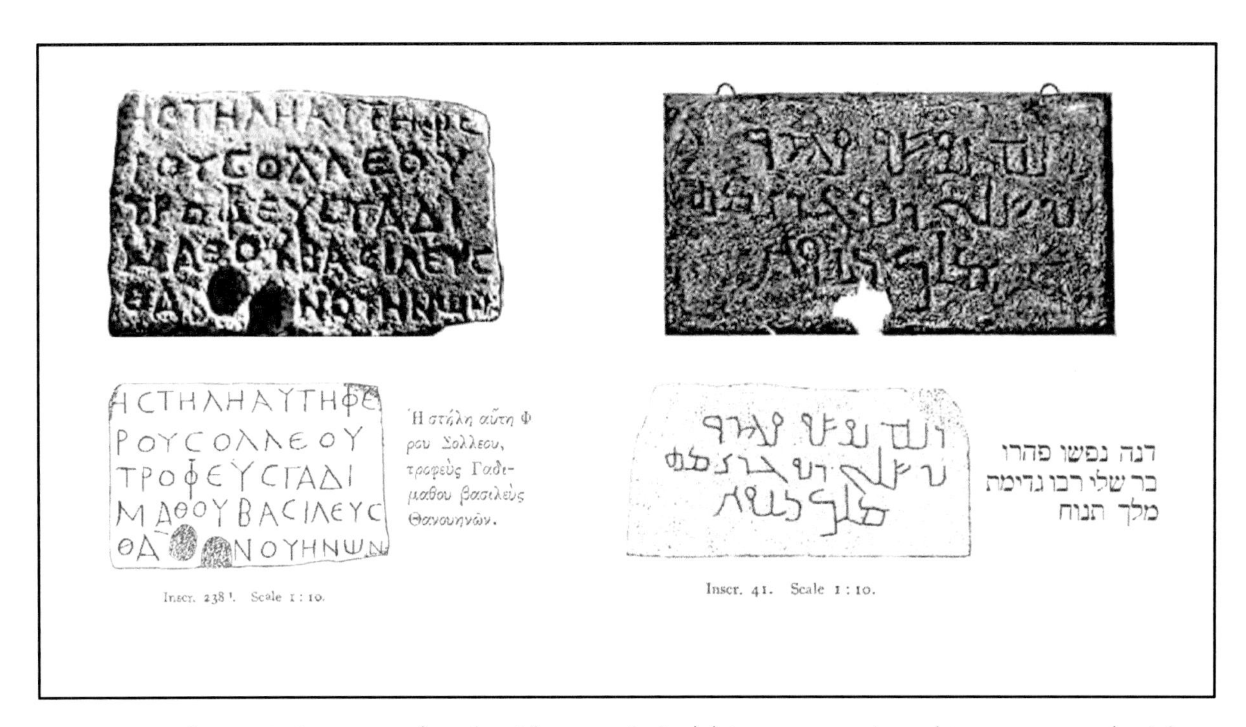

Figure 87 - Left: Inscription nr. 238 (Greek, with transcription) (Littmann, Magie, and Stuart 1930: 138); Right: Inscription nr. 41 (Semitic, with transcription) (Littmann 1930: 38)

Even if the strategy of incorporating local chiefs into the administrative structure appears to be fully developed only in the Byzantine period, much evidence documents the frequent contact that the Roman administration already had with the local groups. It is fairly common, especially in the Hauran, to find mentions of *ethnarchs* and *phylarchs*, probably "specific posts in the provincial administration" that in some circumstances may have also served as intermediaries between the central administration and entirely nomadic groups (Macdonald 1993: 373).[215] One of the most famous inscriptions regarding the military involvement of "Arabs" is the Namara inscription, dated tentatively to AD 328 (see Chapter 2.3). Despite the heated debate around its implications for ethnicity, it testifies a fundamental step not only in the development of the military involvement of "tribes" in the Roman defensive system but also reflects the formation of new identity patterns among segments of the provincial society that became increasingly engaged with the central imperial administration. In particular, contacts established while serving in the army might have made some groups aware of social differences and triggered a process of social self-consciousness. Though not yet on the level of a separate, specified ethnicity, it was enough to make them feeling different from both the *Romanoi* and the other "tribes" that were not directly involved in the Roman administration or under the Roman influence.

One further issue of importance is if these local groups who were involved in the Roman administration can be necessarily described as "nomads". The fact that inscriptions like the one of the tutors of Gadhīmat are also found inside settlements (as opposed to the Safaitic ones) may suggest that the opposite is the case, or at least could be indicative of groups with either settled and nomadic components. In general, Greek inscriptions in the settled areas, especially in the Hauran, suggest that

[215] To this concern, Fisher suggests that the term, either in the historical sources and in the epigraphic evidence, underwent an evolution that first functioned as a "neutral description of local tribal chiefs", and later into a formal administrative role and grade in the military hierarchy over a specific area as suggested by Macdonald (Fisher 2011: 79).

segmented social structures were indeed a shared feature between villages and nomadic pastoral groups (Macdonald 1993: 352-367), and at the same time this "sharing" does not necessarily indicate sedentarisation (Macdonald 1993: 352). Avoiding a clear-cut nomadic-settled dichotomy, a "tribe" (whatever the term describes in each individual occurrence within the epigraphic records) could have included both more nomadic pastoral segments and entirely settled members, being the two possible extremes in a continuous spectrum of economic strategies. This is the impression also given by epigraphic evidence in Umm el-Jimal, where the same group is possibly mentioned in both Safaitic and Nabatean/Aramaic inscriptions (Macdonald 1993: 357-358).[216] In general, Safaitic inscriptions found in the Ḥarra demonstrate that contacts between nomadic pastoral groups and settled communities were frequent. In particular, the use of Greek letters to transcribe Old Arabic (Al-Jallad and al-Manaser 2015: 63) or even properly Greek translations in bilingual inscriptions (Al-Jallad and al-Manaser 2016: in particular 57-60), which share the same "Safaitic narrative" (connected with pasture and surveillance of herds; see also al-Housan 2015) is an important proof of social and cultural exchanges. As mentioned earlier, these types of exchanges could have been also used by nomads who enlisted as Roman troops, where Greek was the only possible medium of communication with the authorities.

There is a distinctive hiatus in the epigraphic records of references to "Arabs" or even to specific "tribes" between the 4th and 5th century, making it difficult to track changes in the identity and social patterns of these groups during the transition into the Byzantine period. The reasons behind this "disappearance" are debated. Apparently, tribal connection or the socio-political role given by serving in the army possibly were not felt sufficiently enough to characterise them as separate social entities as was the case in the preceding period. Nevertheless, even if "absent" in the written or archaeological record, they did not necessarily entirely lose their relevance as a social group. Indeed, the spatial organisation of settlements might be seen as one of the ways in which these social identities continued to be expressed. In particular, the maintenance of the same privacy-systems and the diffusion of similar house and the quarters typologies between periods (especially in the Hauran) suggests a continuity in the importance of family structures, not exclusively on a practical level, but also in terms of identity. In addition, the Byzantine period is markedly characterised by the Christianisation of the society as a whole, closely linked to the involvement of these groups within the Byzantine army, and the process is shown in the epigraphic data. The dedications of churches including the name of representatives of "tribes" involved in the Byzantine army, most notably the Jafnids or individuals connected to them are important examples of these relationship (for instance: Fisher and Wood *et al.* 2015: 320-321 and 323-324). The "intrusion" of religion is ultimately an aspect involving the entire political and social sphere of Late Antiquity, as clearly demonstrated by several sources. Likely, the first groups that were converted to Christianity were more directly connected to the Imperial elite, that being for socio-political benefits or for a sincere assent to the new religion. As far as the rest of the population is concerned, the epigraphic record suggests that the Christianisation of the region as a whole was a slow and gradual process.

Correlating to this gradual Christianisation of the region was the development of the "Arab" identity in which the case of Khirbet es-Samra is a particularly interesting example (Desreumaux *et al.* 2011: 285-304). Here, from the beginning of the occupation of the site the local population had been described prevalently as "Arab", and in general the sedentarisation processes are believed to have played a crucial role in the development of the site. A sort of "cultural resistance" was proposed as a distinguishing

[216] Macdonald (1993: 367) is more cautious in suggesting the presence of tribes with either nomadic and settled groups, since no direct evidence for that is presence. Nonetheless, he admits the possibility that this lack of evidence may depends on the different context the authors of the inscription were at the moment the inscription itself was realised.

feature in the development of the site, resulting not only in the maintenance of strong family traditions as the foundation of the local tradition, but also a later occurring but rapid Christianisation starting from the 5th century onward. This process is reflected by the erection of several churches and private chapels. By the first half of the 7th century, the majority of the churches are also paved with mosaics bearing Greek inscriptions. However, the churches reflect an adaption of pre-Christian traditions to the new religion, which is especially evident in the creation of family chapels. This feature is common to several of the case studies, and possibly testifies a major role of the family in the identarian pattern of the local communities, despite the widespread and dramatic advent of the Christian faith.

The conspicuous epigraphic evidence follows the same trend. Inscriptions on funerary *stelae* in the Roman cemetery, that was likely in use until the 4th century, are mainly in Greek, but the onomastic pattern can apparently be connected to the Nabatean world and other evidence in the Hauran, with a copresence of Roman gentilities and Semitic theophorics. In the 5th century, the Safaitic language disappears from the inscriptions,[217] and is substituted by Aramaic Christo-Palestinian (appearing in the Christian cemetery between AD 550 and 650 and associated with Aramaic and Arab names) and Greek, which was still the official language at that time. The cultural ties related to Hellenization seem to be well-defined within this cemetery, as names of individuals from the period preceding Christianity are never written in Aramaic and later on all of the Christian names are in Greek.

Another important source of evidence for investigating social structures is the exceptional finding of a set of papyri at Nessana in the Negev,[218] all dating from the 6th to the 8th century AD. Even if the papyri refer specifically to Nessana and the Negev context, such evidence offers important insights on the more general socio-economic and political dynamics between the Late Byzantine and the Early Islamic period.[219] In a recent analysis by Stroumsa (2008), she discusses four aspects that are relevant for the present discussion: the absence of a clear social stratification in the settlement; the lack of any reference to tribes or extended family groups; the proud sense of belonging to Nessana (even more marked it was also the place of birth); and finally, the strong continuity of the identity patterns between Byzantine and Early Islamic periods, despite some relevant changes on the administrative level. The absence of a clear-cut social stratification is reflected in the particular emphasis on professions given in the papyri (which can be understood as proper identity groups: Stroumsa 2008: 44-45), in spite of the individual's economic status or tribal association. People bearing Christian or Hellenic names appears to hold more prestigious positions, while more common professions are connected to traditional pre-Christian Semitic onomastics. This is a phenomenon that closely recalls what is seen in the inscriptions in the North Church of Sobata (Shivta), where Semitic names are also absent (Stroumsa 2008: 46).

[217] Interesting for the Safatitic onomastic is that the tribal affiliation often describes the individual as *mawlā*, "servant" of tribe; the same formula applies to orphans in relation to their adoptive father. Such onomastic tradition continues well into the Islamic period (Dirbas 2016: 144-146).
[218] The Papyri were found by the Colt expedition in 1937. They include 13 literary texts and 195 documents, of variable length and state of conservation (Kraemer 1958; Stroumsa 2008; O'Sullivan 2014). 49 documents are datable from 505 and 630 (including contracts, accounts, private letters, and one fiscal register); 35 are dated from 674 and 690 (21 of which are fiscal documents, among which 9 *entagia*: O'Sullivan 2014: 50).
[219] It would be particularly important to compare such general conclusions with the evidence coming from Petra, where the Papyri found in the Room I of the Byzantine Church complex partially overlaps the same chronological span as the Nessana ones (the earliest one is dated to 537, while the last to 593): some differences between the two sets are mentioned later in the discussion, but a more systematic analysis of the social implication of Petra's evidence is required, after the recently concluded phase of publication of the papyri (published by ACOR in six volumes).

Nevertheless, professions were not necessarily dependent on family tradition,[220] and social mobility among the different professional groups was not thwarted by legal or social practices (Stroumsa 2008: 50-51). Such professional divisions in the settlement are most likely connected with economic stratification, but did not necessarily translate into a marked social stratification. The absence of a proper *elite* (either social or economic) is suggested by the fact that despite the mention of "prominent citizens" (for which professional connotation varies), this group never formed a city council. Similarly, it was also noted that there was no reference to farming or pastoralism, two activities in which the majority of the population, which was not only the case for Nessana, was likely involved. . Possibly, these professions were not considered as separate occupations, but constituted a common economic basis for the whole population of the town, and thus not worth being mentioned in the papyri (Stroumsa 2008: 48). In general, it does not seem that belonging to different professional groups triggered any consequences on the organisation of the settlement, such as the creation of physically separated quarters. The papyri even report soldiers and members of the local Church living together with the rest of the population.[221]

A second remark concerns the lack of references to tribes or extended family groups. Clearly, a strong relevance of the family in the daily life of Nessana is well attested in the papyri, especially in terms of a "strong sense of mutual reliance and interdependency within the family" (Stroumsa 2008: 72). One main example is the fact that the tax system was based on households and did not count individuals. Yet, such family groups are apparently quite limited in dimension: in fact, if the relationships between siblings had a major relevance, any other blood ties would not have had any practical importance. The family unit appears to be exclusively considered as nuclear: a single household formed only by siblings-ties, with no regard to the presence of parents. Despite this restricted conception of family, one element has a direct consequence on the spatial organisation of the settlement: references to joint ownership of lands, houses, courtyards, and threshing floors among other issues is extremely frequent and oftentimes the owners are related (Stroumsa 2008: 75). It follows that even if the belonging to more extended family groups is never clearly expressed in the papyri, their importance and presence in daily life is clearly, though indirectly, reflected in these jointly owned properties. In light of such a disparity between legal and ideological levels, it would not be illogical to suggest that a similar relevance of extended families could have been manifested in a settlement's organisation as well.

Nonetheless, the development of a distinct sense of "community" appears to be the main factor in determining both the identity of a group and the morphology of a settlement. The idea of "mutual responsibility" is reflected in the settlement by the lack of separation between public and private spaces (even if some papyri do refer to "public street" while dealing with limits of properties)[222], and was possibly a local response to the difficult environmental conditions in which these communities were faced with.[223] The "communal" identity of the people of Nessana is ultimately the only collective

[220] One exception could have been the clerics of St. Sergius, the church where the Papyri were found: in fact, they appear to have belonged always to the same family, surely from 550 until 690 (O'Sullivan 2015: 50).

[221] One difference between Nessana and Shivta is noted regarding the burial practices of the churchmen: in Shivta, it followed a separate procedure from the one of the laymen. This differentiation may indeed suggest a more rigid social differentiation based on the profession (Stroumsa 2008: 53-55).

[222] Probably in contrast with the proper cities, the distinction was probably more labile and less legally defined: this founds a confirmation when Stroumsa affirms that "in many cases the village relied on extra-judicial, consensual agreements, enforce by the members of the community" (Stroumsa 2008: 76).

[223] The direct consequences of this situation are also the communal efforts to create and maintain water-management systems. This idea closely recalls what Rapaport suggests for the tribal organisation of *badw* in the Mamluk Fayyum, where the more egalitarian structure could offer a better solution to a lower cereals' production of some members of the group following a low level of the Nile: see Rapoport 2004.

identity that is clearly expressed by the papyri. This third remark on the Nessana collection is particularly informative also on the attitude of the local populations towards the imperial institutions, which interestingly did not undergo radical changes over time (and most notably after the "Arab conquest"). The stress on the communal identity of the people of Nessana suggestions a feeling of resistance or weak belonging to the Empire, a simple acceptance of its presence and rules (Stroumsa 2008: 143). In general, the papyri express a clear distinction between the people of Nessana (yet never defined ethnically[224]) and the "outsider" groups, whose people are normally described by the word "Saracen" (Stroumsa 2008: 150-172). The insistency on the provenience from a specific settlement - not even a region - seems to be a reinforcement of a common identity: references to the personal origin from a determined settlement is frequent, even when not required, reflecting a well-established method of self-identification (Stroumsa 2008: 243-246).[225]

The papyri cover more or less two centuries of the settlement's life, from the 6th to the 8th century AD, expressing a striking continuity in the identity and daily life of its inhabitants. Yet, the fourth and last remark on the Nessana collection points out one element of discontinuity, i.e. a progressively more direct involvement of the central Umayyad administration into the local life than what was experienced in earlier periods by the people of Nessana. This does not necessarily mean that there was a forced imposition of a new legal system, since it appears that in several occasions the governor agreed to rely more on the informal status and social conventions (Stroumsa 2008: 79). Still, a more centralised and more insisting supervision was present already in the AD 670s (Stroumsa 2008: 143). This has a direct parallel with the papyrological evidence coming from Egypt, even if with some clear differences.

Despite the fact that Egypt had always had a "different" administrative history, maintaining a particular status through the Roman and Byzantine period, it does show some interesting similarities with the Negev, or at least with the evidence from Nessana, in particular as far as tax collection[226] and justice is

---

[224] The lack of such ethnic determination may also explain the pacific transition under the Muslim control, opposite to other realities in the region where the arrival of an external group determined the "emergence of ethnic consciousness from dormant phase for some groups" (Stroumsa 2008: 182). Not even the linguistic element can help determining an ethnic identity of the community in Nessana, since the change of language do reflect more a switch in the conversation register and an adaptation to the context, than a statement of identity (Stroumsa 2008: 209-211; Stroumsa 2014: 153). The papyrological evidence for Nessana suggests actually that the local population was using comfortably either Greek and Arabic, even if the latter is almost exclusively vernacular, at least until the 690s; the choice of one idioms rather than another was totally dependent on the message to convey (also in terms of power and prestige acquired by its use) and to the context (Stroumsa 2014: 147-152). Ultimately, the shift from Arabic to Greek in the official documents is not to be understood as "a story of a shift from one spoken language to another, but of the change in the connotations each language carried" (Stroumsa 2014: 155).

[225] "This emphasis on one's specific village, town or city as the most salient and most useful marker of man is also found in contemporary inscriptions from the immediate area. The use of local specifications is very common in registers of donations, where people from different communities are included". Further on, "ultimately, the only sense of identity which can be seen to be indisputably operative and important for the people themselves is the local one, in its most limited sense - the identity of the village community" (Stroumsa 2008: 246).

[226] Payment of taxes is still due in kind or gold. Yet, instead of having one single tax as in the Late Roman period, the Umayyad administration split it in two different taxes, the *demosia* (tax paid by landlords) and the *epikephalion* (by adult male) (O'Sullivan 2015: 58-60). The Nessana Papyri also refer to the *rouzikon*, interpreted as a "small extra payment [...] to local Arab soldiers" (O'Sullivan 2015: 60). The main difference between Late Byzantine and Early Islamic periods appears to be a higher fiscal burden on the communities of the Negev during the 7th century, referred by O'Sullivan as matching the theoretical model of "Maximum Exploitation" (O'Sullivan 2015: 59-65). Such fiscal condition is hardly sustainable over a long period, and the evidence of the papyri seems to confirm this hypothesis. For instance, Papyrus 75 clearly refers to tax-protests organised against the *symboulos* (governor) in Gaza by some representants of Negev's town, among which figures also Shivta (O'Sullivan 2015: 62-63).

concerned. The evidence from Egyptian reflects the same pattern following the "Arab" conquest as in the Near East, with a pronounced continuity in social and political structures on almost all levels. Within fifty years, however, a clear change in the administrative regime did take place, with a "more dominant presence [of the central power] at lower administrative levels within the indigenous population" (Sijpesteijn 2013: 260). This is only one manifestation of the Islamisation process that occurred in Egypt. Nonetheless, together with the progressively deeper introduction of new administrative practices, through a more capillary presence of the rulers' emissary on the local level, there was a clear desire and effort to continue local traditions (Sijpesteijn 2013: 262-263). This strategy was not only adopted in Egypt, but was likewise implimented elsewhere,[227] in order to form a new Islamic culture, which is ultimately the result of a continuous and pacific debate with the local traditions where "elements from this local culture were adopted, adjusted, rejected, or replaced by new(ly formed) customs introduced by Arabs" (Sijpesteijn 2013: 264).

In general, several aspects of continuity in the society emerge from the papyrological evidence, despite the establishing of the Islamic rule. Considering these sources together with the epigraphic evidence allows us to delineate a complex pattern of identities, which somehow could be reflected in the "multi-layered" organisation of the settlements. Despite the clear emphasis on the nuclear family in the papyri, there is no evident contradiction with the belonging to a larger social group (like an extended family, a clan, or even a "tribe") as is documented by the epigraphic evidence or to a local community. All of these different "tassels" (family, an extended social group, a town or a village or to other kind of communities) are combined and layered in the "kaleidoscope" of the individual self-perception, and ultimately the formation of an identity. Depending on the specific context (social-economic, geographical and of course political), some of these tassels can become more well-defined and thus reflected more often in the material evidence.[228]

It appears that through all of the different changes in identity patterns, family ties (of variable extensions) always remained more or less important in defining the single person and influencing several aspects of daily life, for instance in the organisation of houses and quarters. This could explain the homogeneity in the architectural typologies across a wide geographical context. Economic, political, and religious factors as delineated in this chapter could have ultimately created further overlaying levels of self-definition, active in specific contexts and moments, determining degrees of inclusion or exclusion of a discrete group into a larger social group, as visible for example on a settlement level or on a trans-regional scale.

---

[227] This finds a further confirmation in Papyrus 77 from Nessana, where Nabr ibn Qays, a high official (possibly the governor himself), reprimands local officials for having oppressed the people of Nessana (O'Sullivan 2015: 66). This evidence is the counterpart of what emerges from other contemporary documents from Nessana, where the population protests against high taxes (see note 226). It interesting to note though, that other evidences suggest that the central authority still levied on Christian officials to rise the payments locally (O'Sullivan 2015: 66).

[228] See for instance al-Salameen and Hazza (2018: 91-92) and al-Otaibi (2015). Both studies underlined as the reference to a Nabatean identity (together with other more precise tribal affiliations) in some Safaitic inscriptions as well as in a 2nd century Palmyrene dedication might reflect the author being a "foreigner" among the community where the text was composed, thus expressing other sort of identities from those normally used. This finds a direct comparison in the use of simple patronymics (and not tribal or ethnic affiliations) in most of the Safaitic inscription from the Ḥarra: these individuals were well-known and easily recognisable because they belonged to the same context. This explains also the reference to local events to date the texts, whose significance was hardly understood from other groups (Al-Housan 2015: 95).

Two examples are particularly relevant for this overlap between family and other sort of identities. One is the ideological development of Islam, where the conception of the *umma* as a supra-tribal group, still framed into the earlier system of true (or created) blood-connections between different social groups, is a clear indication of how religion in Late Antique functioned as unifying factor in an extremely diversified society. Likely, family and kinship relations were one of the most recurring elements of self-identification on the local level, while "Romaness" or being Christian could have been more useful in broader regional or transregional context. The emergence of the "Arab" identity could be read in a similar way. As mentioned earlier in this book (see Chapter 2.3), it is only in the Abbasid period that a proper definition of "Arabness" was developed in terms of a real *ethnicon*. Nonetheless, ethnicity is only one aspect of an identity. Simply stating that there are no "Arabs" present in the region prior to the Abbasid period since no sense of ethnicity or shared identity is established in the sources[229] neglects all other potential spheres of identities where such a perception of alterity by a group could have been felt. The enlistment of "Arab" groups in the army (whether or not this label was a self-definition of these groups themselves) might have generated a distinction among the soldiers, on a geographic, social and/or cultural basis. The reference in the Namara inscription to "King of the Arabs and King of Ma'add" does not necessarily reflect the presence of two distinguished ethnicities, or purely honorific titles, but could for instance distinguish different social "tribal" groups that either were integrated directly into the Roman and later Byzantine armies or were separate social entities.

7.5: The "ruralisation" of the spatial model of the urban space

Parallel to the shifting economic and political climate and the broader religious conversion of the population, there were simultaneously significant changes taking place in the cities in the East from the Byzantine period onwards, which both had an impact on changing identity patterns of various groups and the morphology of settlements in the region as a whole. Moreover, while in this period semi-urban and rural settlements developed and became increasingly autonomous, former urban centres began to assume new features within their built environment. This "rural" settlement boom intensified during the 5th century and a large range of settlements that can be designated as semi-urban developed and absorbed some properly urban functions and possibly acquired a major degree of independence from the internal administration, also thanks to the role of the local Church (Haldon and Kennedy 2012: 333-334). At the same time, some new features found within the *poleis* closely resemble elements characterising typical rural and semi-urban settlements. Perhaps, this shift from an ordered organisation of the urban form to an unplanned and randomly structured urban form could be described as a "ruralisation" of cities and urban settlements. It is evident that some of these aspects of

[229] One of the problematic issues in Webb's argumentation about the "Arab-less" population in the pre-Abbasid period is actually the different destination and context of the different sources (Webb 2016: 60-109). For instance, Webb presents the battle of Dhū Qāras as an excellent example of manipulation of the historical memory in order to create a common myth for Arab ethnic identity: the pre-Islamic epic clash where the Arabs defeated the Persians for the first time. The concept of the Arab identity appears only in the Abbasid texts, while in the local pre-Islamic and Umayyad sources it is absent, although the latter refers to specific tribes in the first case and to the "Ma'add", the latter considered by Webb as the only real ethnic group that is identifiable in the Pre-Islamic and Umayyad Arabic peninsula. In my opinion, one alternative interpretation could be that the context in which the pre-Islamic story of the battle originally circulated was potentially more geographically and culturally more circumscribed than the one of the Umayyad and Abbasid periods. Therefore, the original audience was likely able to recognise the singular smaller tribes named within the story. Similarly, the absence of a connection between the Arabs and Ma'add in the genealogies does not necessarily demonstrate the absence of Arabs, but simply the fact that the term was used to define other kinds of identities than genealogical or ethnic identities.

the new development of the Byzantine cities close the gap to a certain extent between the urban and rural realities (see also Hirschfeld 1997: 71).

Before Late Antiquity, the distinctive character of the *poleis* was clearly expressed through morphological elements, which reflected the administrative status of the city itself, although it was rarely developed from a purely Hippodamic structure. A combination of different architectural knowledge and traditions generated a diversified urban scene, which kept local environmental and topographical conditions in consideration, with an emphasis on "scenography" purposes that was meant to impress the visitor.[230] This panorama began to change radically between the end of the 3rd and the beginning of the 4th century, when several settlements with no clear earlier urban tradition were elected to the "rank" of city. Under such circumstances, one can observe to two different and possibly opposite developments within urban settlements: the maintenance of the former urban structure, despite the construction of new buildings (possibly monumental) required to assert this new status (as seen in Hisban/Esbus) or the complete re-foundation of the settlement itself, according to the Hellenistic urban aesthetics (like Shahba/Philippopolis, which were totally renovated once Emperor Philippus the Arab accessed to the power in AD 244).

In general, it was the common scholarly opinion before the 1980s that Late Antiquity represented a complete disruption of the classical city. Urban settlements were to be seen as unorganised and chaotically developed because of the lack of an authority strong enough to administrate building activities. The population, believed to be a completely new social group who possibly came from non-urban contexts, was free to demise the monumental complexes in the city, indiscriminately occupying public spaces. This was the idea guiding Sauvaget in his description of the encroachment process taking place in the porticoed streets of the Hellenistic *poleis* (Sauvaget 1934; 1941), seeing this as the progressive development of the "suq", one of the principles of what he defined as "Islamic cities". This interpretation of the development of the urban fabric in the Late Antiquity as an disorderly densification of the space was widely accepted, and it was further applied to the social dynamics within that space, where private citizens could take advantage of the lack of a clear civic authority that once regulated the expansion of their own properties. Ultimately, it was viewed that there was a physical rupture of the urban order as reflection of the collapse of the political institution.

Sauvaget's theory triggered a gradual reconsideration of the apocalyptic view of the Middle Eastern Late Antique cities, where a conjunction of political instability (mainly caused by Arab tribes), widespread destruction (because of the Persian and later Arab invasions and earthquakes as well), and plagues depicting the deterioration of once formidable urban centres. Kennedy's paper "From Polis to Medina" (1985) first underscored the strong continuity of the urban development between the Byzantine and the Islamic period that finally led to the transformation of the classical city. This transformation for the first time was viewed as not necessarily negative in its outcome, but rather reflecting a change in the attitude of the population and a different vitality of the city itself (Kennedy 1985: 27).

At the same time, the development of ceramic typologies, resulting in more precise dating of the archaeological remains, led to the reconsideration of the Late Antique period in general. As a consequence this improved chronology began to shed light on when the changes of the urban form in the Byzantine period precisely occurred and later "demonstrate[d] greater continuity, but continuity

---

[230] This is particularly evident in cities like Jerash, Bosra and Palmyra, as noticed by Gros and Tonelli (2010: 463-469). It must also be underscored that some of these urban centres maintained embedded in their new Hellenistic urban grid also earlier nuclei, which appear quite different in their spatial organisation (as in Bosra).

that allows for change" (Walmsley 2007: 37). The major change that took place, already identified by Kennedy, was the inclusion of larger industrial and productive installations into the urban space , as seen in several major centres like Jerash in Jordan,[231] and parallelly the construction of other commercial spaces, materialised in the closing of *portici* on the main *cardus* and occupying the *agora*. Additionally, archaeological evidence from Jarash suggests the development of "irregular" quarters from at least the 4th century.[232] This is a clear discontinuity from the classical urban tradition, where production areas representing potential hygienic or security issues, like tanneries and kilns, were relegated to extra-*muros* areas. These changes likely demonstrate a diverse thriving economy. The "encroachment" process itself, as it was termed, does not appear to be a random development of the lack of administrative control, since in many cases the central paved street was kept free from the new construction and only former sidewalks were obstructed (Walmsley 2007: 38).[233]

Jarash in particular is a useful example, where the involvement of the civic administration, which was still present in the 8th century, can be attested in regimenting the urban development. This involvement is particularly visible in the area of the *Tetrakionia*, where the new congregational Mosque was built (Walmsley 2018: 248-250; Figure 88). Since the mosque orientation did not follow the Roman streets' framework, a series of structures were added to hide the discrepancy, guaranteeing the maintenance of the urban symmetry, yet respecting the limits of the paved street (Walmsley 2007: 84-86). Though, it should be taken into consideration that the different orientation of the mosque followed a similar discrepancy of an earlier constructed bathhouse, which it replaced and that was first built in the 4th century. The construction of the mosque itself experienced a change in the original plan in its First Phase (dated between AD 725 and 735) in order to better fit the new building in the geometric grid formed by the *Cardus* and the *Decumanus* and at the same time with the residential areas developing intensively to its West (Blanke 2018: 47-48; Walmsley 2018: 248-251).

A new attitude towards the organisation of the built environment is expressed in the urban context, where the consuetudinary aspect of the legal system was kept into consideration by the administration (either Byzantine and Islamic). This more closely resembled what attested for the rural sphere, in that the development on the rural built environment appeared to be less framed into rigid urban grids, following more likely the availability of areas to build and a more flexible division between private and public spaces. A good parallel in order to understand the development of private structures in Jerash, mainly the occupation of the sidewalks but not the paved streets, is the Islamic later codification of the *finâ'*, a preferential right of using part of the communal space. In these cases, the guiding principle is the *darar*, *i.e.* the maintenance of an equilibrium between the private and the communal interest. Moreover, if the new expansion of a private building did not determine an obstacle or a violation of other's space and interest, the new development was allowed. This idea not only resembles the Byzantine legal idea offered by Ascalon for the provincial urban context, but also for the development

[231] Despite the fact that the Late Antique period bore witness to the highest development of industrial complexes in Jarash (in particular pottery kilns and tanneries), it is interesting to note that the distribution within the city changed through time. In particular, a marked shift in the location of pottery kilns appears to follow major epidemics between late 6th-7th century: earlier industrial areas were used for mass burials and were never re-opened. Therefore, a new industrial district developed, close to the North Theater and the Artemis complex, and continued to be used well into the Islamic period (Kehrberg-Ostrasz 2018: 119-130).

[232] In particular, the Southwest district, later intensively re-occupied in the Abbasid period (Blanke 2018: 39-58); and the Northwest Quarter, which also appears to have been heavily developed from the 3rd century (also with extensive works of terracing) and to have assist to a consistent Early Islamic residential occupation (Lichtenberger and Raja 2018: 5-36).

[233] Similarly, Di Segni (1995; 1999) demonstrated how civic and religious authorities, despite marked local differences, were very much engaged with building activities in urban centres.

Figure 88 - The Tetrakionia in Jerash with the Mosque (highlighted in blue) (satellite image: Google Earth)

of rural settlements in general.[234] Shivta is a good example for small planned sections that probably followed the same paradigm, but the same could be suggested for the development of "conjunctive" houses in the Hauran. Consequently, the steady evolution of the encroachment process should be understood not in the light of a continuity in legal system between Byzantine and Islamic periods or in the lack of authority or power of the civic institution, but in terms of a continued and persisting application of consolidated social practices.[235]

---

[234] A recent edition of Ascalon was made by Saliou 1996. Avni shows that "one of the most striking aspects of the comparison between the sixth-century Ascalon and eleventh-century Fustat is the similarity in the description of houses and commercial-industrial installations, and their location in the urban context. Large mansions of wealthy people were located near the bazaars and industrial zones, and the basic form of central-court house with the rooms flanking an inner courtyard was continuously maintained in the cities and towns of the Byzantine and Early Islamic period" (Avni 2014: 100).

[235] I would like to thank Prof. Jean-Pierre Van Staevel for discussing this topic with me in person after a lecture held in Bonn in June 2017.

The change of the urban form began around the 4th century and underwent different moments of intensification or abatement according to the regional context. Such local differences could have been determined for instance by the emergence of a new province capital, as was the case for Beth Shean and Tiberia in the Early Islamic period, when the latter became the new capital of the *jund* (Walmsley 2007: 71-112; Avni 2014: 40-106).[236] Nevertheless, the continuity of urban life was maintained in most cases; most importantly, the idea of a city and its ideological centrality in the settlements' hierarchy wasn't undermined by these historical developments. Ultimately, this new spatial and functional reconsideration of the city was the result of a different "balance between monumentality and functionality, and public and private initiatives" (Avni 2014: 106).

Similarly, the Islamic period did not mark the end of large programmes of urban renewal or monumentalising of the city. On the contrary, as the case of Jerash demonstrates, several building projects were realised and these initiatives were enacted either by the local population or the "central administration" that were in continuity with the Late Byzantine patronage structure (Di Segni 1995: 312-332; Walmsley 2007: 90). Although the idea of a "ruralisation" of the urban space during the Byzantine and Early Islamic periods might imply that cities had lost their hierarchical prominence among other settlements, some of the new spatial features of the urban organisation became more similar to the rural organisation of settlements. For instance, the diffusion of clusters of residential buildings and "Complex houses" possibly suggests that shared family ties among their inhabitants is a macroscopic phenomenon that is not limited to the rural sphere. Archaeological data from Jarash clearly demonstrates that such new "rural" domestic architecture is found starting with the Early Islamic period and in particular in the 8th century within newly developed or re-arranged districts.[237] The new dwellings are distinctly different from the pre-Islamic ones, which were normally arranged in geometric rows of five without open courtyards. Simultaneously, a radical change in the water system occurred, with the development of individual cistern providing water to the individual houses instead of a communal hydraulic system (Blanke 2018: 44-51). If these new forms and type(s) of organisation appear to be Early Islamic developments, it is also true that the coexistence of two parallel systems (either urban and hydraulic) could be suggested as well. While the new courtyard complexes do not produce materials earlier than the 7th century, it is also true that the major Early Islamic interventions completely erased earlier structures, making it difficult to reconstruct their form. For example, Jarash had already in the 4th century undergone developments that were morphologically different from the purely Hellenistic tradition. The reasons for such developments are complex and it remains unclear whether being a simple transfer of knowledge or a (considerable) movement of discrete family groups took place from the hinterland to the cities. Nonetheless, the continuity in the material culture of the cities is well-attested and does not suggest any abrupt ethnic or social change in the composition of the urban population, such as the mass exodus of the former Byzantine elite. It is still possible that some groups within the rural population were attracted to the cities following the expansion of various

[236] For instance, Walmsely affirms: "Although there was much continuity, the redrawing of the provincial boundaries was accompanied by a change in status for some towns. Most noticeably the former provincial capitals of Antioch, Apamea and Scythopolis/Baysan lost their privileged status, being replaced by Chalcis/Qinnasrin, Emesa/Hims and Tabriyah/Tiberias respectively, although this change may have predated Islam depending on when the *ainad* were formed. Such an alteration to a town's status clearly had a significant impact on its subsequent history, sometimes favourable, sometimes harmful depending on the nature of that change" (Walmsely 2007: 74).

[237] See the Southwest district (Blanke 2018: 44-48) and the Northwest quarter (Lichtenberger and Raja 2018: 158-161).

productive activities. Likewise, this movement of people was not as dramatic as to lead to the once long supposed abandonment of the hinterland.

The "Arab conquest" had been long considered as the leading cause of the end of "classical cities" in the Near East. The arrival of new groups of people, many of which were thought to have originated from a culture with no formal urban traditions, and the widespread of Islam were seen as the two main factors determining the fate of the previous urban forms. When considering these two factors in more detail, several studies built upon the idea of a distinguished "Islamic" or "Muslim" city, constituting a break with the former "Classical" urban tradition in its morphology and the administration system which managed its respective built environment. Even if in some cases the use of the terms of the "Islamic city" and the "Muslim city" is ambiguous, shifting from being a purely chronological label to a properly cultural or religious one, the debate has been long and heated across several disciplines. The first major contribution to this debate came from historians, art historians, and geographers, together with scholars in architectural-urbanistic studies, who focussed on the Ottoman city in particular. After Sauvaget, who in a way continued an Orientalist perspective of the urban phenomenon in the Islamic world (*i.e.* the "Muslim city" as a result of anarchy and the absence of administration[238]), these scholarly contributions began to focus on the relationship between the urban form and legal and religious texts, finally establishing the cultural independence of the urban phenomenon of the "Islamic city" in respect to former local traditions (see for instance Hakim 2008). As a consequence, a great emphasis was given to the role played by religion in the development of the architectural and urban forms and therefore the presence of buildings that were categorised under specific "Islamic" laws and institutions, like mosques, madrassas, hammams, khans and souks, which had been often considered sufficient to define a city as "Islamic".

Later on, a group of scholars formed who opposed this narrative, and instead argued for the continuity of certain spatial solutions from previous periods. Wirth's idea of "Oriental city" is noteworthy, as he argued that many features considered as typical of the "Islamic city" (irregular network of streets and dead ends, houses with central courtyards, and strict division in quarters) were all borrowed from urban experiences dating to much earlier periods, some of which even dating back to the Assyrian period. In his opinion, only the *suq* was a cultural construction of the Islamic period (Wirth 2000: 103-151). In light of the recent archaeological evidence mentioned above, the idea of the "Islamic" city is misleading for several reasons. Even in the circumstance it is used exclusively as a chronological label, it does not sufficiently acknowledge all of the changes the urban centres underwent during the several centuries of Islamic history. More importantly this term does not reflect the local variation of the many regions where the religion of Islam spread and its various impacts on the respective local culture (see also Raymond 1985). The idea that the urban phenomenon was already brought by the military conquest during the Umayyad and Abbasid expansions and remained substantially the same from the western to the eastern borders of the Caliphate neglects the formation of the Islamic culture itself. As seen in the Egyptian papyri (see Chapt. 7.4), both the establishment of the Islamic administration and the "Islamisation" of the local population were indeed in continuous confrontation with local traditions

[238] Sauvaget, quoted by Raymond (Raymond 1994: 3-18): "'The status of the cities,' wrote J. Sauvaget, in his famous article on Damascus, 'is subject to no particular provision in Islamic law. There are no more municipal institutions ... The city is no longer considered as an entity, as a being in itself, complex and alive: it is just a gathering of individuals with conflicting interests who, each in his own sphere, acts on his own account'. The city is no more than a non-city, Muslim town-planning no planning at all. The Muslim era, Sauvaget was to conclude energetically in his Alep,' is unaccompanied by any positive contribution... the only thing we can credit it with is the dislocation of the urban centre ... The work of Islam is essentially negative'; the town has become 'an inconsistent and inorganic assembly of quarters'; it is as if it were 'the negation of urban order".

through a selection, exchange, change and refusal of different cultural features. Religion, as is much emphasised by many scholars sustaining the idea of the Islamic city, in fact did not represent the guiding principle for the configuration of the urban form in the Islamic period, at least in its early stages. The urban phenomenon continued to be mainly based on the consuetudinary spatial behaviour of the local population and tradition already in place.

At the same time, one's interpretation of the built environment should not be biased by the appearance of new architectural typologies or forms, since the continuity in use of many churches in the Near East does not necessarily imply that social or cultural changes did not take place. Therefore, the construction of mosques within a settlement does not necessarily make it an "Islamic" settlement. Similarly, the caravanserai and the *suq* are not "Islamic" inventions, but simply developments that already began in the Late Antique period. The same functions, though with different architectural features, can be seen in the Byzantine period, when some of such complexes were already embedded in the urban fabric. For instance, House XVII-XVIII in Umm el-Jimal and Building XII in Mampsis can be possibly interpreted as urban caravanserai.

By no means did the urban phenomenon remain completely the same in the Islamic period as it had the Roman and Byzantine periods. In fact, aspects of discontinuity are recognisable and can also be explained by the influence of different cultures in terms of exchange of knowledge and esthetical paradigm, as are the movements of people. As Avni states, "the slow and gradual transformations of cities which included the abandonment of the rigid Roman construction principles, was visible in the sixth century, and it continued to develop for another half millennium, until the eleventh century. The Early Islamic cities of Palestine and Jordan showed a similar orthogonal street grid to their Romana and Byzantine predecessors, evincing a gradual process of change. In fact, the emergence of the medieval medina in many regions of the Near East was visible only in the thirteenth-fourteenth centuries" (Avni 2014: 106). In this period, the urban phenomenon bore witness to a strong regionalism, as the stark differences between the Maghreb and the Syrian cities, for instance, show. This is clearly the result of the different ways that "Islamic culture" interacted with varying local traditions, rarely erasing former customs and building traditions completely. Nonetheless, the ways these changes took place within settlement morphologies seem to follow well established principles, set on a strong consuetudinary foundation as far as the spatial organisation is concerned. This continuity is particularly visible in rural settlements that in turn share several aspects with the Late Antique developments in urban centres.

Chapter 8: Arab settlements or settlements of the "Arabs"?

Is it possible to link the development of specific types of settlement, or at least the adoption of determined spatial patterns within a settlement, with specific demographic phenomena (like the sedentarization of nomads) or specific social structures (like the "tribal" segmented ones)? The short answer is: to a certain extent, yes. Since there is a great degree of variability in settlements, which is highly dependent on the regional socio-economic context a site is located in, a case-by-case perspective was adopted in this study to answer this question at least for the local level. As the sites considered in the present work demonstrate, similar spatial structures can reflect similar social structures, yet may also be determined by different social phenomena that are more difficult to identify. The fact that settled and pastoral communities share several social structures represents the first and main obstacle to our understanding and interpretation of the development of certain settlements. One of the major problems addressed in this study is that pastoralist and settled groups are often considered two clearly defined, separated, and (perhaps even ethnically) different communities, which are exclusively connected by economic or political interests. In an early stage of settlement development, it is possible that small groups did not initially share any social connections (like in the case of newly arrived tribes in the Negev following the Islamic "conquest"), while as time progressed it appears that changing economic and political interests triggered the formation of 'new' social phenomena, such as the integration of the newly arrived group into the 'original' group. In other historical contexts where the arrival of new groups took place, like in the case of some Yemeni tribes following the military campaigns of the first Caliphs, however, such phenomenon could have caused a radical upset in the regional patterns of the 'local' populations and thus also settlement patterns.[239]

Yet, despite the ideologically based dichotomy described in written sources, "nomadic" pastoralist and settled groups are more frequently part of the same society. Changes in the political and economic climate often determines social shifts, but does not imply radical changes in the population in all cases, with movement of discrete groups of people replacing earlier groups. Therefore, changes in the sites' organisation cannot necessarily be seen as determined by "external" groups: the key feature of the populations living on the fringe of arid lands or desert climes is the extremely flexible structure and a constant fluctuation of part of these groups between more settled phases to more mobile phases. This allows a higher degree of resilience to environmental, socio-economic and political stresses. Keeping track of these shifts through the archaeological record is indeed difficult, but in some sites like Tall Hisban, there is evidence to support the validity of these anthropological hypotheses, especially when related to economic strategies employed by the various groups who settled there.

Similarities in the architectural and spatial solutions do reflect the ambiguous nature of the social structures in that they are the physical resultant of pastorialist and stable settled groups inhabiting a space together. Social bonds between both groups may have also led to the transfer of knowledge and in particular the spread of architectural techniques to newly arrived groups in more recent phases. This could have resulted in an accentuated uniformity in building techniques, which can now be seen on the regional level. Therefore, settlements that might have been founded exclusively through

[239] Furthermore, if one considers the nature of borders themselves in ancient times, with few exceptions to be found (Hadrian's wall, for instance), it is evident that horizontal mobility in the liminal regions was constant (see for instance Wells 1999). More stark examples can be seen more frequently in regions where more extreme climatic conditions require a group's socio-economic structures to be flexible and in which frequent mobility plays an important role in the group's survival.

sedentarization of pastoral groups could have become morphologically indistinguishable over the course of time from settlements developed through other evolutionary processes. Although at the same time, sharing similar social structures (in particular the family organisation) or even being part of the same social group, could have determined the adoption of the same spatial solutions either by pastoralists and settled peoples. This partially explains the homogeneity present in residential typologies and even on a more localised level of quarters and blocks.

To further clarify the origin of a settlement, and furthermore the progression of habitation of various pastoralist and settled groups within that settlement, the historical context is of particular importance, especially in light of regional particularisms and micro-histories. For examples, some studies on transitional settlement of pastoral groups gradually becoming more sedentary are available for the Negev, where the main criterion to distinguish temporary structures from the more settled phases is the form of the structures and pens, normally tending to be round or semi-circular in their earlier stages (for instance: Haiman 1995: 29-53; Avni 2014: 271-273). Their proximity to major centres is an important indicator of the role played by the latter in attracting pastoral groups, in light of the socio-economic opportunities they offered. Furthermore, it has been suggested that it takes two generations before the process of sedentarisation is complete, which in terms of building activities is proceeded by a progressive addition of new structures to the "original" ones that belonged to the first phase of sedentarisation. A different trend can be seen in the Hauran region, where it appears that sections of mobile pastoral groups were already settled inside existing villages or around them, as is suggested by the epigraphic documentation present in sites like Umm el-Jimal. Nonetheless, the architecture is highly homogeneous in both the spatial organisation and technique, and therefore it is almost impossible to distinguish between the two social groups eventually both settled there.

Nevertheless, we can try to distinguish between the two groups by taking a closer look at the process of the addition of structures to earlier buildings (a process that specifically characterises the conjunctive houses), which is common to almost all case studies that have been analysed. This solution, which was adopted either by already settled family groups or sedentarising nomadic pastorals, was fitting to their needs for both social and economic flexibility. In general, even when detailed reconstructions of the development of single residential units are available, it is difficult to state with certainty whether the expansion from "simple house" to more complex types of houses (like the "L-Shape" or the "U-Shape" houses) reflects a progressive sedentarisation of people or simply the growth of a family. In the latter case, such an expansion can be the outcome of an improvement in the economic condition and/or the growth in size of the family (from nuclear to extended). The same interpretation can be suggested for the progressive densification of the residential spaces in the houses in the Negev.

While recognising sedentarisation, as a main factor determining the development of a settlement, and its related processes are extremely difficult, establishing the degree of "spontaneity" of such processes (even where plausibly recognisable) represents an even more challenging task. Some observations of the way different houses are organised within a block or a quarter may be helpful to this end. In particular, the presence of markedly separated quarters (as in Umm es-Surab and Umm el-Jimal) or more scattered complexes (clustered or concentrating in determined areas of a settlement as in Sharah) might reflect different types of sedentarisation of pastoralist groups. In the circumstance that the development of site like Umm el-Jimal was undoubtedly related to sedentarisation processes, the structure of the quarters could potentially reflect the presence of at least three larger social groups (one for each quarter), being either extended families or clans. The inner compactness of these quarters and their physical separation could suggest that the settling down of the group occurred in a limited time span and involved a larger group of people (and perhaps the group in its entirety). Later development of the houses within the quarter (the addition of new room in a complex, the creation of entirely enclosed courtyards, and creation of pens and other facilities) could then reflect the growth of the single families in size and/or economic wealth. If this hypothesis could be confirmed, being that these quarters belonged to individual families, one could wonder if sedentarisation took place because it was encouraged or forced by a central authority. In the particular case of Umm el-Jimal, it is plausible that the Byzantine administration of Syria or Arabia could have encouraged specific pastoralist tribes (or some of their sections) to settle, possibly in order to secure the border region, as was the case for the Jafnids. On the contrary, a looser urban structure without properly defined quarters and characterised by concentrations of single residential units, like in Mseikeh and Sharah, might reflect more spontaneous sedentarization processes that were undertaken by small groups, especially single nuclear families. These small groups or families were attracted by economic opportunities or newly established social ties with the settled community (Fisher 2011: 112). Similarly, the presence of differently organised blocks within the same village, as seen in the case of some clusters of houses organised around a larger courtyard in Shivta, could be associated with the arrival of a group maintaining a sort of separation or distinction from the rest of the settlement, but yet wanting to remain a part of the same settlement and community.

While these remain hypotheses at this stage that will require more study and further confirmation through material evidence, they offer some important considerations when reconstructing both social structures and the spaces they built and inhabited. Furthermore, a single site can hardly be informative on its own and only when it is considered in the wider regional and temporal contexts, we can understand its complex development. In fact, within the same region there may be important changes between different periods, such as the spontaneity of the sedentarisation of nomadic people. For instance, according to Haiman, in the Negev the Byzantine period represented a phase of spontaneous sedentarisation processes, while in the Early Islamic period there was a greater state initiative that influenced sedentarisation (Haiman 1995: 45-48).

What is also of particular importance to the understanding of sedentarisation processes of specific groups is considering the evidence, both archaeological and historical, that we may have on the identity of one or more of those groups. Identity (both assumed by a specific group itself and how a group is perceived by others) plays a key role within this interpretative framework, and here lies the reason for the recurring label of the "Arab settlement". While being aware of the danger of applying the term "Arab" too broadly for earlier periods than the Abbasid period, when an official Arab identity was formalised (see Chapters 2.3 and 7.4 for a discussion), it is likewise important to consider that the term "Arab" is not a term that is exclusive to a single ethnicity. Much of the scientific literature often implies this and discusses the "Arabs" as a single ethnic population completely exclusive from those that were previously integrated into the Roman and Byzantine empires. In my opinion, the element of the Arab identity that is now associated with its "ethnicity" is one of the last stages in the development of another formation of a common identity, which gradually tied together several groups that were integrated into the imperial entities ruling over the Near East with those who were not. This common social identity, even if not formalised as to be designated with a specific term, might had found expression in a similar familial structure and/or economic strategies, which manifested itself into similar ways of organising the domestic space and possibly even larger portions of the settlement. It is a possibility that these kinship groups and their related social structures could allow for a common basis of recognition of these groups and at the same time distinguish them from the rest of the provincial population.

Another important stage for the development of these not-ethnical proto-Arab identity was the inclusion of some sections of tribes into the Roman army. When serving under the same banner, these tribes were confronted with groups who did not share the same social structure and might have felt internal pressure (or the desire) to concretely (re-)define their identity features in respect to the *Romanoi*. At the same time, their integration in the Roman army distinguished them from other tribes (or segmented groups) that were not yet under the direct influence or control of the Romans. It is possible that the label *Ma'aad*, considered by Webb as the first term for an identity designating proto-Arab groups, designated such groups that were included in the Roman army (see Chapter 7.2).

The change in the defensive strategy and have affected the visibility of proto-Arab groups in the historical and archaeological records. This has led to the debate on the so-called "absence of Arabs" in the Byzantine period in this region. The processes of sedentarisation and the mass conversion of nomadic pastoralists were already well underway by this period, resulting in well-integrated tribes throughout Syria and the Transjordan (see Chapter 7.3). This would have already reduced the amount of clearly defined tribes and segmented groups from the previous period. It is likewise possible that the absence of "Arabs" and specific tribes that are distinguished in the epigraphic finds and historical sources in the Byzantine territories after the 4th century reflect this new context, where other kinds of identities began to have a greater influence in the region (see Chapter 7.4). In particular, the increasing territorial control over some regions in the provinces of *Siria* and *Arabia* by the Jafnids during the 5th-6th centuries, facilitated and reinforced by their involvement in the religious disputes, could have affected relevant changes in the relationship between the *Romanoi* and the other groups. This does not necessarily imply that the former socio-economic identities and ties lost their importance entirely. The maintenance of the same settlement organisation, for instance, suggests that family ties and kin-belonging were still the guiding principle for the spatial behaviour. The increased building of churches and chapels that were often embedded in quarters and blocks, which is hardly a (settlement-wide) communal destination for such complexes (see also Shboul and Walmsley 1998: 278), seems to reflect familial/kinship relationships, and therefore represents a proof for such family-based spatial behaviour, as well as a further testimony of the increasing "conversion" to Christianity, as argued by Fowden (see Chapter 7.3).

The re-emergence of an "official" or distinguishable Arab identity, which occurred during the Early Islamic period as attested by written sources and archaeological evidence, occurred for precisely the opposite reasons why such an identity was "invisible" during the Byzantine period. Namely, the new ruling entity as "Arabs" deliberately distinguished themselves from the local groups, whether or not they too defined themselves as (Christian) "Arabs", and they developed a completely new political and administrative system that set them apart as a group of their own. Ultimately, the Muslim identity of the newly established "Arab" ruling elite is not the most essential factor to understand the process leading to the development of an "Arab ethnicity" in the 9th century, despite the over-emphasis of its importance in historical discussions. Christian tribes (who also defined themselves as "Arabs") joined the Muslim armies to participate in the uprising, which underlines that other factors than religion were involved in both shaping the new society and more importantly the new ruling elite social class. Likewise, copresence of Muslim and Christian sections within the same tribe suggests that different religious orientations within the same kinship group did not undermine their inner social stability.

Another related and important point that should be made on the formation of the "Arab" identity during the 9th century is the deliberate contextual modifications of genealogies in order to gain or maintain political power and influence through real or invented familial relations, according to the changing political and economic contexts. Becoming a part of the "Arab" elite during the earliest stages of the Islamic era guaranteed political benefits. Likewise, this was the case much later in the medieval period as Rapoport describes for some settled communities in the Fayyum during the Mamluk period, who claimed familial ties with the first "Arabs" come to Egypt with the "Islamic conquest" (see Chapter 7.4). The chronological recurrence and geographic diffusion of such choices demonstrates how strongly kinship relations played a role in the daily life of these communities. In fact, kin-based logic, kinship relations, and their "segmented" social structures in general should be considered key elements that distinguished these "Arab" groups from other groups in the Byzantine and Early Islamic periods. What Webb suggests for the development of the Arab ethnicity can also be similar for other emerging social identities: one discrete group seeks to "contrast" itself with others within a determined and well-defined context, being within a local community, within an army, or within an empire as a whole.

The same principles on the formation of the "Arab" identity can also apply to the built environment of the "Arab settlement". Despite all of these shifts and overlaps of identities, which can at times appear contradictory, the principle behind the spatial organisation of the settlements, especially visible in the houses, appears to remain the same: they are firmly based on kinship or familial identity. The segmented kin-based social structure (shared by the settled and pastoral communities) remained far more important than economic and administrative changes in terms of the evolving built environment, especially in the peripheral and semi-desert or desert areas along the eastern borders of the Roman and Byzantine empires, which later became Umayyad Caliphate. In spite of the regional particularisms, this feature is shared by all sites considered for the present analysis. The differences between the sites' built environments is ultimately related to the influence of the groups present, both with similar and differing social structures and identities, which interacted within that specific sub-region or site and local factors (e.g. geographic, environmental). In the Trans-Jordan and Syria, the impact of the kinship relations in shaping the built-environment is even more evident on the quarter and settlement levels because of a marked centrality of extended family-groups within the society. Similarly, a more direct role of involvement of sedentarised nomadic pastoralists in the development of the settlements might be suggested. In turn, the kin or family-based organisation is less evident in the Negev area, because of an extremely restricted idea of family and a more developed local communal identity, as is expressed in the Nessana papyri (see Chapter. 7.4).

In conclusion, the present work demonstrated how the built environment, or more specifically the spatial behaviour shaping the settlement on different levels including the settlement as a whole, the quarters and the dwellings and residential blocks, represents crucial evidence for social dynamics within a society. The persistence of the same spatial behaviour over several centuries and across different regions are telling indications that extended family groups are at the core of the evolving "Arab" identity, despite continuous and variable shifts in the political, economic and religious institutions. These close-knit familial groups and their resilient social structures are ultimately additional reasons that supported the continuity within the settlements' organisation also following the "Islamic conquest" both in rural sites as well as in the urban context. Changes in the shape of the villages, towns, and cities were already taking place well before the 7th and 8th centuries as a result of a combination of socio-economic and political factors (see Chapter 7.5). The most evident consequence of such changes was not only the morphological and functional development of the Late Antique urban centres, but also the emergence of more "intermediate" semi-urban settlements, both sharing a number of features. Consequently, the religious impact of Islam on already existing settlements, which has been previously strongly asserted to have resulted in the formation of the *medinas*, a distinguished urban phenomenon, appears to be overestimated, as is also the case for many other aspects of the daily life of the local communities. This is most likely due to the preceptive nature of the Quran and Sharia law, where several mentions on how to build and administrate the development of a settlement can be found. Nonetheless, these rules closely resemble the building habits documented for the Late Byzantine periods, either through written sources and more importantly though the study of the architectural, remains themselves.

Are there then "Arab settlements" or settlements of the "Arabs"? I would suggest that the term the "Arab settlement", despite the problematic definition of the label "Arab", presents a series of benefits for this highly debated topic. Mostly, it underlines the continuity in the settlement patterns during the pre-Islamic and the Islamic period, and at the same time acknowledges the continuous importance of the kinship relations and their social structures in shaping the morphology of the sites and the identity patterns. At the same time, this term underscores the limited influence of the purely religious sphere in determining the transformations undergone by settlements. Furthermore, "Arab" itself helps geographically delimit the argumentation to a specific region (the Near East) and a precise chronologic horizon (Late Antiquity). In these terms, the "Arab settlement" represents a sub-category of what Wirth described as the "Oriental city", underlining the geographical specificity of its urban morphology and emphasizing its differences from other regions that were also incorporated in the Umayyad and Abbasid Empires in light of their respective regional traditions.

It is clear from this study that the term "Arabs" does not unequivocally identify nomadic pastoral groups, nor does the "Arab settlement" define one fixed site typology. The formulation of the "settlement of the 'Arabs'", rather than suggesting that a settlement was first built up by sedentarised nomadic pastoral peoples, is meant to encompass a wide range of sedentarisation processes that organically evolved based on a plethora of factors, including regional patterns and economic and political conditions. The social engagement of such groups in the development of sites are very difficult to identify based on the settlement morphology alone, because they are often not distinguishable from episodes of "horizontal mobility" and are impossible to categorise within a single typology, especially in light of the different sedentarisation forms that were possible for such groups. As Fisher has stated succinctly: "Arguments for and against the sedentarisation and settlement of Arabs in the Late Antiquity will always be hampered by insufficient evidence. However, sedentarisation is ultimately a part of the larger process of integration into the Empire" (Fisher 2011: 115-116). Luckily, however, the similarities between the organisational forms expressed by both settled communities and sedentarised groups are indeed numerous, especially when contact between the two was frequent over a long period of time. A case-by-case approach, which considers the local and regional context in its entirety, as this study tested and successfully implemented, could offer some crucial information for future studies on the social actors in play and the kind of processes involved. More importantly, employing an interdisciplinary approach that integrates archaeological, historical and anthropological disciplines, as this study attempts, is a means to help shed more light on both the complex and fascinating phenomena of settlement formation in the Near East during a key period of transition and more importantly on the people who built them.

References

Abbreviations

CIS: Corpus Inscriptionum Latinarum
IG: Inscriptiones Graecae

Primary Sources

Diodorus Siculus
Bibl. Hist. *Bibliotheca Historica*

Eusebius of Caesarea
On. *Onomasticon*

Flavius Josephus
JW *Bellum Judaicum*
Ant *Antiquitates Judaicae*

Ibn Khaldūn
2015 *The Muqaddimah: An introduction to history*. Trans. by F. Rosenthal, edited by N.J. Dawood. Princeton classics. Princeton: Princeton University Press.

Notitia Dignitatum

Pliny the Elder
Nat. Hist. *Naturalis Historia*

Secondary Sources

al-Housan, A.-Q.
2015 A selection of Safaitic inscriptions from the Mafraq Antiquities Office and Museum. *Arabian Epigraphic Notes* 1: 77-102.
2017 A selection of Safaitic inscriptions from Al-Mafraq, Jordan: II. *Arabian Epigraphic Notes* 3: 19-46.

Al-Jallad, A., and A. al-Manaser
2015 New Epigraphica from Jordan I: a pre-Islamic Arabic inscription in Greek letters and a Greek inscription from north-eastern Jordan. *Arabian Epigraphic Notes* 1: 51-70.
2016 New Epigraphica from Jordan II: three Safaitic-Greek partial bilingual inscriptions. *Arabian Epigraphic Notes* 2: 55-66.

al-Manaser, A., and L. Ellis
2018 New Islamic inscriptions from the Jordanian Badia region. *Arabian Epigraphic Notes* 4: 69-86.

Al-Otaibi, F. M.
2015 Nabatean Ethnicity: Emic Perspective. *Mediterranean Archaeology and Archaeometry* 15: 293-303.

al-Salameen, Z., and M. Hazza
2018 New Nabatean Inscriptions from Umm al-Jimāl. *Arabian Epigraphic Notes* 4: 87-106.

Anastasio, S., P. Gilento, and R. Parenti
2016 Ancient Buildings and Masonry Techniques in the Southern Hauran, Jordan. *Journal of Eastern Mediterranean Archaeology and Heritage Studies* 4.4: 299-320.

Antoun, R. T.
1972 *Arab village: A social structural study of a Transjordanian peasant community* (Indiana University social science series, v. 29). Bloomington: Indiana University Press.

Appadurai, A.
1986 Theory in Anthropology: Center and Periphery. *Comparative Studies in Society and History* 28: 356-361.

Ashkenazi, E., Y. Avni, and G. Avni
2012 A comprehensive characterization of ancient desert agricultural systems in the Negev Highlands of Israel. *Journal of Arid Environments* 86: 55-64.

Avni, G.
2014 *The Byzantine-Islamic transition in Palestine: An archaeological approach.* Oxford: Oxford University Press.

Avni, G., N. Porat, and Y. Avni
2013 Byzantine–Early Islamic agricultural systems in the Negev Highlands: Stages of development as interpreted through OSL dating. *Journal of Field Archaeology* 38: 332-346.

Ball, W.
2000 *Rome in the East: The Transformation of an Empire.* London and New York: Routledge.

Bar, D.
2003 The Christianisation of Rural Palestine during Late Antiquity. *The Journal of Ecclesiastical History* 54: 401-421.
2005 Rural Monasticism as a Key Element in the Christianization of Byzantine Palestine. *Harvard Theological Review* 98: 49-65.

Baumgarten, Y.
2004 *Archaeological survey of Israel: Map of Shivta (166).* Jerusalem: Israel Antiquities Authority.

Blanke, L.
2018 Abbasid Jerash Reconsidered: Suburban Life in Jerash's Southwest District over the Longue Durée, in A. Lichtenberger and R. Raja (eds) *The Archaeology and History of Jerash. 110 Years of Excavations* (Jerash Papers 1): 39-58. Turnhout: Brepols.

Bloom, J. M.
1991 Creswell and the Origin of the Minarets, in O. Grabar (ed.) *Muqarnas VIII: An Annual on Islamic Art and Architecture*: 55-58. Leiden: E.J. Brill.

Bonte, P.
2006 Sistema segmentario, in P. Bonte, M. Aime, and M. Izard (eds) *Dizionario di antropologia e etnologia*: 730. Torino: Einaudi.

Bourdieu, P.
1977 *Outline of a Theory of Practice*. Cambridge: Cambridge University Press.

Bowersock, G. W.
1994 *Roman Arabia*. Cambridge (Massachusetts) and London: Harvard University Press.

Brimer, B.
1981 Shivta - An Aerial Photographic Interpretation. *Israel Exploration Journal* 31: 227-229.

Brown, R. M.
1998a A large residence (House XVIII), in B. de Vries (ed.) *Umm el-Jimal: A frontier town and its landscape in northern Jordan* (Journal of Roman Archaeology, Supplementary Series Number 26): 195-204. Portsmouth R.I.: Journal of Roman Archaeology.
1998b The Roman Praetorium and its later domestic re-use, in *Umm el-Jimal: A frontier town and its landscape in northern Jordan* (Journal of Roman Archaeology, Supplementary Series Number 26): 161-194. Portsmouth R.I.: Journal of Roman Archaeology.

Bruant, J.
2010 Premiers sondages archéologiques sur le rampart oriental du village antique de Sharah (Syrie du Sud, Leja), in M. Al-Maqdissi, F. Braemer, J.-M. Dentzer, J. Dentzer-Feydy, and M. Vallerin (eds) *Hauran V: La Syrie du Sud du Néolithique à l'Antiquité tardive: recherches récentes: actes du colloque de Damas 2007* (Bibliothèque archéologique et historique 191): 215-222. Beyrouth: Institut français du Proche-Orient.

Büntgen, U., V. S. Myglan, F. C. Ljungqvist, M. McCormick, N. Di Cosmo, M. Sigl, J. Jungclaus, S. Wagner, P. J. Krusic, J. Esper, J. O. Kaplan, M. A. C. de Vaan, J. Luterbacher, L. Wacker, W. Tegel, and A. V. Kirdyanov
2016 Cooling and societal change during the Late Antique Little Ice Age from 536 to around 660 AD. *Nature Geoscience* 9: 231-236.

Büntgen, U., W. Tegel, K. Nicolussi, M. McCormick, D. Frank, V. Trouet, J. O. Kaplan, F. Herzig, K.-U. Heussner, H. Wanner, J. Luterbacher, and J. Esper
2011 2500 Years of European Climate Variability and Human Susceptibility. *Science (New York, N.Y.)* 331: 578-582.

Butler, H. C.
1930 Division 2, Ancient Architecture in Syria, in H. C. Butler and E. Littmann (eds) *Syria: publications of the Princeton University Archaeological Expeditions to Syria in 1904-5 and 1909*. Leyden: E. J. Brill

Clauss-Balty, P.
2008 Maisons romano-byzantines dans les villages de Batanée: missions 2002-2004, in P. Clauss-Balty and J.-M. Dentzer (eds) *Hauran III: L'habitat dans les campagnes de Syrie du Sud aux époques classique et médiévale* (Bibliothèque archéologique et historique 181): 41-103. Beyrouth: Institut français du Proche-Orient.
2010 Les villages et l'habitat rural à l'époque romano-byzantine: le cas de Sharah, sur le rebord nord-ouest du Leja, in M. Al-Maqdissi, F. Braemer, J.-M. Dentzer, J. Dentzer-

Feydy, and M. Vallerin (eds) *Hauran V: La Syrie du Sud du Néolithique à l'Antiquité tardive: recherches récentes: actes du colloque de Damas 2007* (Bibliothèque archéologique et historique 191): 199-214. Beyrouth: Institut français du Proche-Orient.

Creswell, K. A. C.
1958 *A Short Account of Early Muslim Architecture*. Harmondsworth: Penguin Books.

Cribb, R.
1991 Mobile Villagers: The structure and Organisation of Nomadic Pastoral Campsites in the Near East, in C. Gamble and W. A. Boismier (eds) *Ethnoarchaeological approaches to mobile campsites: Hunter-gatherer and pastoralist case studies* (Ethnoarchaeological series 1): 371-393. Ann Arbor (Michigan): International Monographs in Prehistory.

de Veaux, L., and S. T. Parker
1998 The 'Nabatean' Temple: a re-examination, in B. de Vries (ed.) *Umm el-Jimal: A frontier town and its landscape in northern Jordan* (Journal of Roman Archaeology: Supplementary Series Number 26): 153-160. Portsmouth R.I.: Journal of Roman Archaeology.

deMenocal, P. B.
2001 Cultural Responses to Climate Change During the Late Holocene. *Science (New York, N.Y.)* 292: 667-673.

Dentzer, J.-M.
1999 L'espace des tribus arabes à l'époque hellénistique et romaine: nomadisme, sédentarisation, urbanisation. *Comptes-rendus des séances de l'Académie des Inscriptions et Belles-Lettres* 143: 231-261.

Desreumaux, A., J.-B. Humbert, G. Thébault, and T. Bauzou
2011 Des Romains, des Araméens et des Arabes dans le Balqa' Jordanien: le cas de Hadeitha – Khirbet es-Samra, in A. Borrut, M. Debié, D. Pieri, A. Papaconstantinou, and J. P. Sodini (eds) *Le Proche-Orient de Justinien aux Abbassides: Peuplement et dynamiques spatiales (Actes du Colloque Continuités de l'occupation entre les périodes byzantine et abbasside au Proche-Orient, 7.-9. siècle, Paris, 18-20 Octobre 2007)*: 285-304. Turnhout: Brepols.

Dever, W. G.
1995 Ceramics, Ethnicity, and the Question of Israel's Origins. *The Biblical Archaeologist* 58: 200-213.

de Vries, B.
1998a Location, plan and design of the late-antique town, in B. de Vries (ed.) *Umm el-Jimal: A frontier town and its landscape in northern Jordan* (Journal of Roman Archaeology: Supplementary Series Number 26): 91-127. Portsmouth R.I.: Journal of Roman archaeology.
1998b Towards a history of Umm el-Jimal in late antiquity, in B. de Vries (ed.) *Umm el-Jimal: A frontier town and its landscape in northern Jordan* (Journal of Roman Archaeology: Supplementary Series Number 26): 229-241. Portsmouth R.I.: Journal of Roman archaeology.
1998c Previous research, in B. de Vries (ed.) *Umm el-Jimal: A frontier town and its landscape in northern Jordan* (Journal of Roman Archaeology: Supplementary Series Number 26): 27-38. Portsmouth R.I.: Journal of Roman archaeology.

2000 Continuity and Change in the Urban Character of the S. Hauran from the 5th and 9th c. AD: The archaeological Evidence at Umm el-Jimal, *Mediterranean Archaeology. Australian and New Zealand Journal for the Archaeology of the Mediterranean World* 13: 39-45.

Di Segni, L.
1995 The Involvement of Local, Municipal and Provincial Authorities in Urban Building in Late Antique Palestine and Arabia, in J. H. Humphrey (ed.) *The Roman and Byzantine Near East I: Some Recent Archaeological Research* (Journal of Roman Archaeology: Supplementary Series Number 14): 312-332. Ann Arbor (Michigan): Journal of Roman Archaeology.
1999 Epigraphic Documentation on Building in the Provinces of Palaestina and Arabia, 4th-7th C., in J. H. Humphrey (ed.) *The Roman and Byzantine Near East II: Some Recent Archaeological Research* (Journal of Roman Archaeology: Supplementary Series Number 31): 149-178. Portsmouth R.I.: Journal of Roman Archaeology

Dirbas, H.
2016 'Abd al-Asad and the Question of a Lion-God in the pre-Islamic Tradition: An Onomastic Study. *Arabian Epigraphic Notes* 2: 141-150.

Donner, F. M.
1981 *The early Islamic conquests*. Princeton: Princeton University Press.

Dumont, L.
1971 *Introduction a deux theories d'anthropologie sociale: Groupes de filiation et alliance de mariage* (Les textes sociologiques no 6). Paris: Mouton.

Durkheim, É.
1893 *De la division du Travail Social: Études sur l'organisation des sociétés supérieures*.

Dussaud, R. and F. Macler
1903 *Mission dans les régions désertiques de la Syrie moyenne* (Nouvelles Archives des Missions scientifiques, Tome X). Paris: Imprimerie Nationale

Erickson-Gini, T.
2013 Shivta. *Hadashot Arkheologiyot* 125.

Esders, S.
2012 'Faithful believers': Oaths of Allegiance in the Post-Roman Societies as Evidence for Eastern and Western 'Visions of Community', in W. Pohl, C. Gantner, and R. E. Payne (eds) *Visions of community in the post-Roman world: The West, Byzantium and the Islamic world, 300-1100*: 357-374. Farnham, Burlington VT: Ashgate.

Evans-Pritchard, E.
1940 *The Nuer: A Description of the Modes of Livelihood and Political Institutions of a Nilotic People*. Oxford: Clarendon Press.

Fabietti, U.
2010 *Elementi di antropologia culturale*. Milano: Mondadori.
2011 *Culture in bilico: Antropologia del Medio Oriente*. Milano: Bruno Mondadori.

Faust, A., and S. Bunimovitz
2003 The four-room house. Embodying Iron Age Israelite Society. *Near Eastern Archaeology* 66: 22-31.

Ferch, A. J., L. T. Geraty, L. A. Haynes, L. E. Hubbard, L. G. Running, M. B. Russell, and W. K. Vyhmeister
1989 *Hesban 3: Historical foundations: Studies of literary references to Hesban and vicinity*. Berrien Springs (Michigan): Joint publication of the Institute of Archaeology and Andrews University Press.

Finkelstein, I., and A. Perevolotsky
1990 Processes of Sedentarization and Nomadization in the History of Sinai and the Negev. *Bulletin of the American Schools of Oriental Research*: 67-88.

Fisher, G.
2011 *Between Empires: Arabs, Romans, and Sasanians in Late Antiquity* (Oxford classical monographs). Oxford and New York: Oxford University Press.
2015 *Arabs and Empires Before Islam*. New York: Oxford University Press.

Fisher, G., P. Wood, G. Bevan, G. Greatrex, B. Hamarneh, P. Schadler, and W. Ward
2015 Arabs and Christianity, in G. Fisher (ed.) *Arabs and empires before Islam*: 276-372. New York: Oxford University Press.

Fowden, E. K.
2015 Rural converters among the Arabs, in A. Papaconstantinou, N. McLynn, and D. L. Schwartz (eds) *Conversion in late antiquity: Christianity, Islam, and beyond: papers from the Andrew W. Mellon Foundation Sawyer Seminar, University of Oxford, 2009-2010*: 175-196. Burlington, VT: Ashgate.

Franz, K.
2005 Resources and Organizational Power: Some Thoughts on Nomadism in History, in S. Leder and B. Streck (eds) *Nomaden und Sesshafte Bd. 2: Shifts and drifts in nomad-sedentary relations*: 55-78. Wiesbaden: L. Reichert.

Fuks, D., O. Ackermann, A. Ayalon, M. Bar-Matthews, G. Bar-Oz, Y. Levi, A. M. Maeir, E. Weiss, T. Zilberman, and Z. Safrai
2017 Dust clouds, climate change and coins: consiliences of palaeoclimate and economy in the Late Antique southern Levant. *Levant* 49: 205-223.

Gatier, P.-L.
2011 Inscriptions grecques, mosaïques et églises des débuts de l'époque islamique au Proche-Orient, in A. Borrut, M. Debié, D. Pieri, A. Papaconstantinou, and J. P. Sodini (eds) *Le Proche-Orient de Justinien aux Abbassides: Peuplement et dynamiques spatiales (Actes du Colloque Continuités de l'occupation entre les périodes byzantine et abbasside au Proche-Orient, 7.-9. siècle, Paris, 18-20 Octobre 2007)*: 7-28. Turnhout: Brepols.

Gawlikowski, M.
1986 A residential area by the South Decumanus, in F. Zayadine (ed.) *Jerash archaeological project, 1981-1983*:107-136. Amman: Department of Antiquities of Jordan.
1995 Les Arabes de Syrie dans l'Antiquité, in K. van Lerberghe and A. Schoors (eds) *Immigration and emigration within the ancient Near East: Festschrift E. Lipiński* (Orientalia

Lovaniensia analecta 65): 83-92. Leuven: Uitgeverij Peeters en Departement Oriëntalistiek.

Genequand, D.
2015 The Archaeological Evidence for the Jafnids and the Nasrids, in G. Fisher (ed.) *Arabs and Empires before Islam*: 172-213. New York: Oxford University Press.

Gilento, P.
2013 Stratigrafia e tipologie costruttive dello Hawran giordano: Verifica delle potenzialità di un 'antico' strumento archeologico: la registrazione degli edifici. Unpublished PhD dissertation, Scuola di Dottorato in Preistoria e Protostoria, Storia e Archeologia del mondo antico, Università degli Studi di Siena.
2014 La chiesa dei Santi Sergio e Bacco, Umm as-Surab (Giordania): Risultati storico-costruttivi dall'analisi archeologica degli elevati. *Arqueología de la Arquitectura* 11: e013.
2015 Ancient architecture in the village of Umm al-Surab, Northern Jordan: Construction process and building techniques, a case study. *Syria* 92: 329-360.

Gros, P., and M. Torelli
2010 *Storia dell'urbanistica: Il mondo romano*. Roma and Bari: GLF editori Laterza [2nd edition].

Guérin, A.
2008 Le village de Mseikeh et le Léjà à la période islamique (VIIe – XVe siècle): Archéologie du peuplement et histoire des territoires, in P. Clauss-Balty and J.-M. Dentzer (eds) *Hauran III: L'habitat dans les campagnes de Syrie du Sud aux époques classique et médiévale* (Bibliothèque archéologique et historique 181): 233-310. Beyrouth: Institut français du Proche-Orient.

Haiman, M.
1995 Agriculture and Nomad-State Relations in the Negev Desert in the Byzantine and Early Islamic Periods. *Bulletin of the American Schools of Oriental Research*: 29-53.

Hakim, B. S.
2008 *Arabic-Islamic cities: Building and planning principles*. London: Kegan Paul International.

Haldon, J., and H. Kennedy
2012 Regional Identities and Military Power: Byzantium and Islam ca. 600-750, in W. Pohl, C. Gantner, and R. E. Payne (eds) *Visions of community in the post-Roman world: The West, Byzantium and the Islamic world, 300-1100*: 317-353. Farnham, Burlington VT: Ashgate.

Hall, J. M.
1997 *Ethnic identity in Greek antiquity*. Cambridge and New York: Cambridge University Press.

Harper, K.
2017 *The Fate of Rome: Climate, Disease, and the End of an Empire*. Princeton: Princeton University Press.

Hirschfeld, Y.
1995 *The Palestinian dwelling in the Roman-Byzantine period* (Studium Biblicum Franciscanum collectio minor 34). Jerusalem: Franciscan Printing Press.
1997 Farms and Villages in Byzantine Palestine. *Dumbarton Oaks Papers* 51: 33-71.

2003 Social aspects of the late-antique village of Shivta. *Journal of Roman Archaeology* 16: 395-408.

Hirschfeld, Y., and Y. Tepper
2006 Columbarium towers and other structures in the environs of Shivta. *Journal of the Institute of Archaeology of Tel Aviv University* 33: 83-116.

Honigman, S.
2002 Les divers sens de l'ethniquè Ἄραψ dans les sources documentaires grecques d'Égypte. *Ancient Society* 32: 43-72.

Hoyland, R. G.
2001 *Arabia and the Arabs: From the bronze age to the coming of Islam*. London and New York: Routledge.
2007 Epigraphy and the emergence of Arab identity, in P. Sijpesteijn (ed.) *From al-Andalus to Khurasan: Documents from the medieval Muslim world* (Islamic history and civilization. Studies and texts, v. 66): 217-242. Leiden: Brill.
2009 Late Roman Provincia Arabia, Monophysite Monks and Arab Tribes. *Semitica et Classica International Journal of Oriental and Mediterranean Studies:* 117-139.

Irwin, R.
2018 *Ibn Khaldun: An intellectual biography*. Princeton and Oxford: Princeton University Press.

Jones, S.
1997 *The archaeology of ethnicity: Constructing identities in the past and present*. London: Routledge.

Kalaitzoglou, G.
2018 A Middle Islamic Hamlet in Jerash: Its Architectural Development, in A. Lichtenberger and R. Raja (eds) *Middle Islamic Jerash (9th Century- 15th Century)* (Jerash Papers 3): 97-116. Turnhout: Brepols.

Kennedy, D. L.
2004 *The Roman army in Jordan*. London: Council for British Research in the Levant.

Kennedy, H. N.
1985 From polis to medina: urban change in the late antique and early Islamic Syria. *Present & Past* 106: 3-27.

Kehrberg-Ostrasz, I.
2018 Jerash Seen from Below. Part Two: Aspects of Urban Living in Late Antiquity, in A. Lichtenberger and R. Raja (eds) *The Archaeology and History of Jerash. 110 Years of Excavations* (Jerash Papers 1): 119-130. Turnhout: Brepols.

King, G. R.D.
1983a Byzantine and Islamic Sites in Northern and Eastern Jordan, in *Proceedings of the Seminar for Arabian Studies, Vol. 13, Proceedings of the Sixteenth SEMINAR FOR ARABIAN STUDIES, Oxford, 20th - 22nd July 1982*: 79-91
1983b Two Byzantine Churches in Northern Jordan and their Re-Use in the Islamic Period. *Deutsches Archäologisches Institut - Station Damaskus: Damaszener Mitteilungen* 1: 3-36.

Kraemer, C. J.
1958 *Excavations at Nessana 3: Non-Literary Papyri.* Princeton: Princeton University Press.

Kressel, G. M.
1996 Being Tribal and Being Pastoralist, in U. Fabietti and P. C. Salzman (eds) *The anthropology of tribal and peasant pastoral societies: The dialectics of social cohesion and fragmentation*: 129-138. Pavia and Como: Collegio Ghislieri and Ibis.

LaBianca, Ø. S., L. E. Hubbard, and L. G. Running
1990 *Hesban I: Sedentarization and nomadization: Food system cycles at Hesban and vicinity in Transjordan.* Berrien Springs (Michigan): Institute of Archaeology and Andrews University Press.

Lapidus, I. M.
1988 *A History of Islamic Societies.* Cambridge, New York, Melbourne: Cambridge University Press.
1990 Tribes and State Formation in Islamic History, in P. S. Khoury and J. Kostiner (eds) *Tribes and state formation in the Middle East*: 25-47. Berkeley: University of California Press.

Leppin, H.
2012 Roman Identity in a Border Region: Evagrius and the Defence of the Roman Empire, in W. Pohl, C. Gantner, and R. E. Payne (eds) *Visions of community in the post-Roman world: The West, Byzantium and the Islamic world, 300-1100*: 241-258. Farnham, Burlington VT: Ashgate.

Lewis, B.
1993 *The Arabs in history.* Oxford: Oxford University Press.

Lichtenberger, A., and R. Raja
2015 New Archaeological Research in the Northwest Quarter of Jerash and Its Implications for the Urban Development of Roman Gerasa. *American Journal of Archaeology* 119: 483-500.
2018 A View of Gerasa/Jerash from its Urban Periphery: the Northwest Quarter and its Significance for the Understanding of the Urban Development of Gerasa from the Roman to the Early Islamic Period, in A. Lichtenberger and R. Raja (eds) *The Archaeology and History of Jerash. 110 Years of Excavations* (Jerash Papers 1): 143-166. Turnhout: Brepols.

Little, D. P.
1984 *A Catalogue of the Islamic Documents from Al-Haram as-Šărīf in Jerusalem* (Beiruter Texte und Studien 29). Wiesbaden and Beirut: F. Steiner, Orient-Institut der Deutschen Morgenländischen Gesellschaft.

Littmann, E.
1930 Division 4: Semitic inscriptions, in H. C. Butler and E. Littmann (eds) *Syria: publications of the Princeton University Archaeological Expeditions to Syria in 1904-5 and 1909.* Leyden: E. J. Brill

Littmann, E., D. Magie, and D. R. Stuart
1930 Division 3: Greek and Latin inscriptions; Section A: Southern Syria, in H. C. Butler and E. Littmann (eds) *Syria: publications of the Princeton University Archaeological Expeditions to Syria in 1904-5 and 1909.* Leyden: E. J. Brill

Macdonald, M. C. A.
1993 Nomads and the Hawran in the late Hellenistic and Roman Periods: a reassessment of the epigraphic evidence. *Syria* 70: 303-403.
2009a Arabians, Arabias, and the Greeks: contact and perceptions, in M. C. A. Macdonald (ed.) *Literacy and identity in pre-Islamic Arabia* (Variorum collected studies series 906): 1-33. Surrey and Burlington: Ashgate Pub.
2009b Arabs, Arabias, and Arabic before late antiquity. *Topoi* 16: 277-332.
2009c On Saracens, the Rawwafah inscription, and the Roman army, in M. C. A. Macdonald (ed.) *Literacy and identity in pre-Islamic Arabia.* Variorum collected studies series 906: 1-26. Surrey and Burlington: Ashgate Pub.

Macdonald, M. C. A., A. Corcella, T. Daryaee, G. Fisher, M. Gibbs, A. Lewin, D. Violante, and C. Whately
2015 Arabs and Empires before the Sixth Century, in G. Fisher (ed.) *Arabs and empires before Islam*: 12-89. New York: Oxford University Press.

Magness, J.
2003 *The Archaeology of the Early Islamic Settlement in Palestine.* Winona Lake Ind.: Eisenbrauns.

Margalit, S.
1987 The North Church of Shivta: the discovery of the first church. *Palestine Exploration Quarterly* 119: 106-121.

Marx, E.
1996 Are there Pastoral Nomads in the Arab Middle East?, in U. Fabietti and P. C. Salzman (eds) *The anthropology of tribal and peasant pastoral societies: The dialectics of social cohesion and fragmentation*: 101-115. Pavia and Como: Collegio Ghislieri and Ibis.
2005 Nomads and Cities: The Development of a Conception, in S. Leder and B. Streck (eds) *Nomaden und Sesshafte Bd. 2: Shifts and drifts in nomad-sedentary relations*: 3-15. Wiesbaden: L. Reichert.

McCormick, M., U. Büntgen, M. A. Cane, E. R. Cook, K. Harper, P. Huybers, T. Litt, S. W. Manning, P. A. Mayewski, A. F. M. More, K. Nicolussi, and W. Tegel.
2012 Climate Change during and after the Roman Empire: Reconstructing the Past from Scientific and Historical Evidence. *Journal of Interdisciplinary History* 43: 169-220.

Meeker, M. E.
2005 Magritte on the Bedouins: *Ce n'est pas une société segmentaire*, in S. Leder and B. Streck (eds) *Nomaden und Sesshafte Bd. 2: Shifts and drifts in nomad-sedentary relations*: 79-97. Wiesbaden: L. Reichert.

Meir, A.
1996 Territoriality among the Negev Bedouin in Transition from Nomadism to Sedentarism, in U. Fabietti and P. C. Salzman (eds) *The anthropology of tribal and peasant pastoral societies: The dialectics of social cohesion and fragmentation*: 187 207. Pavia and Como: Collegio Ghislieri and Ibis.

Michel, A.
2001 *Les églises d'époque byzantine et umayyade de la Jordanie. V-VIII siècle.* Turnhout: Brepols Publishers.

2011 Le devenir des lieux de culte chrétiens sur le territoire jordanien entre le VIIe et le IXe siècle: un état de la question, in A. Borrut, M. Debié, D. Pieri, A. Papaconstantinou, and J. P. Sodini (eds) *Le Proche-Orient de Justinien aux Abbassides: Peuplement et dynamiques spatiales (Actes du Colloque Continuités de l'occupation entre les périodes byzantine et abbasside au Proche-Orient, 7.-9. siècle, Paris, 18-20 Octobre 2007)*: 233-269. Turnhout: Brepols.

Millar, F.
1998 The Theodosian Empire (408-450) and the Rabs: Saracens or Ishmaelites. *Mediterranean Archaeology. Australian and New Zealand Journal for the Archaeology of the Mediterranean World* 11: 296-313.

Mitchel, L. A.
1992 *Hesban 7: Hellenistic and Roman strata: A study of the stratigraphy of Tell Hesban from the 2d century B.C. to the 4th century A.D.* Berrien Springs MI: Institute of Archaeology and Andrews University Press.

Morgan, L. H.
1871 *Systems of Consanguinity and Affinity of the Human Family, by Lewis H. Morgan*. Washington: Smithsonian institution.

Negev, A.
1980 House and city planning in the Ancient Negev and the Provincia Arabia, in G. Golany (ed.) *Housing in arid lands: Design and planning*: 3-32. London: Architectural Press.
1988a *The Architecture of Mampsis Negev: Vol. 1: the Middle and Late Nabatean Periods*. Jerusalem: The Hebrew Institute of archaeology.
1988b *The Architecture of Mampsis Negev: Vol. 2: the Late Roman and Byzantine Periods*. Jerusalem: The Hebrew Institute of Jerusalem.

Nehmé, L.
2017 New dated inscriptions (Nabatean andpre-Islamic Arabic) from a site near al-Jawf, ancient Dūmah, Saudi Arabia. *Arabian Epigraphic Notes* 3: 121-164.

Nöldeke, T.
1875 Zur Topographie und Geschichte des Damaszenischen Gebietes und der Hauran Gegend. *Zeitschrift der Deutschen Morgenländischen Gesellschaft* 29: 419-443.

Norris, J., and A. Al-Manaser
2018 The Nabateans against the Ḥwlt – once again. An edition of new Safaitic inscriptions from the Jordanian Ḥarrah desert. *Arabian Epigraphic Notes* 4: 1-24.

Osinga, E.
2017 The Countryside in Context: stratigraphic and ceramic analysis at Umm el-Jimal and environs in northeastern Jordan (c. 1st–20th century). Unpublished PhD dissertation, Southampton University.

O'Sullivan, S.
2014 Fiscal Evidence from the Nessana Papyri, in P. Sijpesteijn and A. T. Schubert (eds) *Documents and the History of the Early Islamic World*: 50-74. Leiden and Boston: Brill.

Parker, S. T.

1998a The 'Nabatean' Temple, in B. de Vries (ed.) *Umm el-Jimal: A frontier town and its landscape in northern Jordan* (Journal of Roman archaeology: Supplementary Series Number 26): 149-152. Portsmouth R.I.: Journal of Roman archaeology.

1998b The defences of the Roman and Byzantine town, in B. de Vries (ed.) *Umm el-Jimal: A frontier town and its landscape in northern Jordan* (Journal of Roman archaeology: Supplementary Series Number 26): 143-148. Portsmouth R.I.: Journal of Roman archaeology.

1998c The later castellum ('barracks'), in B. de Vries (ed.) *Umm el-Jimal: A frontier town and its landscape in northern Jordan* (Journal of Roman archaeology: Supplementary Series Number 26): 131-142. Portsmouth R.I.: Journal of Roman archaeology.

Peters, E. L.

1967 Some Structural Aspects of the Feud among the Camel-Herding Bedouin of Cyrenaica. *Africa: Journal of the International African Institute* 37: 261-282.

Pini, N.

2019 Semi-urban or semi-rural settlements: a new definition of urban centres required?, in Proceedings of the 19th International Congress of Classical Archaeology (Cologne/Bonn, 22-26 May 2018) (in press).

Pouillon, F.

1996 Bédouins des Lumières, Bédouins romantiques: mouvement littéraire et enquête sociologique dans le voyage en Orient (XVIIIe-XIXe siècles), in U. Fabietti and P. C. Salzman (eds) *The anthropology of tribal and peasant pastoral societies: The dialectics of social cohesion and fragmentation*: 57-79. Pavia and Como: Collegio Ghislieri and Ibis.

Rapoport, Y.

2004 Invisible Peasants, Marauding Nomads: Taxation, Tribalism, and Rebellion in Mamluk Egypt. *Mamluk Studies Review* VIII: 1-22.

Raymond, A.

1985 *Grandes villes arabes a l'epoque ottomane*. Paris: Sindbad.

1994 Islamic City, Arab City: Orientalist Myths and recent views. *British Journal of Middle Eastern Studies* V. 21: 3-18.

Renfrew, C., and P. G. Bahn

2004 *Archaeology: Theories, methods, and practice*. London: Thames & Hudson.

Retsö, J.

2003 *The Arabs in antiquity: Their history from the Assyrians to the Umayyads*. London and New York: Routledge.

Röhl, C.

2010 *Shivta: Architektur und Gesellschaft einer byzantinischen Siedlung im Negev*. PhD dissertation, Philosophischen Fakultät, Universität zu Köln.

Safrai, Z.

1994 *The Economy of Roman Palestine*. London: Routledge.

Said, E. W.
2003 *Orientalism*. London: Penguin Group.

Saidel, B. A.
2009 Pitching camp: Ethnoarchaeological investigations of inhabited tent camps in the Wadi Hisma, Jordan, in J. Szuchman (ed.) *Nomads, tribes, and the state in the ancient Near East: Cross-discipilinary perspectives* (Oriental Institute seminars 5): 87–105. Chicago: Oriental Institute of the University of Chicago.

Saliou, C.
1996 *Le traité d'urbanisme de Julien d'Ascalon (VI e siècle). Droit et architecture en Palestine au VIe siècle* (Travaux et Mémoires du Centre de Recherche d'Histoire et Civilisation de Byzance, Collège de France Monographies 8). Paris: Editions De Boccard.

Salzman, P. C.
1978a Does Complementary Opposition Exist?. *American Anthropologist* 80: 53-70.
1978b The Study of 'Complex Society' in the Middle East: A Review Essay. *International Journal of Middle East Studies* 9: 539-557.
1996a Introduction: Varieties of pastoral Societies, in U. Fabietti and P. C. Salzman (eds) *The anthropology of tribal and peasant pastoral societies: The dialectics of social cohesion and fragmentation*: 21-37. Pavia and Como: Collegio Ghislieri and Ibis.
1996b Peasant Pastoralism, in U. Fabietti and P. C. Salzman (eds) *The anthropology of tribal and peasant pastoral societies: The dialectics of social cohesion and fragmentation*: 149-166. Pavia and Como: Collegio Ghislieri and Ibis.

Sartre, M.
1999 Les metrokomiai de Syrie du Sud. *Syria* 76: 197-222.
2005 *The Middle East Under Rome*. Cambridge (Mass.) and London: Harvard University Press.

Sauvaget, J.
1934 Le plan de Laodicée-sur-Mer. *Bulletin d'études orientales* 4: 81-116.
1941 *Alep: Essai sur le développement d'une grande ville syrienne, des origines au milieu du XIXe siècle*. Paris: P. Geuthner.

Schlee, G.
2005 Forms of Pastoralism, in S. Leder and B. Streck (eds) *Nomaden und Sesshafte Bd. 2: Shifts and drifts in nomad-sedentary relations*: 17-53. Wiesbaden: L. Reichert.

Schmitt, O.
2005 Rome and the Bedouins of the Near East from 70 BC to 630 AD: 700 Years of Confrontation and Coexistence, in S. Leder and B. Streck (eds) *Nomaden und Sesshafte Bd. 2: Shifts and drifts in nomad-sedentary relations*: 271-288. Wiesbaden: L. Reichert.

Segal, A.
1983 *The Byzantine city of Shivta (Esbeita), Negev Desert, Israel* (BAR international series 179). Oxford: BAR.
1985 Shivta-A Byzantine Town in the Negev Desert. *Journal of the Society of Architectural Historians* 44: 317-328

Seligman, J.
2011 The Rural Hinterland of Jerusalem in the Byzantine Period. PhD dissertation, University of Haifa.

Shahîd, I.
2006 *Byzantium and the Arabs in the Fourth Century*. Washington: Dumbarton Oaks.
2009 *Byzantium and the Arabs in the Sixth Century: 2.2: Economic, Social, and Cultural History*. Washington: Dumbarton Oaks.

Shboul, A., and A. Walmsley
1998 Identity and Self-image in Syria-Palestine in the Transition from Byzantine to Early Islamic Rule: Arab Christians and Muslims. *Mediterranean Archaeology* 11: 255-287.

Sijpesteijn, P. M.
2013 *Shaping a Muslim State: The World of a Mid-English-Century Egyptian Official*. Oxford: University Press.

Storfjell, J. B.
1983 The stratigraphy of Tell Hesban, Jordan in the Byzantine period. Unpublished PhD dissertation, Andrews University.

Stroumsa, R.
2008 *People and Identities in Nessana*, PhD dissertation, Duke University (published in 2012 via Proquest, Umi Dissertation Publishing).
2014 Greek and Arabic in Nessana, in P. Sijpesteijn and A.T. Schubert (eds) *Documents and the History of the Early Islamic World*: 143-157. Leiden and Boston: Brill.

Szuchman, J.
2009 Integrating approaches to Nomads, tribes, and the State in the Ancient Near East, in J. Szuchman (ed.) *Nomads, tribes, and the state in the ancient Near East: Cross-discipilinary perspectives* (Oriental Institute seminars 5): 1-14. Chicago: Oriental Institute of the University of Chicago.

Tapper, R.
1990 Anthropologist, Historian, and Tribespeople on Tribe and State Formation in the Middle East, in P. S. Khoury and J. Kostiner (eds) *Tribes and state formation in the Middle East*: 48-73. Berkeley: University of California Press.

Tepper, Y., L. Weissbrod, and G. Bar-Oz
2015 Behind sealed doors: unravelling abandonment dynamics at the Byzantine site of Shivta in the Negev Desert. *Antiquity* 89:348

ter Haar Romeny, B.
2012 Ethnicity, Ethnogenesis and the Identity of Syriac Orthodox Christians, in W. Pohl (ed.) *Visions of community in the post-Roman world: The West, Byzantium and the Islamic world, 300-1100*: 183-204. Farnham, Burlington VT: Ashgate.

Walker, B. J.
2013 Planned Villages and Rural Resilience on the Mamluk Frontier: A Preliminary Report on the 2013 Excavation Season at Tall Hisban. Annemarie Schimmel Kolleg Working Paper 11

Walker, B. J., R. D. Bates, J. P. Hudon, and Ø. S. La Bianca
2014 Tall Hisban 2013 and 2014 Excavation Seasons: Exploration of the Medieval Village and Long-Term Water Systems. *Annual of the Department of Antiquities of Jordan* 58 (2014-2015): 483-523.

Walker, B, S. Laparidou, A. Hansen, C. Corbino
2017 Did the Mamluks Have an Environmental Sense? Natural Resource Management in Syrian Villages. *Mamluk Studies Review* XX: 167-245.

Walker, B. J., and Ø. S. LaBianca
2003 The Islamic Qusur of Tall Hisban: Preliminary Report on the 1998 and 2001 Seasons. *Annual of the Department of Antiquites of Jordan* 47: 443-471.
2018 Hisban Cultural Heritage Project. *Archaeology in Jordan Newsletter: 2016 and 2017 Seasons:* 47-48.

Walker, B., R. Bates, S. Polla, A. Springer, and S. Weihe
2017 Residue Analysis as Evidence of Activity Areas and Phased Abandonment in a Medieval Jordanian Village. *Journal of Islamic Archaeology* 4: 217-248.

Walmsley, A.
2007 *Early Islamic Syria: An Archaeological Assessment.* London: Duckworth.
2011 Trends in the urban history of Eastern Palaestina Secunda, in A. Borrut, M. Debié, D. Pieri, A. Papaconstantinou, and J. P. Sodini (eds) *Le Proche-Orient de Justinien aux Abbassides: Peuplement et dynamiques spatiales (Actes du Colloque Continuités de l'occupation entre les périodes byzantine et abbasside au Proche-Orient, 7.-9. siècle, Paris, 18-20 Octobre 2007)*: 271-284. Turnhout: Brepols.
2018 Urbanism at Islamic Jerash: New Readings from Archaeology and History, in A. Lichtenberger and R. Raja (eds) *The Archaeology and History of Jerash. 110 Years of Excavations* (Jerash Papers 1): 241-256. Turnhout: Brepols.

Watson, A. M.
1974 The Arab Agricultural Revolution and Its Diffusion, 700-1100. *The Journal of Economic History* 34: 8-35.

Webb, P.
2016 *Imagining the Arabs: Arab Identity and the Rise of Islam.* Edinburgh: Edinburgh University Press.

Weiß, A.
2007 Nomaden jenseits der Topoi - anstelle einer Einleitung, in A. Weiß (ed.) *Nomaden und Sesshafte Bd.8: Der imaginierte Nomade: Formel und Realitätsbezug bei antiken, mittelalterlichen und arabischen Autoren*: 3-15. Wiesbaden: Dr. Ludwig Reichert Verlag.

Wells, P. S.
1999 *The barbarians speak. How the conquered peoples shaped Roman Europe.* Princeton and Oxford: Princeton University Press.

Whitcomb, D.
1994 Amsar in Syria?: Syrian cities after the conquest. *ARAM Periodical* 6: 13-33.
2009 From pastoral peasantry to tribal urbanites: Arab tribes and the foundation of the Islamic State in Syria, in J. Szuchman (ed.) *Nomads, tribes, and the state in the ancient Near*

East: Cross-discipilinary perspectives (Oriental Institute seminars 5): 241-259. Chicago: Oriental Institute of the University of Chicago.

Wirth, E.
2000 *Die orientalische Stadt im islamischen Vorderasien und Nordafrika.* Mainz: von Zabern.

Wolf, E.
1951 The social organization of Mecca and the Origins of Islam. *Southwestern Journal of Anthropology* 7: 329-356.

Plates

Plate 1
Size comparison of the case studies (drawing by the author)

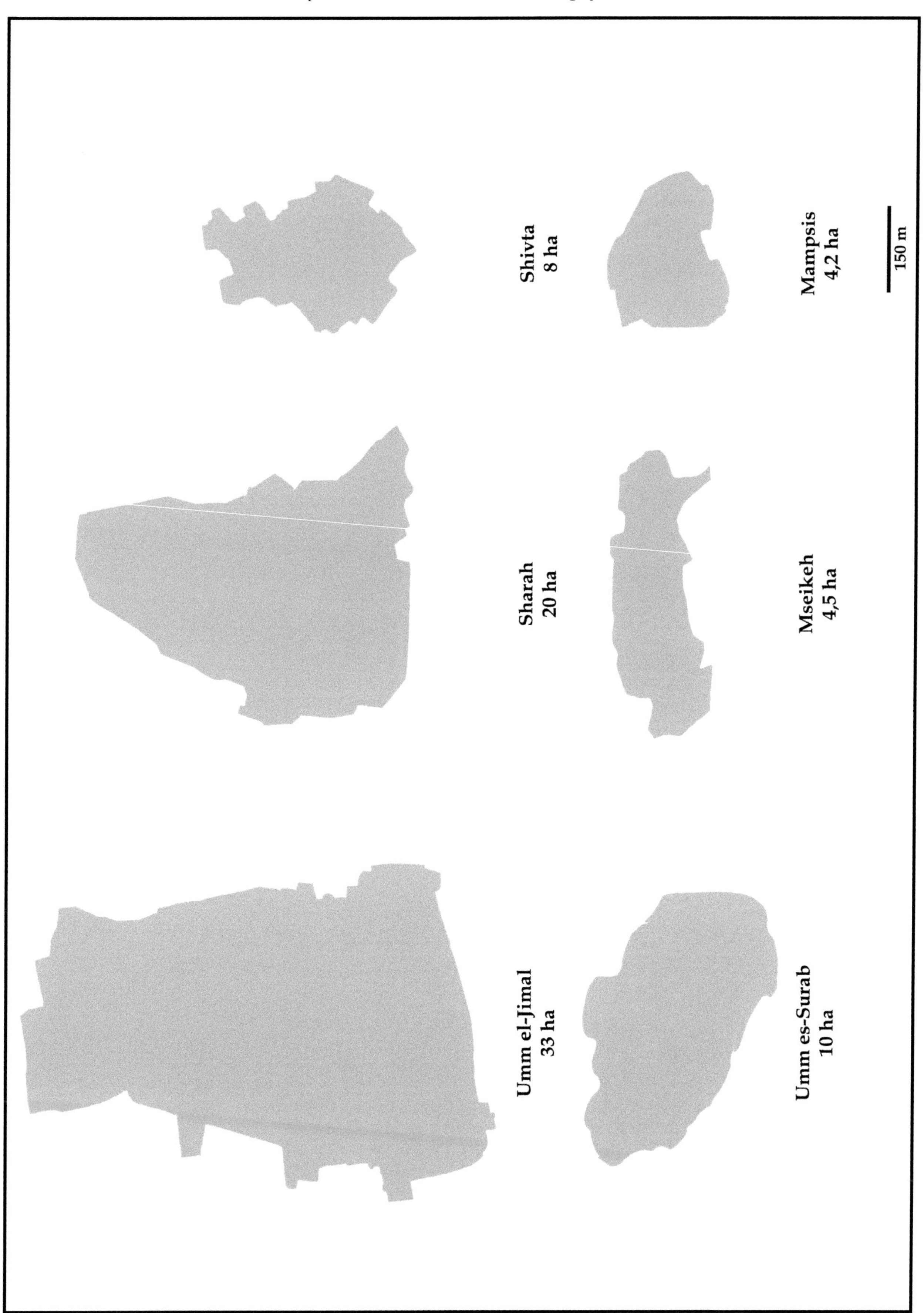

Plate 2a

"Cluster-Quarter" in Umm el-Jimal (drawing by the author)

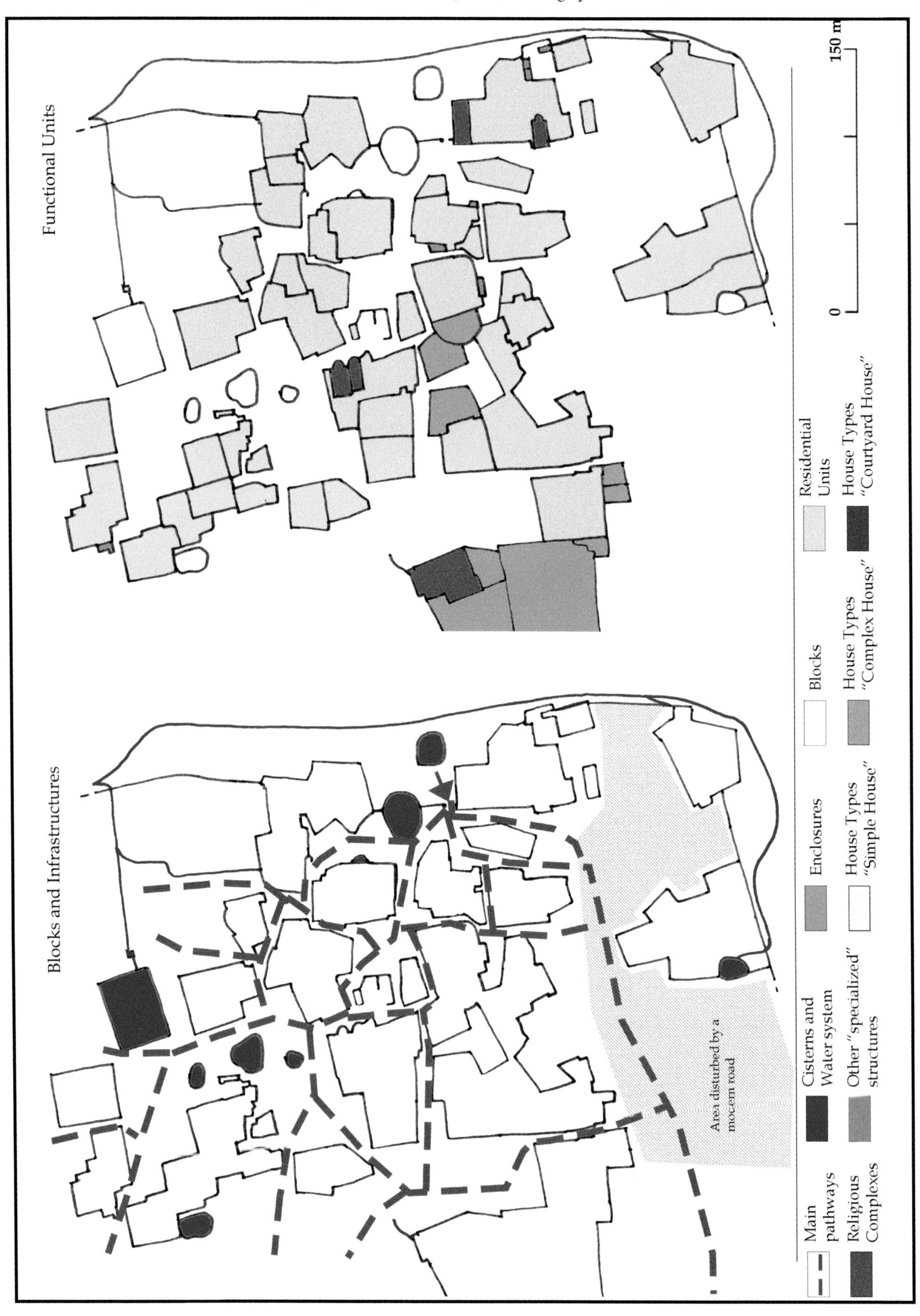

Plate 2b

"Cluster-Quarter" in Umm el-Jimal (drawing by the author)

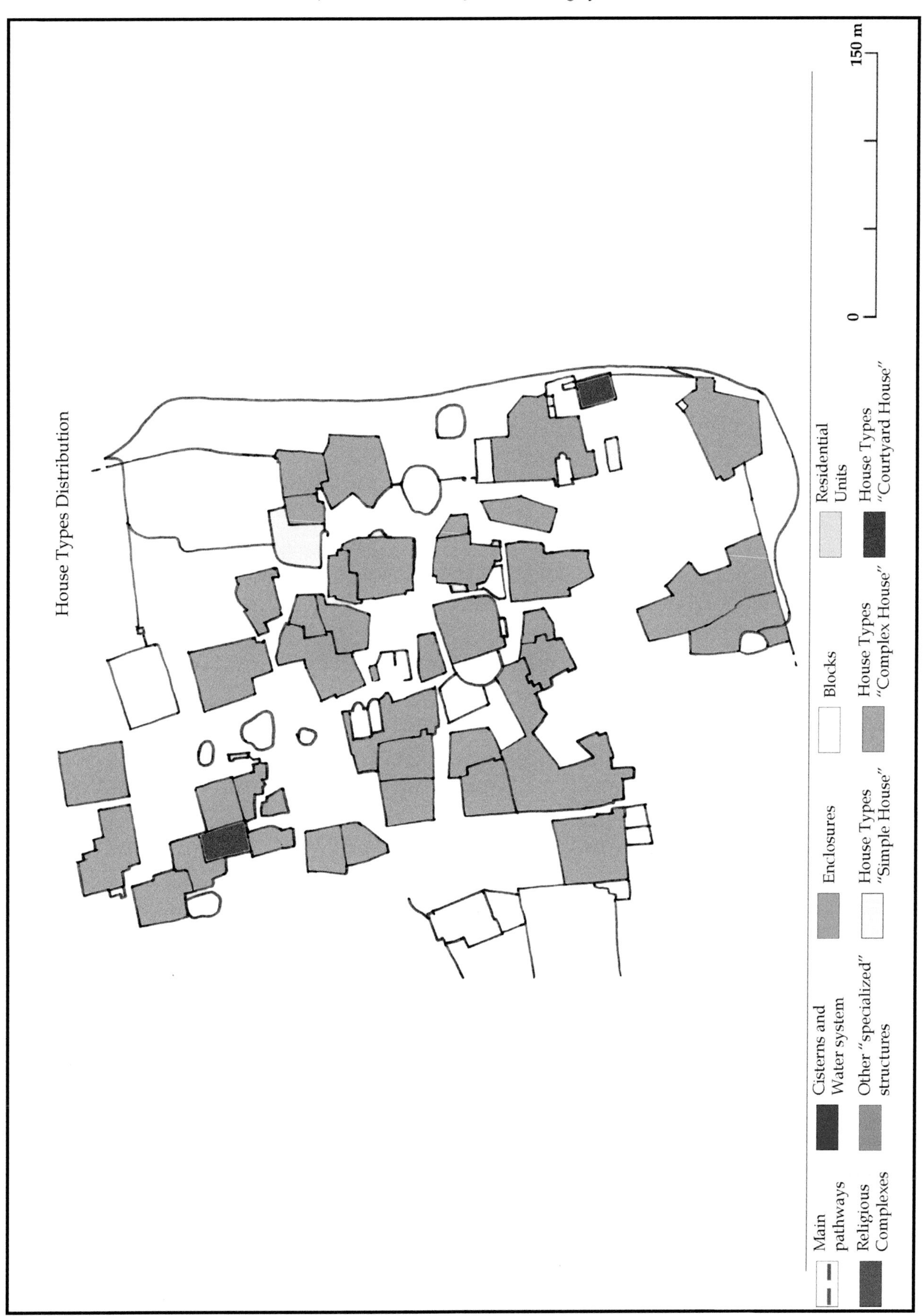

Plate 3

External "Cluster-Quarter" in Umm el-Jimal (drawing by the

Plate 4

Above: Quarter developed by a former Roman fort (Umm er-Rasas). Below: enclosed quarters (Mampsis) (drawings by the author)

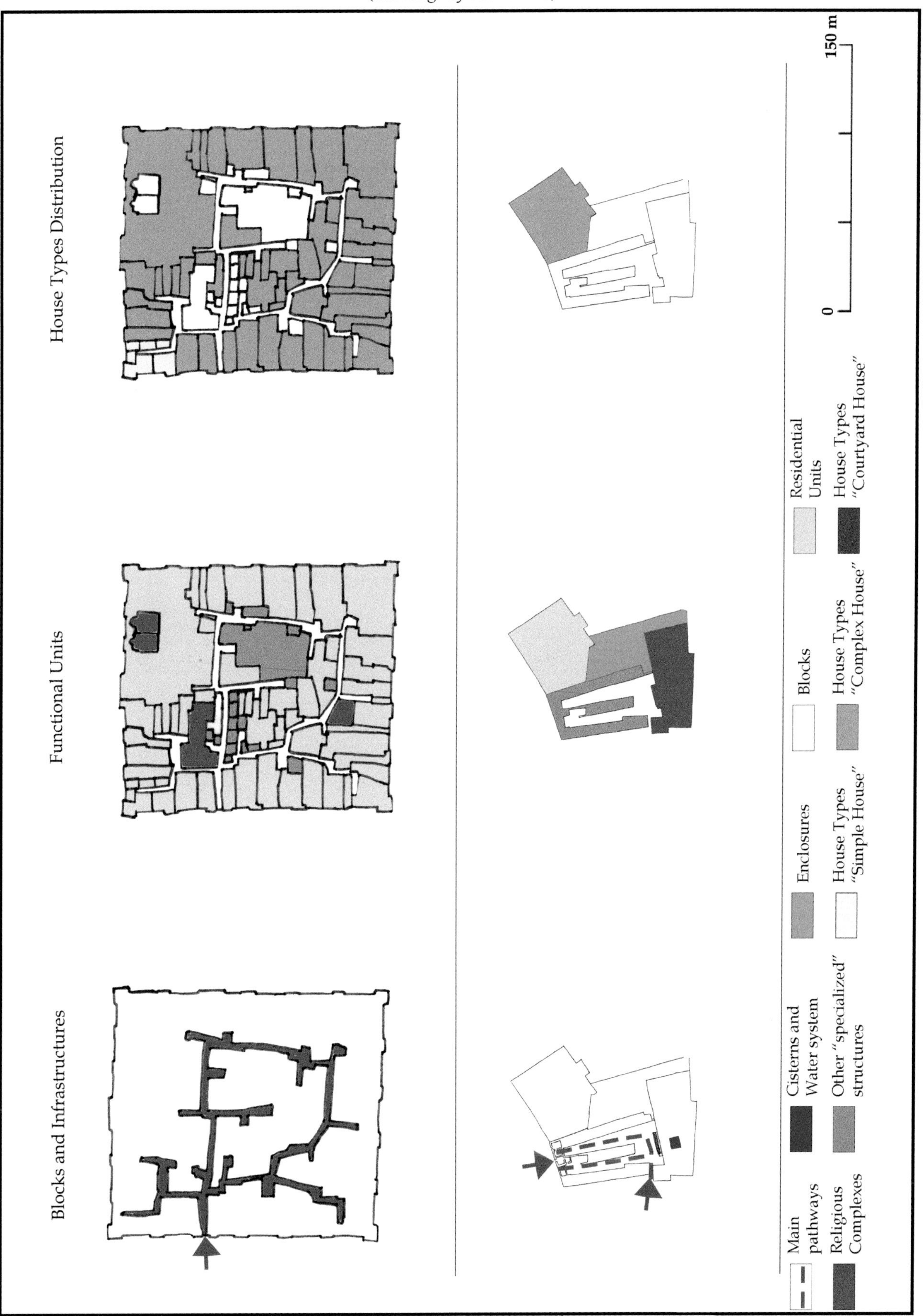

Plate 5

"Cluster-block" in Umm el-Jimal, with direct access to each unit (drawing by the author)

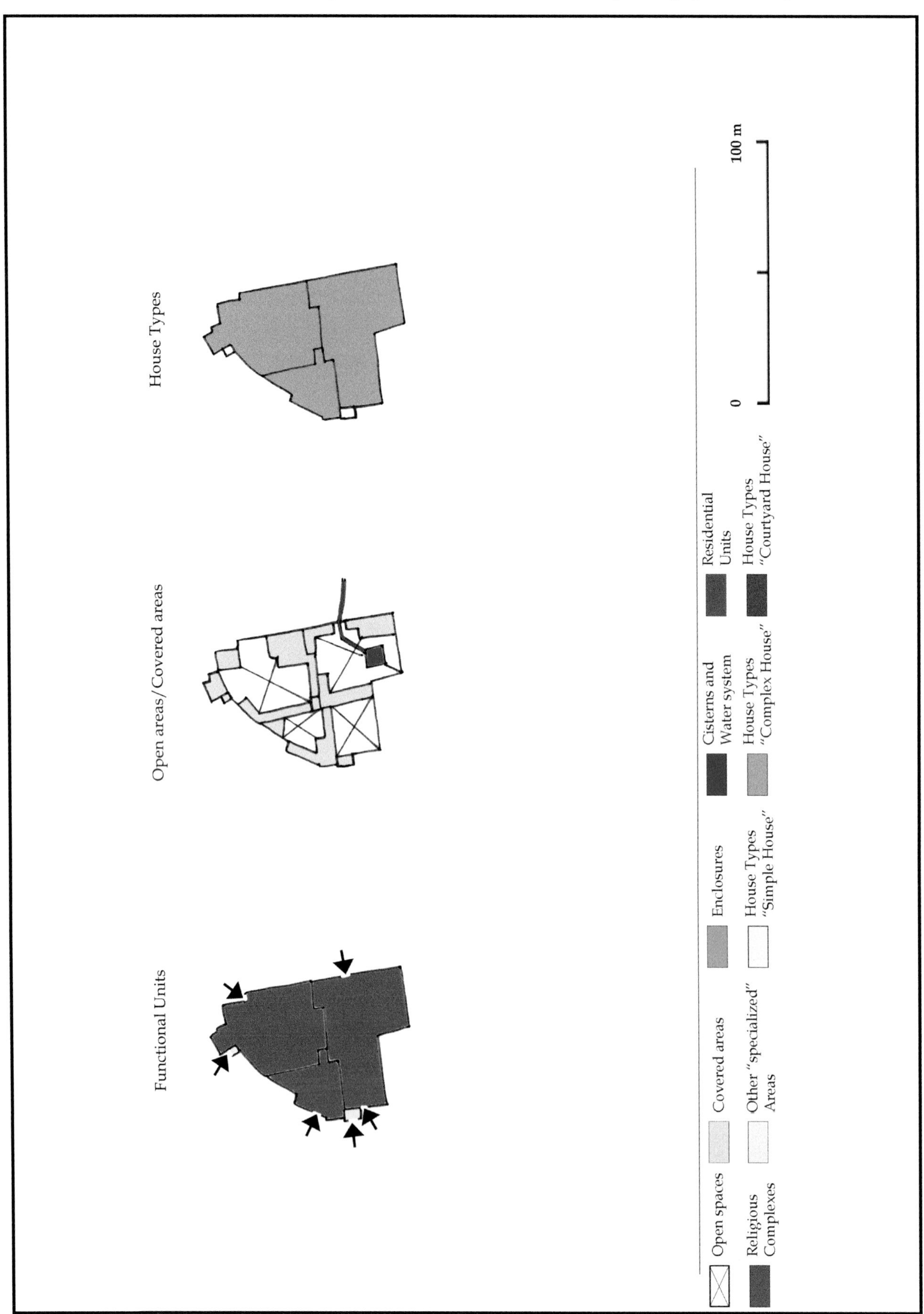

Plate 6

"Cluster-block" in Umm el-Jimal, with "buffer" zones to access each unit (drawing by the author)

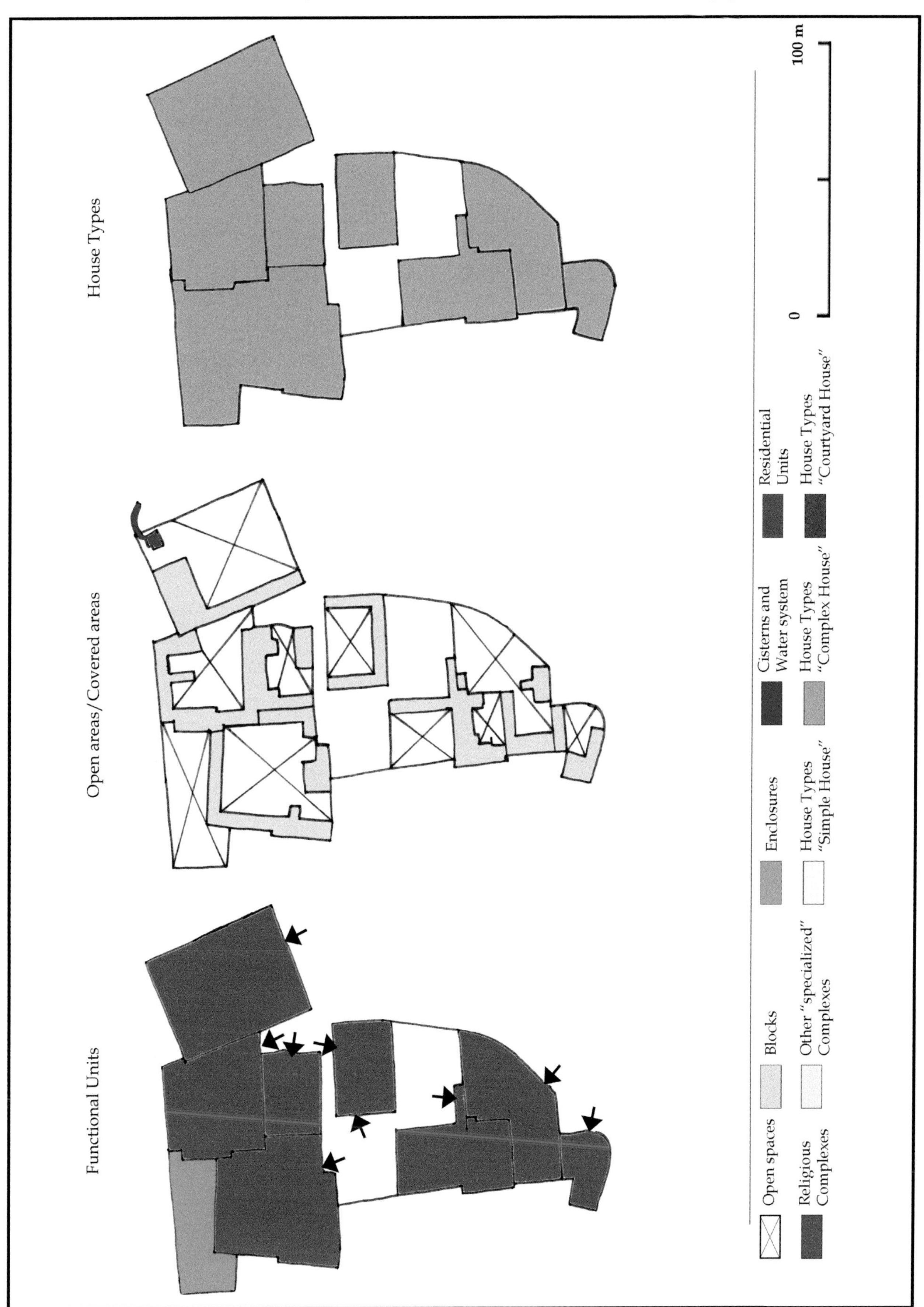

Plate 7

"Insulae" in Shivta, in the Negev (drawing by the author)

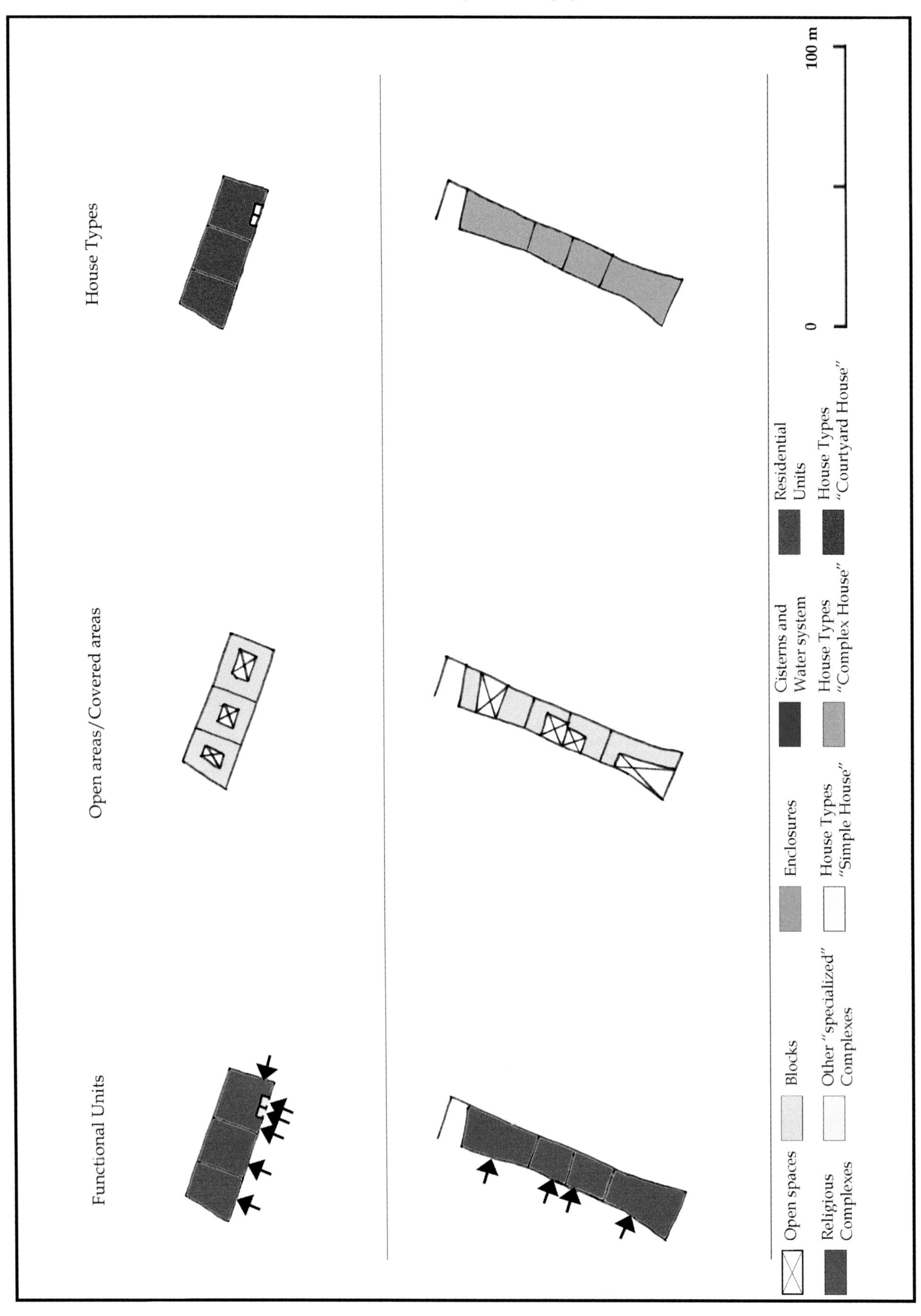

Plate 8
Two examples of "Simple" houses (drawing by the author)

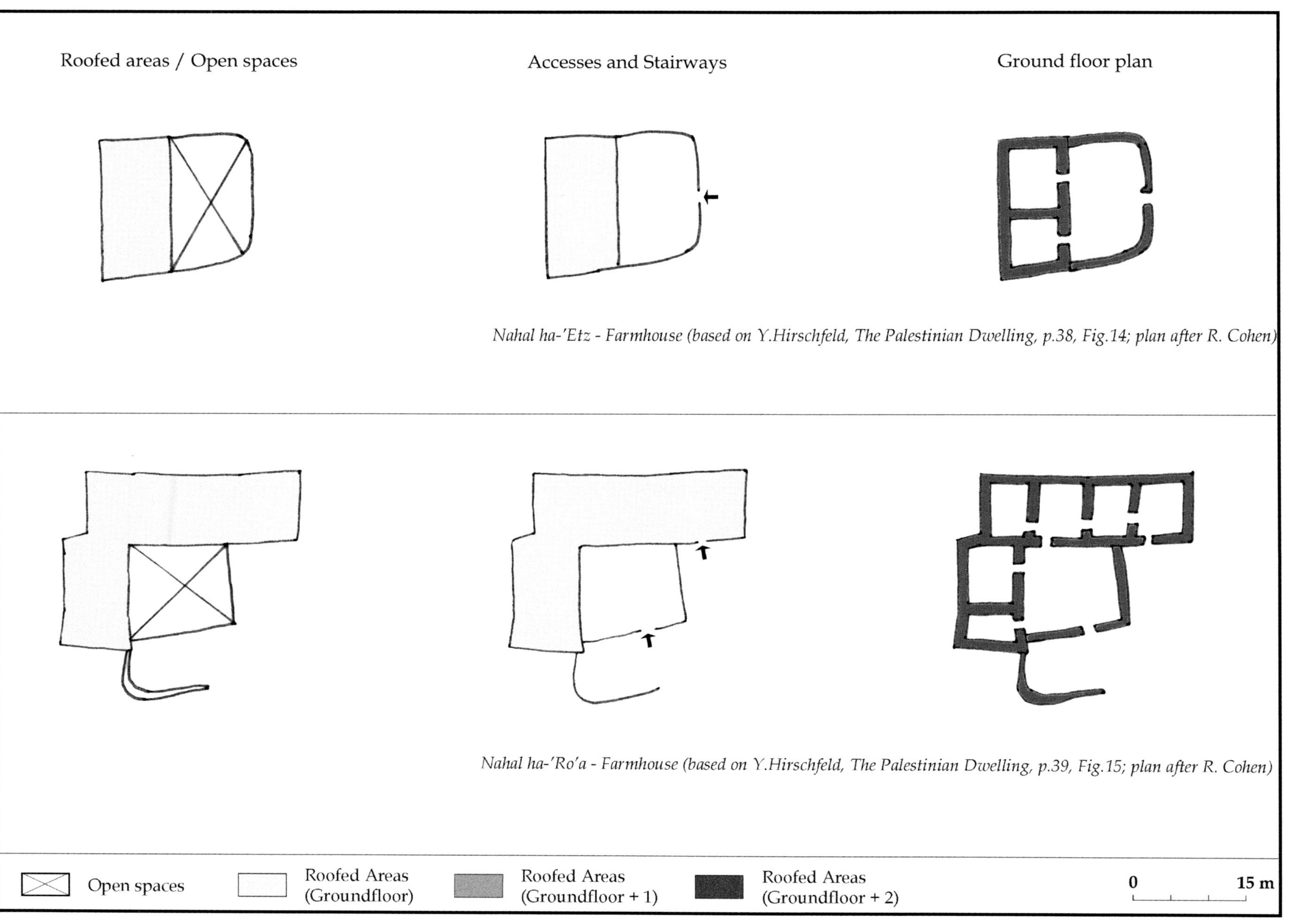

Plate 9a

Example of "Complex" house (drawing by the author)

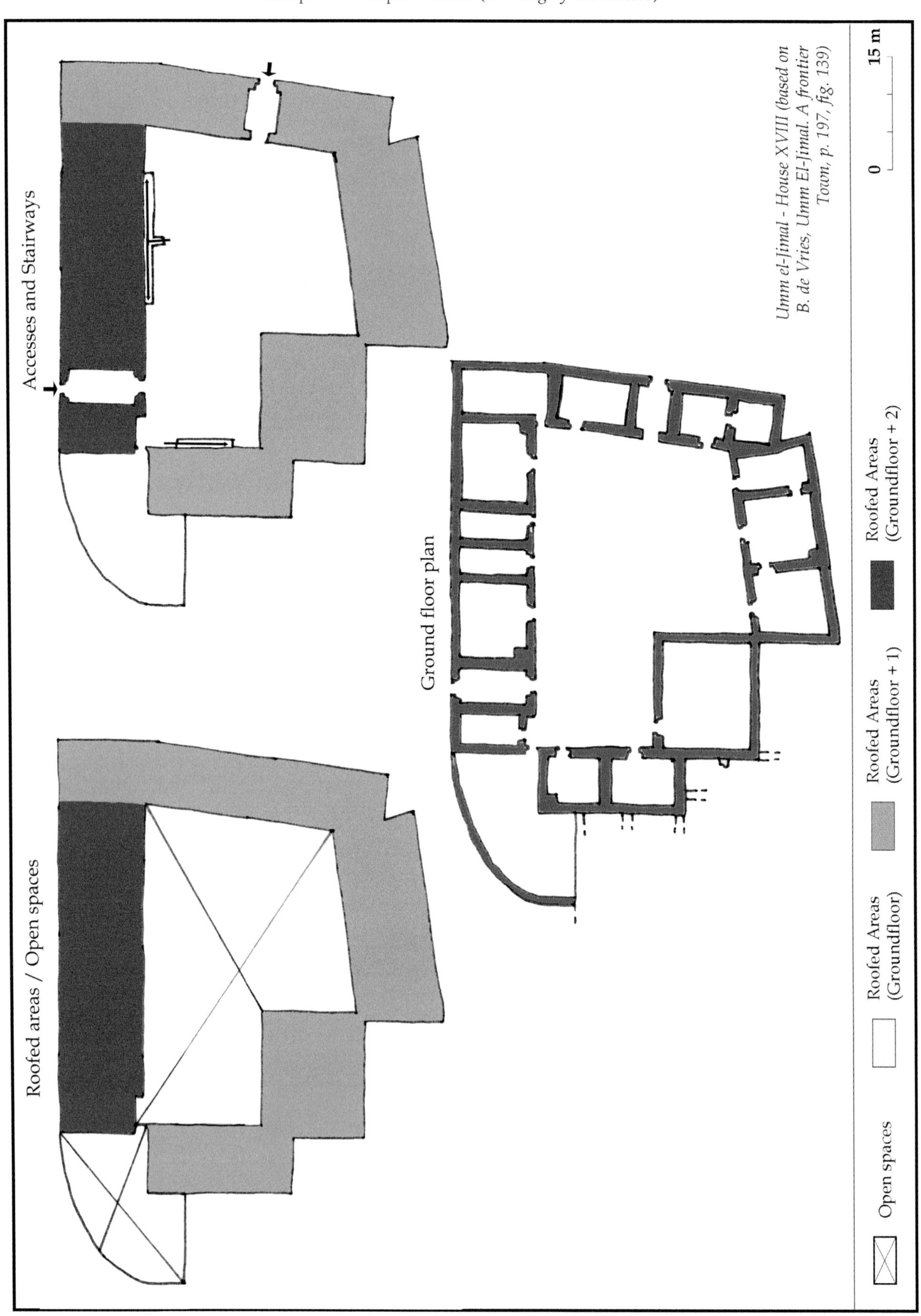

Plate 9b

Example of "Complex" house (drawing by the author)

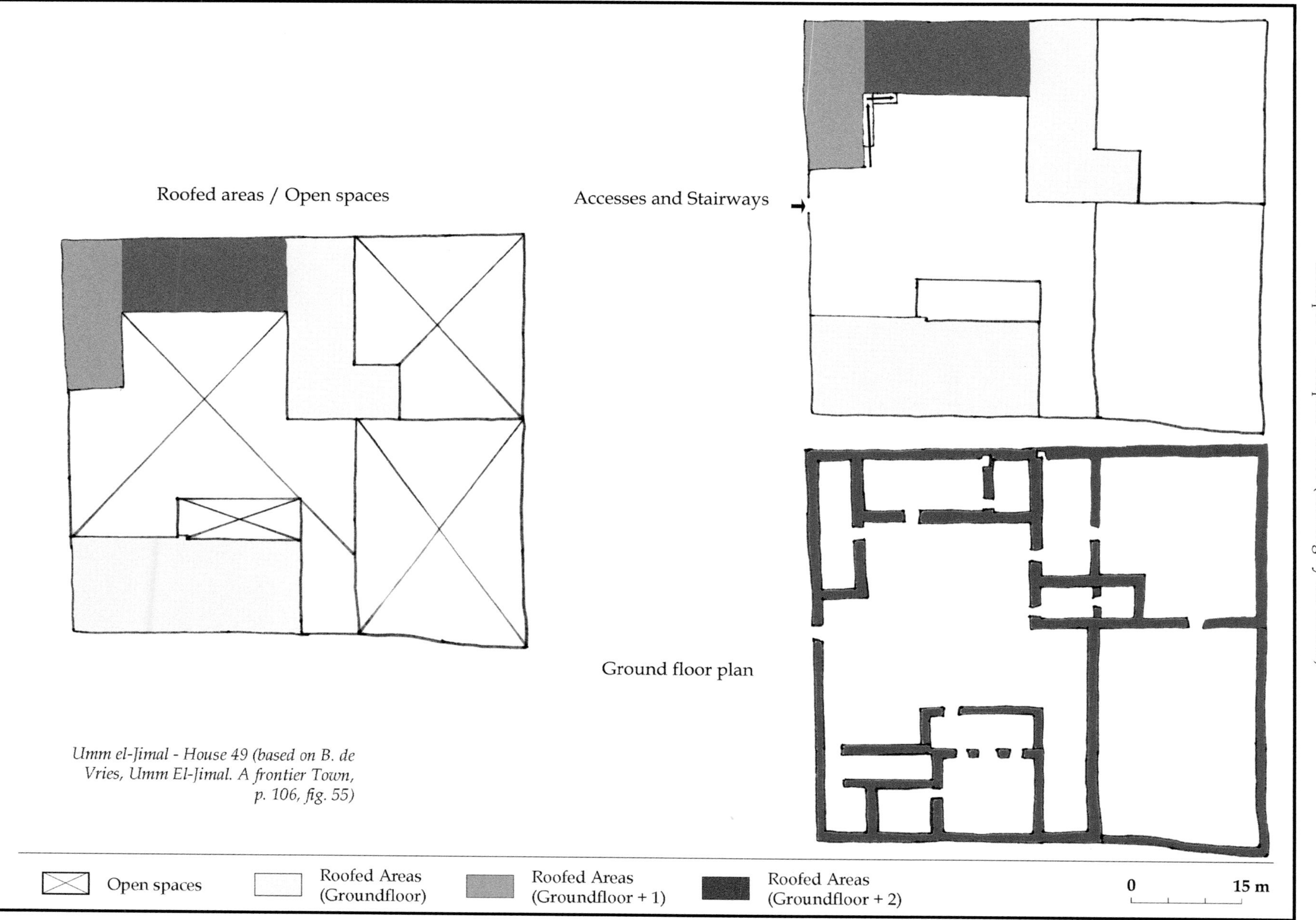

Plate 10a

Example of "Courtyard" house (drawing by the author)

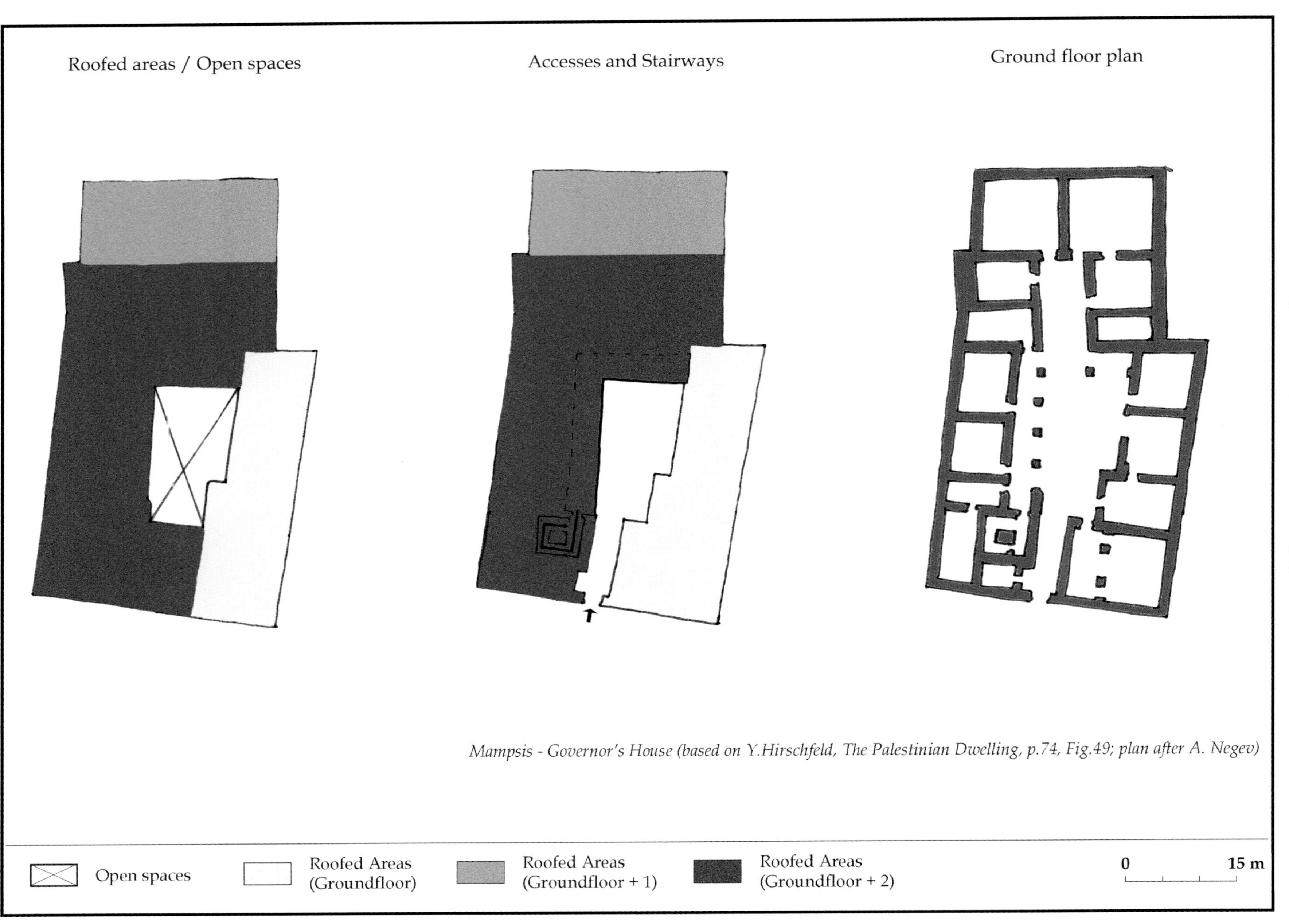

Plate 10b

Example of "Courtyard" house (drawing by the author)

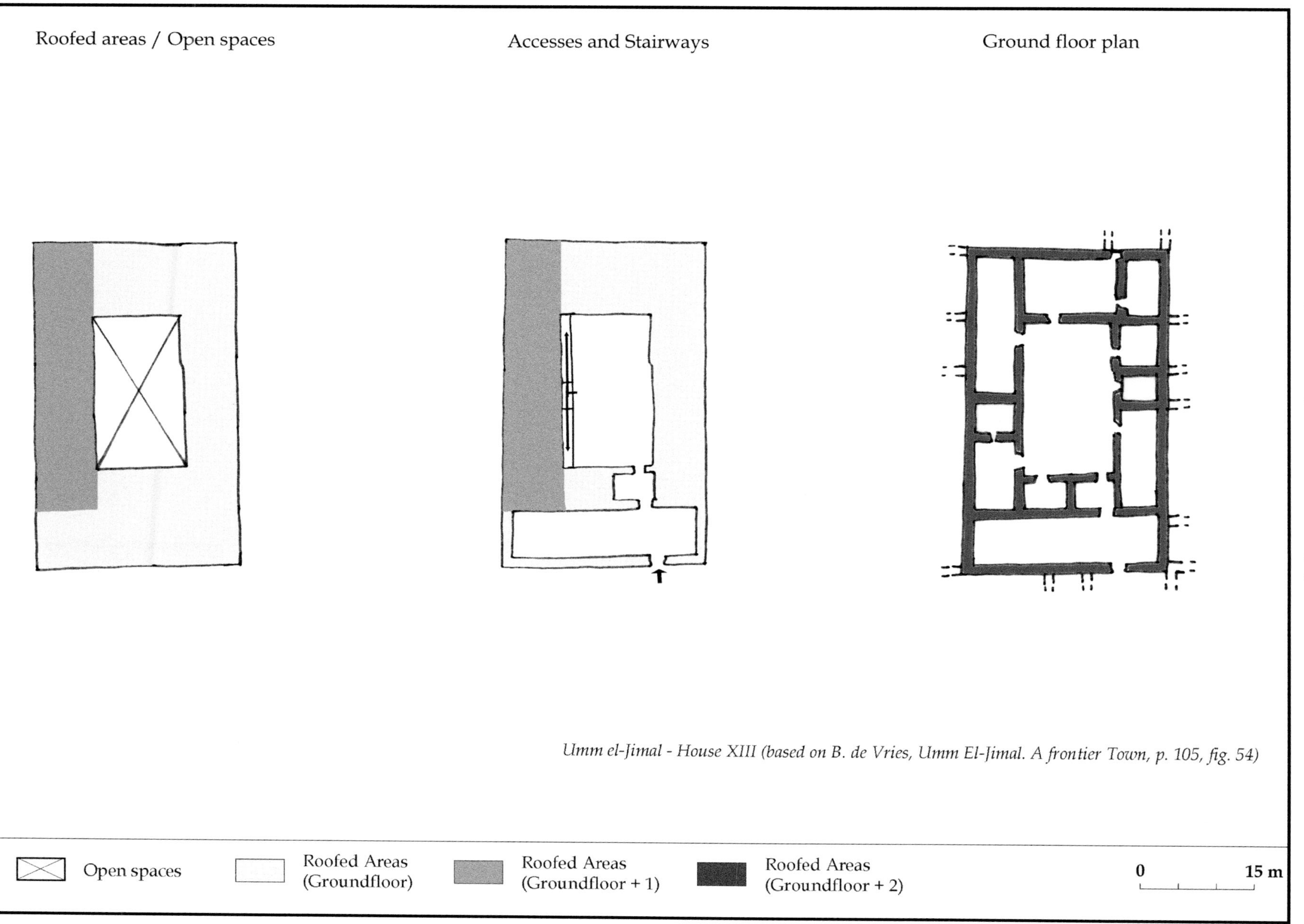